Genetherapy

Genetherapy

Edited by:

Preethi Kartan

Delve
PUBLISHING

www.delvepublishing.com

Genetherapy

Preethi Kartan

Delve Publishing
224 Shoreacres Road
Burlington, ON L7L 2H2
Canada
www.delvepublishing.com
Email: orders@arclereducation.com

ISBN: 978-1-77407-787-0 (Hardcover)

DECLARATION

ABOUT THE EDITOR

Preethi is a postgraduate in Biotechnology from University of Leeds, UK. She is currently working as a Scientific Associate at one of the CRO's in India and her interest lie in life sciences related writing.

TABLE OF CONTENTS

LIST OF CONTRIBUTORS

YK Sung
Department of Pharmaceutics and Pharmaceutical Chemistry, University of Utah, Salt Lake City, UT 84112, USA.
Center for Controlled Chemical Delivery, University of Utah, Salt Lake City, UT 84112, USA.
Department of Chemistry, Dongguk University, Chung-gu, Seoul 04620, Korea
Center for Controlled Chemical Delivery (CCCD), Department of Pharmaceutics and Pharmaceutical Chemistry, University of Utah, BPRB, Room 205, Salt Lake City, UT 84112, USA.

SW Kim
Department of Pharmaceutics and Pharmaceutical Chemistry, University of Utah, Salt Lake City, UT 84112, USA.
Center for Controlled Chemical Delivery, University of Utah, Salt Lake City, UT 84112, USA.

Rebecca E. Taylor
Mechanical Engineering, Biomedical Engineering and Electrical and Computer Engineering, Carnegie Mellon University, Pittsburgh, PA 15213, USA

Maliha Zahid
Department of Developmental Biology, University of Pittsburgh School of Medicine, Pittsburgh, PA 15201, USA

Lyse A. Norian
Department of Urology, University of Iowa Carver College of Medicine, Iowa City, IA 52242, USA

Britnie R. James
Interdisciplinary Graduate Program in Immunology, University of Iowa Carver College of Medicine, Iowa City, IA 52242, USA

Thomas S. Griffith
Department of Urology, University of Iowa Carver College of Medicine, Iowa City, IA 52242, USA

Interdisciplinary Graduate Program in Immunology, University of Iowa Carver College of Medicine, Iowa City, IA 52242, USA

Yuen Yi C. Tam
Department of Biochemistry and Molecular Biology, University of British Columbia, 2350 Health Sciences Mall, Vancouver, B.C. V6T 1Z3, Canada

Sam Chen
Department of Biochemistry and Molecular Biology, University of British Columbia, 2350 Health Sciences Mall, Vancouver, B.C. V6T 1Z3, Canada

Pieter R. Cullis
Department of Biochemistry and Molecular Biology, University of British Columbia, 2350 Health Sciences Mall, Vancouver, B.C. V6T 1Z3, Canada

Nuria Vilaboa
Unidad de Investigación, Hospital Universitario La Paz-IdiPAZ, 28046 Madrid, Spain
CIBER de Bioingeniería, Biomateriales y Nanomedicina (CIBER-BBN), Madrid, Spain

Frank Boellmann
Pacific Biosciences, Menlo Park, CA 9402

Richard Voellmy
Dept. of Physiological Sciences, University of Florida, Gainesville, FL 32611
HSF Pharmaceuticals S.A., 1814 La Tour-de-Peilz, Switzerland

Alejandra Meza-Ríos
Institute for Molecular Biology in Medicine, Department of Molecular Biology and Genomics, Health Sciences University Center, University of Guadalajara, Guadalajara, Jalisco, Mexico

Leonel García-Benavides
Unit of Cardiovascular Investigation, Department of Physiology, Health Sciences University Center, University of Guadalajara, Guadalajara, Jalisco, Mexico

Jesus García-Bañuelos
Institute for Molecular Biology in Medicine, Department of Molecular Biology and Genomics, Health Sciences University Center, University of Guadalajara, Guadalajara, Jalisco, Mexico

Adriana Salazar-Montes
Institute for Molecular Biology in Medicine, Department of Molecular Biology and Genomics, Health Sciences University Center, University of Guadalajara, Guadalajara, Jalisco, Mexico

Juan Armendáriz-Borunda
Institute for Molecular Biology in Medicine, Department of Molecular Biology and Genomics, Health Sciences University Center, University of Guadalajara, Guadalajara, Jalisco, Mexico
INNOVARE, Guadalajara, Jalisco, Mexico

Ana Sandoval-Rodríguez
Institute for Molecular Biology in Medicine, Department of Molecular Biology and Genomics, Health Sciences University Center, University of Guadalajara, Guadalajara, Jalisco, Mexico

Giulliana Augusta Rangel Gonçalves
Hospital Israelita Albert Einstein, São Paulo, SP, Brazil

Raquel de Melo Alves Paiva
Hospital Israelita Albert Einstein, São Paulo, SP, Brazil

Yoon-ha Jang
Department of Chemical and Biological Engineering, Sookmyung Women's University, Seoul 04310, Korea

Kwang-il Lim
Department of Chemical and Biological Engineering, Sookmyung Women's University, Seoul 04310, Korea
Institute of Advanced Materials and Systems, Sookmyung Women's University, Seoul 04310, Korea
Research Institute of ICT Convergence, Sookmyung Women's University, Seoul 04310, Korea

Magid H Amer
Department of Medicine, St Rita's Medical Center, 825 West Market Street, Suite #203, Lima, OH 45805, USA

Carolyn M. Yrigollen
The Raymond G. Perelman Center of Cellular and Molecular Therapeutics, Children's Hospital of Philadelphia, Philadelphia, PA 19104, USA

Beverly L. Davidson
The Raymond G. Perelman Center of Cellular and Molecular Therapeutics, Children's Hospital of Philadelphia, Philadelphia, PA 19104, USA
Department of Pathology and Laboratory Medicine, University of Pennsylvania, Philadelphia, PA 19104, USA

Jesse D. Sengillo
Jonas Children's Vision Care, and Bernard & Shirlee Brown Glaucoma Laboratory, Department of Ophthalmology, Columbia University Medical Center, New York, NY, 10032 USA
Edward S. Harkness Eye Institute, New York-Presbyterian Hospital, New York, NY, 10032 USA
State University of New York Downstate Medical Center, Brooklyn, NY, 11203 USA

Sally Justus
Jonas Children's Vision Care, and Bernard & Shirlee Brown Glaucoma Laboratory, Department of Ophthalmology, Columbia University Medical Center, New York, NY, 10032 USA
Edward S. Harkness Eye Institute, New York-Presbyterian Hospital, New York, NY, 10032 USA

Thiago Cabral
Jonas Children's Vision Care, and Bernard & Shirlee Brown Glaucoma Laboratory, Department of Ophthalmology, Columbia University Medical Center, New York, NY, 10032 USA
Edward S. Harkness Eye Institute, New York-Presbyterian Hospital, New York, NY, 10032 USA
Department of Ophthalmology, Federal University of Espírito Santo, Vitória, 29075-910 Brazil
Department of Ophthalmology, Federal University of São Paulo, São Paulo, 04021-001 Brazil

Stephen H. Tsang
Jonas Children's Vision Care, and Bernard & Shirlee Brown Glaucoma Laboratory, Department of Ophthalmology, Columbia University Medical Center, New York, NY, 10032 USA
Edward S. Harkness Eye Institute, New York-Presbyterian Hospital, New York, NY, 10032 USA
Department of Pathology & Cell Biology, Institute of Human Nutrition, College of Physicians and Surgeons, Columbia University, New York, NY, 10032 USA

LIST OF ABBREVIATIONS

OVs	Oncolytic viruses
HG	homo-galacturonan
HIV	Human Immunodeficiency Virus
CPPs	Cell penetrating peptides
PTDs	protein transduction domains
TRAIL	tumor necrosis factor–related apoptosis-inducing ligand
DISC	death-inducing signaling complex
FADD	Fas-associated death domain protein
TTR	target transthyretin
CCI	chronic constrictive injury
MSCs	Mesenchymal stem cells
CAR-T	Chimeric antigen recipient T
LS	Leigh syndrome
RNA	ribonucleic acid
RNAi	RNA interference
FXADs	Fragile X-Associated Disorders
LHON	Leber hereditary optic neuropathy
CHM	choroideremia
CMV	cytomegalovirus
MSC	Mesenchymal stem cell
AAV	adeno-associated virus
SNPs	single-nucleotide polymorphisms
KUN	Kunjin virus
SNHL	sensorineural hearing loss
VECs	vascular endothelial cells
VSMCs	vascular smooth muscle cells
ITR	inverted terminal repeats

PREFACE

Gene therapy is the technique used to treat disease by altering gene inside the cell. Many diseases like muscular dystrophy, haemophilia and cancer are caused due to malfunctioning of the genes in the cell. Gene therapy helps in correcting mutated gene by replacing mutated gene with healthy gene or by knocking out mutated genes or introducing healthy copy of gene in the cell to fight against disease or by making diseased cell more prominent to the body's immune system. The advantage of gene therapy is it helps in correcting underlying cause of the disease instead of the symptoms of the disease.

Gene therapy requires knowledge of identifying the genetic basis of disease and cloning correct copy of sequence.

DNA being negatively charged cannot easily cross the plasma membrane; this is a potential issue. To address this issue lot of efforts are going on to produce suitable and functional vectors for transporting genes into cell.

Various types of vectors are being used in gene therapy such as Replication deficient viruses and non-viral gene transfer systems using cationic liposome/DNA complexes.

This book describes some of the applications and recent advancements in the field of gene therapy.

1

Recent Advances in the Development of Gene Delivery Systems

YK Sung[1,2,3,4] and SW Kim[1,2]

[1] Department of Pharmaceutics and Pharmaceutical Chemistry, University of Utah, Salt Lake City, UT 84112, USA.
[2] Center for Controlled Chemical Delivery, University of Utah, Salt Lake City, UT 84112, USA.
[3] Department of Chemistry, Dongguk University, Chung-gu, Seoul 04620, Korea
[4] Center for Controlled Chemical Delivery (CCCD), Department of Pharmaceutics and Pharmaceutical Chemistry, University of Utah, BPRB, Room 205, Salt Lake City, UT 84112, USA.

ABSTRACT

Background

Gene delivery systems are essentially necessary for the gene therapy of human genetic diseases. Gene therapy is the unique way that is able to use

Citation: Sung, Y., Kim, S. Recent advances in the development of gene delivery systems. Biomater Res 23, 8 (2019). https://doi.org/10.1186/s40824-019-0156-z.

the adjustable gene to cure any disease. The gene therapy is one of promising therapies for a number of diseases such as inherited disorders, viral infection and cancers. The useful results of gene delivery systems depend open the adjustable targeting gene delivery systems. Some of successful gene delivery systems have recently reported for the practical application of gene therapy.

Main Body

The recent developments of viral gene delivery systems and non-viral gene delivery systems for gene therapy have briefly reviewed. The viral gene delivery systems have discussed for the viral vectors based on DNA, RNA and oncolytic viral vectors. The non-viral gene delivery systems have also treated for the physicochemical approaches such as physical methods and chemical methods. Several kinds of successful gene delivery systems have briefly discussed on the bases of the gene delivery systems such as cationic polymers, poly(L-lysine), polysaccharides, and poly(ethylenimine)s.

Conclusion

The goal of the research for gene delivery system is to develop the clinically relevant vectors such as viral and non-viral vectors that use to combat elusive diseases such as AIDS, cancer, Alzheimer, etc. Next step research will focus on advancing DNA and RNA molecular technologies to become the standard treatment options in the clinical area of biomedical application.

BACKGROUND

Gene therapy is the unique technique that uses gene to prevent or recover any diseases. The technique of gene therapy may allow doctors to treat a disorder by inserting a gene into patient's cell instead of using drugs or surgery. Some researchers and doctors are examining several approaches to gene therapy, including: i) replacing a mutated gene that causes disease with a healthy gene; ii) 'knocking out' or inactivating, a mutated gene that is functioning improperly; and iii) introducing new genes into the cells to protect from any diseases.

Although gene therapy is a promising treatment option for a number of diseases such as cancer, inherited disorders and certain viral infections, the technique remains risky and is still under examination to make safe until it will be effective. The gene therapy has currently examined only for diseases that have no other cure techniques. Genetic molecules reach to the nuclei of host cells to induce gene expression [1,2,3]. Gene therapy has derived to provide a patient's somatic cells with genetic information for producing specific therapeutic proteins to modulate genetic diseases. In order to get a successful design of gene delivery system, it must be required the complete understanding of interaction between targeting cell and gene delivery system. The gene delivery systems consist of the three components such as a plasmid-based gene expression system that controls the function of a gene within the targeting cell, a gene that encodes a specific therapeutic protein, and a gene delivery system that controls the delivery of the gene expression plasmid to specific location within the body [4, 5]. The successful gene delivery system requires the foreign genetic molecule to remain stable within the host cells [6,7,8]. As a tool of transgene expression, viral-based vectors had firstly emerged in the 1980's [9]. The virus as a vaccine vector was used the vaccinia virus in 1984 as a way to protect chimpanzees against Hepatitis B [9]. On the other hand, non-viral gene delivery system had firstly reported on showing cellular phenotype change via DNA exposures in the body [10, 11].

In this article, the recent advances in the development of gene delivery systems have reviewed, including the progress of viral gene delivery systems and non-viral gene delivery systems. The viral gene delivery systems have discussed for the viral vectors based on DNA, RNA, and oncolytic viral vectors, respectively. The non-viral gene delivery systems have also treated for the physicochemical approaches such as physical method and chemical method. Several kinds of gene delivery systems have discussed on the gene delivery systems of cationic biomedical polymers such as polysaccharides, polyethylenimine (PEI), and poly(L-lysine)(PLL) derivatives.

Development of Gene Delivery Systems

Viral Vector Gene Delivery Systems

Viruses mediated gene delivery systems utilize the ability of virus to inject their DNA inside host cells. Those take the advantage of the virus ability to replicate their own genetic materials. Virus is an effective form of gene delivery because of the virus structure preventing degradation via the lyposome of DNA [12,13,14]. In gene delivery systems, the genes open to intend for delivery system package into the replication. The deficient viral particles form the viral vectors [15, 16]. Viruses used for gene delivery systems include retrovirus, adenovirus, adeno-associated viruses, and herpes simplex viruses. There are two categories of gene delivery systems such as the germline gene delivery systems and the somatic gene delivery systems. Although the germline gene delivery systems may have great potential, the germline gene therapy cannot use ethically [17, 18]. In practice, gene delivery system for human has been only limited to somatic cell alteration. In somatic gene delivery system, there are three types such as in vivo delivery, ex vivo delivery and in situ delivery. In vivo delivery systems, the genetic materials transfer directly into the targeting tissue. In ex vivo delivery system, the genetic materials implant into the targeting tissue bone marrow, cultivated, and manipulated in vitro. The genetic materials transfer into the targeting tissues. There are no immunologic responses in the way but only the techniques use in the various cases. In addition, only a small percentage of implanted cells remain viable at present [19, 20]. In situ delivery system, the administration of the genetic material goes directly into the target tissue. As most of the current delivery systems need no effective targeting, the way is proper [21,22,23]. The technique is the least advanced strategy at present, but it might potentially be useful method.

Viral Vectors Based on DNA for Gene Delivery Systems

Viral vectors based on DNA for gene delivery are usually longer lasting and integrating into the genomes. DNA-based viral vectors are lentivirus, poxvirus, adenovirus, adeno-associated virus, retrovirus, human foamy virus (HFV) and herpes virus [24]. DNA-based viral vectors for gene delivery systems utilize the viral vectors to deliver genetic materials to the host cells. The viral vectors are efficient for delivering the genetic materials to the host cells [25]. The DNA-based viral vectors for the gene delivery systems include plasmids containing transgenes for gene therapy [26]. Although the DNA-based viral vectors are in early stages of clinical trials, the class of materials has been emerged to yield extremely promising candidate of gene delivery systems for gene therapy in the wide range of diseases including cancer, AIDS, neurological disorders such as Parkinson's and Alzheimer's diseases, and cardiovascular disorders [27,28,29].

Viral Vectors Based on RNA for Gene Delivery Systems

Viral vectors based on RNA for gene delivery have developed for the ability to transcribe directly for infectious RNA transcripts. RNA-based gene delivery is not permanent and usually transient. In the gene delivery systems based on RNA, human foamy virus, oncoretro-viral vectors, and lenti-viral vectors are included in the gene therapy. The sophisticated system provides the RNA-dependent polymerase complexes coupled with negative-strand RNA templates [30]. RNA-based gene delivery systems have carried out for HIV with lenti-viral vectors-modified CD 34(+) cells in patients undergoing transplantation for AIDS-related lymphoma [31].

Oncolytic Viral Vectors for Gene Delivery Systems

Oncolytic viruses (OVs) are an emerging treatment option for cancer type diseases. They have recently focused on cancer research aiming to develop their therapeutic treatment potential [32]. The positive and negative effects of the multitude for their modification have discussed to increase their infectivity, anti-tumor immunity and treatment safety for the interaction of OVs and tumor cells. Oncolytic adenovirus co-expressing IL-12 and IL-18 improves tumor-specific immunity via differentiation of T cells expressing IL-12Rβ_2 or IL-18Rα [33]. The ultimate aim is to design a virus that is able to replicate effectively within the host, special target and lyse tumor cells. Adenovirus-mediated decorin expression induces cancer cell death through activation of p53 and mitochondrial apoptosis [34]. Oncolytic adenovirus (Ad) vectors present a promising modality to treat cancer by using the gene therapy [35]. Oncolytic adenovirus expressing IL-23 and p35 elicits IFN-ɤ-and TNF-α-Co-producing T cell-mediated antitumor immunity. It has reported that cytokine immune-gene therapy is one of promising strategies for cancer treatments [36,37,38,39].

Non-Viral Vector Gene Delivery Systems

The physicochemical approaches have recently proposed for non-viral based delivery systems [40]. Non-viral vectors of delivery systems for gene therapy are less likely to induce an immune response in the system of biocompatible materials such as lipid, naked DNA, chromosomes, plasmid, cationic polymers, and conjugate complexes [41].

Non-viral gene delivery mediated artificially by physical method uses to introduce genetic materials through the cell membranes. The physical method of gene delivery includes needle injection, ballistic DNA injection, sono-poration, photo-poration, magneto-fection, and hydro-poration. Needle injection is the direct inject of genetic materials using a needle. On the other hand, ballistic DNA injection is the gold coated DNA articles that force into cells. Electroporation is the electric pulse that creates pores in a cell membrane to allow entry of genetic materials. Sono-poration is the sound wave that creates pores in a cell membrane to allow entry of genetic materials. Photo-poration is the Laser pulse that creates pores in a cell membrane to allow entry of genetic materials. Magneto-fection is the magnetic particles that are complexed with DNA and an external magnetic field, and then they concentrate nucleic acid particles into target cells. Finally, the hydro-poration is the hydrodynamic capillary effect that manipulates cell permeability. The non-viral chemical methods for gene delivery systems use synthetic or natural compounds to form some particles that facilitate the transfer of genes into the cells. The synthetic vector has an ability to interact with RNA or DNA electrostatically and bind to compact the genetic information in the accommodation of larger genetic transfers [42]. Then, non-viral chemical vectors are able to enter cells by endocytosis. In general, there are two non-viral vectors such as liposomes and polymers. Liposome based non-viral vectors use liposomes to facilitate gene delivery by the formation of lipoplexes. The lipoplexes form spontaneously when the negatively charged DNA contacts with positively charged liposomes. The polymer-based non-viral vectors use the polymers to interact with DNA to form polyplexes.

Practical Application of Gene Delivery Systems

A successful gene therapy needs a necessary achievement to develop the proper gene delivery system. The gene delivery system in therapeutic setting utilizes non-immunogenic vectors capable of cell specificity that is able to deliver an adequate amount of transgene expression to cause the desired effect [43]. DNA or RNA microarrays used in a variety of gene sequencing can identify the thousands of genes with analytical software looking at the gene expression patterns [44]. The gene delivery system has utilized to generate hybrid biosynthetic vector to deliver a possible vaccine as an application method. The vector overcomes the traditional barriers to gene deliveries by combining *E. coli* with synthetic polymers to create some vectors [45]. The gene therapy system gives a great opportunity for treating some diseases from genetic disorder, cancer, and other infections. The non-viral vector has a merit in its low immunogenicity and rather low cost reproducibility [46]. The non-viral vector has no limitation in DNA size for packing and modification with ligands to specific cell targeting [47].

The gene delivery systems based on synthetic polymeric materials have tested and evaluated for gene transfer to animals [48]. One of the candidates is the plasmid DNA that is able to carry the gene into the nucleus of the desired cells safely. The gene delivery systems based on non-viral vectors have also risen as a new promising therapeutic modality [49].

Gene Delivery Systems Based On Cationic Polymers

In the research of gene delivery systems, the potential benefits of polymeric carriers for genes have evaluated by various investigators for cationic biopolymers such as liposomes and chitosan derivatives [50, 51]. Regarding to cationic polymers for the gene delivery, the polymeric systems can mask the negative DNA charges and condense the large genes into the small molecular structures. The cationic polymer-based nucleic acid is a complex called as polyplex. The cationic non-viral lipid-based gene carriers, 'lipoplexes' have been currently evaluated clinically [52,53,54,55,56]. The gene complex is one of key targeting delivery for gene therapy. Most researches focus on the effect of targeting ligands that are coordinated covalently to the DNA complex. The cationic polymers can conjugate to the targeting ligands. One of candidate is poly(L-lactide), which has widely utilized as a polymer for attaching the targeting ligands [57,58,59,60]. The plasmid DNA condensation of cationic polymer undergoes a compaction under the wide variety of conditions. The multivalent cationic polymers have usually used as the condensing agents [61, 62]. The tertiary structures similar to the non-condensing plasmid DNA have been investigated on the complexed with low hydrophobized stearyl-poly(L-lactide) [63]. The DNA condensation has still allowed for the systems of attaching hydrophilic segments such as polyethylene glycol, dextran, hyaluronic acid or hydrophobic stearyl chains [64,65,66,67].

Gene Delivery Systems Based On Polysaccharides

A range of biodegradable cationic polymers based on grafted oligo-amine residues for natural polysaccharides was reported [68]. The grafting idea, in which side chain oligomers have attached to either linear or branched hydrophilic polysaccharides, allows three-dimensional interaction with anionic surfaces of double strand DNA chains. The gene delivery system based on polysaccharide indicates that the structure of the cationic polysaccharide has a significant role in the transfection activity for gene delivery. The colloidal polysaccharide particles had investigated for the mucosal delivery of complexes such as peptides, proteins, oligonucleotides, and plasmids [69]. Natural polysaccharide-based nano-particles had also studied as gene and drug delivery systems [70]. The mechanism was proposed to prepare polysaccharide-based nanoparticles, including covalent and ionic crosslinking, electrolyte complexing, and the self-assembly of hydrophobic polysaccharides. The natural polysaccharides such as chitosan and pectin had studied as gene delivery or drug delivery carriers [71]. Chitosan families have strongly recommended poly-cationic derivatives of poly-N-acetyl-D-glucosamine, which is the positive charge molecule with important muco-adhesion. Pectin is also made of several structural segments, which are homo-galacturonan (HG) rhamnogalaturonan (RG-I) as smooth and hairy regions.

Gene Delivery Systems Based On Poly(Ethylenimine)S

Cationic polymers used for gene complexation are mainly polyamines that become cationic at physiologic conditions. Poly(ethylenimine)(PEI) consists of many nitrogen atoms including primary, secondary, and tertiary amine groups. Those amine groups are able to enhance their buffering capacities. Those cause the subsequent endosomal or lysosomal rupture, and escape into the cytoplasm via a proton sponge effect [72, 73]. PEI is able to provide some buffering capacity enhanced to the other functional molecules. PEI conjugated poly(crystamine bis(acrylamide)-diaminohexane) [poly(CBA-DAH)] had been synthesized to decrease weight ratio and increase transfection efficiency. Poly(CBA-DAH) is composed of multiple disulfides, which can be cleaved in the cytoplasm by intracellular reducing agent such as glutathione(GSH) [74,75,76,77]. GSH is composed of triple-peptide, which has synthesized in the cytosol from precursor amino acids [78]. The design of PEI-conjugated bio-reducible polymer for efficient gene delivery had investigated to confirm the successful vectors and polyplexes formation with pDNA. The other studies had also shown the enhanced buffering capacity such as imidazole, histidine, and PEI [79,80,81,82,83]. Biodegradable Poly(ethylenimine)s for plasmid DNA Delivery had been investigated [84]. Branched poly(Ethylenimine)-cholesterol water-soluble lipo-polymers for gene delivery was also developed [85]. Modified Linear Polyethylenimine-Cholesterol Conjugates for DNA Complexation had made for gene delivery [86]. Tumor efficacy and biodistribution of linear polyethylenimine-cholesterol/DNA complexes had tested for gene delivery system [87]. Polyethylenimine with acid-labile linkages as a biodegradable gene carrier had synthesized and tested for gene delivery system [88]. Reducible poly(amidoethylenimine)s designed for gene delivery systems had been made and tested for gene delivery [89,90,91,92].

Gene Delivery Systems Based On Poly(L-Lysine)

Poly(L-lysine)(PLL) is one of the cationic polymers for gene transfers. PLL is a homo-polypeptide belonging to the group of cationic polymers at pH 7. PLL contains a positively charged hydrophilic amino group. Replica particles (RP) based on PLL cross-linked via a homo-bi-functional linker to support co-adsorption of a plasmid DNA has designed for gene delivery system [93]. The stabilization of poly-L-lysine/DNA polyplexes for gene delivery in vivo to the liver has been developed [94]. The synthesis and characterization of poly-L-lysine-based gene delivery systems have been reported [95]. N-terminal modified poly(L-lysine)-antibody conjugate had been used as a carrier for targeted gene delivery in mouse lung endothelial cells [96]. Biodegradable poly(ethylene glycol)-*co*-poly(L-lysine)-*g*-histidine multi-block copolymers had been used for non-viral gene delivery systems [97]. Poly(L-lysine) and copolymers for gene delivery had been treated and summarized in polymeric gene delivery [98]. Poly-L-lysine-graft-PEG-comb-type polycation copolymers had also developed for gene delivery systems [99]. Nano-particle DNA carrier with poly(L-lysine) grafted polysaccharide copolymers and terplex DNA delivery system as a gene carrier were firstly developed and characterized [100, 101]. A new non-viral DNA delivery vectors of the terplex systems and polyethylene glycol-grafted poly(L-lysine) as polymeric gene carriers were also developed for gene delivery systems [102,103,104,105]. The clinical evaluations have currently made in the field of biomedical applications [106,107,108,109,110,111,112,113,114,115,116,117].

CONCLUSION

An important goal for the research of gene delivery system is to develop clinically relevant vectors that use to combat elusive diseases such as AIDS, cancer, Alzheimer, etc. DNA-based viral vectors for gene delivery systems utilize the viral vectors to deliver genetic materials to the host cells. The DNA viral vectors are efficient for delivering the genetic materials to the host cells. On the other hand, one of promising RNA vectors has established employing naked syn-mRNA, viral or non-viral RNA embedded within viral particles, as well as self-amplifying RNA replicons. RNA-based systems appear superior to common DNA-based modalities. RNA technologies have especially matured in terms of immunogenicity, RNA stability, their formulation and production. Improvement of these parameters has created promising gene transfer systems suitable for in vivo treatment. Next step research will probably focus on advancing DNA and RNA molecular technologies to become standard treatment options in the biomedical applications.

REFERENCES

1. Guo X, Huang L. Recent advances in non-viral vectors for gene delivery. Acc Chem Research. 2012;45(7):971–9.
2. Jones CH, Chen CK, Ravikrishnan A, Rane S, Pfeifer BA. Overcoming nonviral gene delivery barriers: perspective and future. Mol. Pharmaceutics. 2013;10:4082–98.
3. Friedmann T, Roblin R. Gene therapy for human genetic disease. Science. 1972;175:949–55.
4. Han S-O, Mahato RI, Sung YK, Kim SW. Development of biomaterials for gene therapy. Mol Therapy. 2000;2:302–17.
5. Mahato RI, Smith LC, Rolland A. Pharmaceutical perspectives of nonviral gene therapy. Adv Genet. 1999;41:95–156.
6. Park, D W, Hoffman, A S, Pu, S, Stayton, P S, Design and development of polymers for gene delivery Nature Rev Drug Discovery 2005; 4:581–593.
7. Nayerossadat N, Maedeh T, Ali PA. Viral and nonviral delivery systems for gene delivery. Adv Biomed Res. 2012;1:27–31.
8. Mali S. Delivery systems for gene therapy. Indian J Human Genet. 2013;19:3–8.
9. Moss B, Smith GL, Gerin JL, Purcell RH. Live recombinant vaccinia virus protects chimpanzees against hepatitis B. Nature. 1984;311(5981):67–9.
10. Yellepeddi V K, Vectors for Non-viral Gene Delivery- Clinical and Biomedical Applications. Austin Therapeutics, ISSN: 2472–3673. 2015; 2(1):1014.
11. Kim S W, Biomaterials to gene delivery. J Control Release, 2011; 155(2):116–118. Doi:https://doi.org/10.1016/j.jconrel.2011.03.023. PMID:21457735
12. Kamimura K, Suda T, Zhang G, Liu D. Advances in gene delivery systems. Pharm Med. 2011;25(5):293–306.
13. van Nies P, Westerlaken I, Blanken D, Salas M, Mencia M, Danelon C. Self-replication of DNA by its encoded proteins in liposome-based synthetic cells. Nat Commun. 2018;9:1583–5.
14. Singh BN, Prateeksha A, Gupta VK, Chen J, Atanasov AG. Organic nanoparticle-based combinatory approaches for gene therapy. Trends Biotechnol. 2017;35(12):1121–4.

15. Nimesh S, Halappanavar S, Kaushik N K, Kumar P, Advances in gene delivery systems. Biomed Res Int. 2015;610342 doi: https://doi.org/10.1155/2015/610342 PMID: 25918717.
16. Lodish H, Berk A, Zipursky S L, Matsudaira P, Baltimore D, Darnell J, Molecular Cell Biology, 4th ed., New York, W. H. Freeman; 2000. ISBN-10: 0–7167–3136-3. www.ncbi.nlm.nih. gov/books/NBK21599.
17. Resnik B, Langer PJ. Human germline gene therapy reconsidered. Hum. Gene Ther. 2001;12:1449–58.
18. McDonough PG. The ethics of somatic and germline gene therapy. Ann N Y Acad Sci. 1997;816:378–82.
19. Suhonen J, Ray J, Bloemer U, Gage F H, Kaspar B, *Exo vivo* and *in vivo* gene delivery to the brain. Curr Protoc Hum Genet. 2006; Chapter 13: Unit 13.3. doi:https://doi.org/10.1002/0471142905.hg1303s51. PMID: 18428389.
20. Herrero MJ, Sabater L, Guenechea G, Sendra L, Montilla AI, Abargues R, Navarro V, Alino SF. DNA delivery to *ex vivo* human liver segments. Gene Ther. 2012;19:504–12.
21. Takefumi S, Akira I, Shin E, Shiro B. *In situ* gene therapy for prostate cancer. Curr Gene Ther. 2005;5:111–9.
22. Hu WW, Wang Z, Hollister SJ, Krebsbach PH. Localized viral vector delivery to enhance in situ regenerative gene therapy. Gene Ther. 2007;14:891–901.
23. Davis PB, Cooper MJ. Vectors for airway gene delivery. AAPS J. 2007;9:2–5.
24. Augusta G, Gonçalves R, de Melo R, Paiva A. Gene therapy: advances, challenges and perspectives. Einstein. 2017;15(3):369–75.
25. Wivel NA, Wilson JM. Method of gene delivery. Oncology Clinics North America. 1998;12(3):483–501.
26. Crooke ST. An overview of progress in antisense therapeutics. Antisense Nucleic Acid Drug Dev. 1998;8:115–22.
27. Stull RA, Szoka FC Jr. Antigene, ribozyme and aptamer nucleic acid drugs: Progress and prospects. Pharm Res. 1995;12:465–83.
28. Patil SD, Rhodes DG, Burgess DJ. DNA-based therapeutics and DNA delivery systems: a comprehensive review. AAPS J. 2005;7(1):E61–77.

29. Siddhesh DP, David GR, Diane JB. DNA-based therapeutics and DNA delivery systems: a comprehensive review. AAPS J. 2005;7(1):61–77.
30. Mogler, MA, Kamrud, KI, RNA-based viral vectors. Expert Rev Vaccines 2015; 14(2):283–312. doi: https://doi.org/10.1586/14760584.2015.979798. PMID: 25382613.
31. DiGiusto DL, Krishnan A, Li H, Li S, Rao A, Mi S, Yam P, Stinson S, Kalos M, Simon J, Lacey F, Yee JK, Li M, Couture L, Hsu D, Forman SJ, Rossi JJ, Zaia JA. RNA-based gene therapy for HIV with lentiviral vector-modified CD34(+) cells in patients undergoing transplantation for AIDS-related lymphoma. Sci Transl Med. 2010;2(36):36–43.
32. Howells A, Marelli G, Lemoine NR, Wang Y, Oncolytic virus-interaction of virus and tumor cells in the battle to eliminate cancer. Front Oncol 2017; http://doi.org/https://doi.org/10.3389/fonc.2017.00195.
33. Choi I-K, Lee J-S, Zhang S-N, Park J, Lee K-M, Sonn CH, Yun C-O. Oncolytic adenovirus co-expressing IL-12 and IL-18 improves tumor-specific immunity via differentiation of T cells expressing IL-12Rß$_2$ or IL-18Rα. Gene Ther. 2011;18(9):942–7.
34. Yoon A-R, Hong JW, Yun C-O. Adenovirus-mediated decorin expression induces cancer cell death through activation of p53 and mitochondrial apoptosis. Oncotarget. 2017;8(44):1–20.
35. Choi JW, Lee YS, Yun C-O, Kim SW. Polymeric oncolytic adenovirus for cancer gene therapy. J Control Release. 2015;10(219):181–91. https://doi.org/10.1016/jconrel.2015.10.009 Epub 2015 Oct 23.
36. Choi I-K, Li Y, Oh E, Kim J, Yun C-O, Oncolytic adenovirus expressing IL-23 and p35 elicits IFN-ɤ-and TNF-α-co-producing T cell-mediated antitumor immunity, PLoS One, 2013; 8(7):e67512. https://doi.org/https://doi.org/10.1371/journal.pone.0067512.
37. Hernandez-Gea V, Toffanin S, Friedman SL, Llovet JM. Role of the microenvironment in the pathogenesis and treatment of hepatocellular carcinoma. Gastroenterology. 2013;144:512–27.
38. El-Aneed A. Current strategies in cancer gene therapy. Europ J Pharmacol. 2004;498:1–8.
39. Baban CK, Cronin M, O'Hanlon D, O'SullivanG C, Tangney M. Bacteria as vectors for gene therapy of cancer. Bioengineered Bugs. 2010;1(6):385–94.

40. Uddin SN, Islam KK. Cationic polymers and its uses in non-viral gene delivery systems: a conceptual research. Trends Med Res. 2006;1(2):86–99. https://doi.org/10.3923/tmr.2006.86.99.
41. 4Remes A, Williams DF. Immune response in biocompatibility. Biomaterials. 1992;13(11):731–43.
42. Sung YK, Kim SW. The practical application of gene vectors in cancer therapy. Integrative Cancer Sci Therap. 2018;5(5):1–5. https://doi.org/10.15761/ICST.1000287.
43. Katz M G, Fargnoli A S, Williams R D, Brigges C R, Gene therapy delivery systems for enhancing virial and non-virial vectors for cardiac diseases:current concepts and future applications. Human Gene Ther 2013; 24(11):1–7, https://doi.org/https://doi.org/10.1089/hum.2013.2517.
44. Guyon I, Weston J, Barnhill S, Vapnik V. Gene selection for cancer classification using support vector machines. Mach Learn. 2002;46:389–422.
45. Jones CH, Ravikrishnan A, Chen M, Reddinger R, Kamal Ahmadi M, Rane S, Hakansson AP, Pfeifer BA. Hybrid biosynthetic gene therapy for vector development and dual engineering capacity. Proc Nat Academy Sci USA. 2014;111(34):12360–5.
46. Lee M, Kim SW. Polyethylene glycol-conjugated copolymers for plasmid DNA delivery. Pharm Res. 2005;22:1–10.
47. Glover DJ, Lipps HJ, Jans DA. Towards safe, non-viral therapeutic gene expression in humans. Nat. Rev. Genet. 2005;6:299–310.
48. Lostale-Seijo I, Montenegro J. Synthetic materials at the forefront of gene delivery. Nature Rev Chem. 2018;2:258–77.
49. Kim SW, Nam J-P, Kim S, Sung YK. Recent development of bio-reducible polymers for efficient gene delivery system. J Cancer Treat Diag. 2018;2(5):17–23.
50. Ginn SL, Amaya AK, Alexander IE, Edelstein M, Abedi MR. Gene therapy: clinical trials worldwide to 2017: an update. J Gene Med. 2018;20(e3015):1–16.
51. Huli-Curtis SL, Uusi-Kerttula H, Jones R, Hanna L, Chester JD, Parker AL. Evaluation of CD46 re-targeted adenoviral vectors for clinical ovarian cancer intraperitoneal therapy. Cancer Gene Ther. 2016;23:229–34.

52. Blaese RM, Culver KW, Miller AD, Carter CS, Fleisher T, et al. T lymphocytedirected gene therapy for ADA-SCID: initial trial results after 4 years. Science. 1995;270:475–80.
53. Brenner MK, Rill DR, Moen RC, Krance RA, Mirro J Jr, et al. Gene-marking to trace origin of relapse after autologous bone-marrow transplantation. Lancet. 1993;341:85–6.
54. Hacein-Bey-Abina S, Von Kalle C, Schmidt M, McCormack MP, Wulffraat N, et al. LMO2-associated clonal T cell proliferation in two patients after gene therapy for SCID-X1. Science. 2003;302:415–9.
55. Gomez CE, Najera JL, Krupa M, Esteban M. The poxvirus vectors MVA and NYVAC as gene delivery systems for vaccination against infectious diseases and cancer. Curr Gene Ther. 2008;8:97–120.
56. Berges BK, Wolfe JH, Fraser NW. Transduction of brain by herpes simplex virus vectors. Mol Ther. 2007;15:20–9.
57. Stone D. Novel viral vector systems for gene therapy. Viruses. 2010;2010(2):1002–7.
58. Manno CS, Pierce GF, Arruda VR, Glader B, Ragni M, et al. Successful transduction of liver in hemophilia by AAV-factor IX and limitations imposed by the host immune response. Nat Med. 2006;12:342–7.
59. Kabanov AV, Kabanov VA. DNA complexes with polycations for the delivery of genetic material into cells. Bioconjug Chem. 1995;6:7–20.
60. Buwalda SJ, Dijkstra PJ, Feijen J. Poly(ethylene glycol)-poly(L -lactide) star block copolymer hydrogels cross-linked by metal-ligand coordination. J Polym Sci Part A. 2012;50(9):1783–93.
61. Bloomfield V. A condensation of DNA by multivalent cations: considerations on mechanism. Biopolymers. 1991;31:1471–81.
62. Bloomfield VA. DNA condensation. Curr Opin Struct Biol. 1996;6:334–41.
63. Kim JS, Maruyama A, Akaike T, Kim SW. Terplex DNA delivery system as a gene carrier. Pharm. Res. 1998;5:116–21.
64. Toncheva V, Wolfert MA, Dash PR, Oupicky D, Ulbrich K, et al. Novel vectors for gene delivery formed by self-assembly of DNA with poly(L-lysine) grafted with hydrophilic polymers. Biochim Biophys Acta Gen Rev. 1998;1380:354–68.
65. Maruyama A, Katoh M, Ishihara T, Akaike T. Comb-type polycations effectively stabilize DNA triplex. Bioconjug Chem. 1997;8:3–6.

66. Katayose S, Kataoka K. Water-soluble polyion complex associates of DNA and poly(ethylene glycol) poly(L-lysine) block copolymer. Bioconjug Chem. 1997;8:702–7.
67. Kim JS, Maruyama A, Akaike T, Kim SW. In vitro gene expression on smooth muscle cells using a terplex delivery system. J Control Release. 1997;47:51–9.
68. Azzam T, Eliyahu H, Shapira L, Linial M, Barenholz Y, Domb AJ. Polysaccharide-oligoamine based conjugates for gene delivery. J Med Chem. 2002;45:1817–24.
69. Janes KA, Calvo P, Alonso MJ. Polysaccharide colloidal particles as delivery systems for macromolecules. Adv. Drug Del. Rev. 2001;47:83–97.
70. Liu Z, Jiao Y, Wang Y, Zhou C, Zhang Z. Polysaccharides-based nanoparticles as drug delivery systems. Adv Drug Del Rev. 2008;60:1650–62.
71. Morris GA, Samil M, Ernest KS, Adams HG. Polysaccharide drug delivery systems based on pectin and chitosan. Biotech Genet Eng Rev. 2010;27(1):257–84.
72. Benjaminsen RV, Mattebjerg MA, Henriksen JR, Moghimi SM, Andersen TL. The possible "proton sponge" effect of Polyethylenimine (PEI) does not include change in lysosomal pH. Mol Ther. 2013;21(1):149–57.
73. Luu QP, Shin JY, Kim YK, Islam MA, Kang SK, Cho MH, Cho CS. High gene transfer by the osmotic polysorbitol-mediated transporter through the selective caveolae endocytic pathway. Mol Pharm. 2012;9(80):2206–18.
74. Doss CG, Debottam S, Debajyoti C. Glutathione-responsive nano-transporter-mediated siRNA delivery: silencing the mRNA expression of Ras. Protoplasma. 2013;250(3):787–92.
75. Hong R, Han G, Fernandez JM, Kim BJ, Forbes NS, Rotello VM. Glutathione-mediated delivery and release using monolayer protected nanoparticle carriers. J Amer Chem Soc. 2006;128(4):1078–9.
76. Oupicky D, Li J. Bio-reducible polycations in nucleic acid delivery: past, present, future trends. Macromol Biosci. 2014;14(7):908–22.
77. Wen HY, Dong HQ, Xie WJ, Li YY, Wang K, Pauletti GM, Shi DL. Rapidly disassembling nano-micelles with disulfide-linked PEG shells for glutathione-mediated intracellular drug delivery. Chem. Commun.

(Camb). 2011;47(12):3550–2.

78. Chakravarthi S, Jessop CE, Bulleid NJ. The role of glutathione in disulphide bond formation and endoplasmic-reticulum-generated oxidative stress. EMBO Rep. 2006;7(3):271–5.
79. Bello Roufai M, Midoux P. Histidylated poly-lysine as DNA vector: elevation of the imidazole protonation and reduced cellular uptake without change in the polyfection efficiency of serum stabilized negative polyplexes. Bioconjug Chem. 2001;12(1):92–9.
80. Pack DW, Putnam D, Langer R. Design of imidazole-containing endo-somolytic biopolymers for gene delivery. Biotechnol Bioeng. 2000;67(2):217–23.
81. Pires LR, Oliveira H, Barrias CC, Sampaio P, Pereira AJ, Maiato H, Pego AP. Imidazole-grafted chitosan-mediated gene delivery: *in vitro* study on transfection, intracellular trafficking and degradation. Nano-medicine (Lond). 2011;6(9):1499–512.
82. Yang Y, Xu Z, Chen S, Gao Y, Gu W, Chen L, Li Y. Histidylated cationic polyorganophosphazene/DNA self-assembled nanoparticles for gene delivery. Int J Pharm. 2008;353(1–2):277–82.
83. Zhang X, Duan Y, Wang D, Bian F. Preparation of arginine modified PEI-conjugated chitosan copolymer for DNA delivery. Carbohydrate Polymer. 2015;122:53–9.
84. Ahn CH, Chae SY, Bae YH, Kim SW. Biodegradable poly(ethylenimine) for plasmid DNA delivery. J. Control. Rel. 2002;80:273–8.
85. Wang DA, Narang AS, Kotb M, Gaber AO, Miller DD, Kim SW, Mahato RI. Novel branched poly(Ethylenimine)-cholesterol water-soluble lipopolymers for gene delivery. Biomacromolecules. 2002;3:1197–202.
86. Furgeson DY, Chan WS, Yockman JW, Kim SW. Modified linear polyethylenimine-cholesterol conjugates for DNA complexation. Bioconjug Chem. 2003;14:840–5.
87. Furgason DY, Yockman JW, Janat MM, Kim SW. Tumor efficacy and biodistribution of linear polyethylenimine-cholesterol/DNA complexes. Molecular Ther. 2004;9:837–42.
88. Kim YH, Park JH, Lee M, Park TG, Kim SW. Polyethylenimine with acid-labile linkages as a biodegradable gene carrier. J. Control. Rel. 2005;103:209–13.

89. Christensen LV, Chang C-W, Kim WJ, Kim SW. Reducible poly(amido ethylenimine)s designed for triggered intracellular gene delivery. Bioconjug Chem. 2006;17:1233–40.

90. Christensen LV, Chang C-W, Yockman J, Conners WR, Jackson RH, Zhong Z, Feijen J, Bull DA, Kim SW. Reducible Poly(amido ethylenediamine) for Hypoxia-Inducible VEGF Delivery. J. Control. Rel. 2007;118:254–61.

91. Jeong JH, Christensen LV, Yockman JW, Zhong Z, Engbersen JFJ, Kim WJ, Feijen J, Kim SW. Reducible poly(amido ethylenimine) direct to enhance RNA interference. Biomaterials. 2007;28:1912–7.

92. Jeong JH, Kim SH, Christensen LV, Feijen J, Kim SW. Reducible poly(amido ethylenimine)-based gene delivey system for improved nucleus trafficking of plasmid DNA. Bioconjug Chem. 2010;1(2):296–301.

93. Zhang X, Oulad-Abdelghani M, Zelkin AN, Wang Y, Haîkel Y, Mainard D, Voegel JC, Caruso F, Benkirane-Jessel N. Poly(L-lysine) nanostructured particles for gene delivery and hormone stimulation. Biomaterials. 2010;31(7):1699–706.

94. Kwoh DY, Coffin CC, Lollo CP, Jovenal J, Banaszczyk MG, Mullen P, Phillips A, Amini A, Fabrycki J, Bartholomew RM, Brostoff SW, Carlo DJ. Stabilization of poly-L-lysine/DNA polyplexes for in vivo gene delivery to the liver. Biochim. Biophys. Acta. 1999;1444:171–90.

95. Choi JS, Joo DK, Kim CH, Kim K, Park JS. Synthesis of a Barbell-like triblock copolymer, poly(L-lysine) dendrimer-*block*-poly(ethylene-glycol)-*block*-poly(L-lysine) dendrimer, and its self-assembly with plasmid DNA. J. Am. Chem. Soc. 2000;122:474–80.

96. Trubetskoy VS, Torchilin VP, Kennel SJ, Huang L. Use of N-terminal modified poly(L-lysine)-antibody conjugate as a carrier for targeted gene delivery in mouse lung endothelial cells. Bioconjug Chem. 1992;3(4):323–7.

97. Bikram M, Cheol-Hee Ahn C-H, Chae S-Y, Lee M, Yockman JW, Kim SW. Biodegradable poly(ethylene glycol)-co-poly(l-lysine)-g-histidine multi-block copolymers for non-viral gene delivery. Macromolecules. 2004;37(5):1903–16.

98. Martinez-Fong D, Mullersman JE, Purchio AF, Armendariz-Borunda J, Martinez-Hernandez A. Non-enzymatic glycosylation of poly-L-lysine: a new tool for targeted gene delivery. Hepatology. 1994;20(6):1602–8.

99. Banaszczyk MG, Kwoh DY, Carlo DJ. Poly-L-lysine-graft-PEG-comb-type polycation copolymers for gene delivery. J Macromol Sci Part A. 1999;36(7–8):1061–84.
100. Maruyama A, Ishihara T, Kim J-S, Kim SW, Akaike T. Nanoparticle DNA carrier with poly(L-lysine) grafted polysaccharide copolymer and poly(D,L-lactic acid). Bioconjug Chem. 1997;8:735–40.
101. Kim J-S, Maruyama A, Akaike T, Kim SW. Terplex DNA delivery system as a gene carrier, pharm. Research. 1998;15:116–20.
102. Kim J-S, Kim BI, Maruyama A, Akaike T, Kim SW. A new non-viral DNA delivery vector: the terplex system. J ControlRel. 1998;53:175–80.
103. Choi YH, Liu F, Kim JS, Choi YK, Park JS, Kim SW. Polyethylene glycol-grafted poly-L-lysine as polymeric gene carrier. J Control Release. 1998;54:39–43.
104. Choi YH, Liu F, Park JS, Kim SW. Lactose-poly(ethylene glycol)-grafted poly-L-lysine as hepatoma cell-targeted gene carrier. Bioconjug Chem. 1998;9:708–12.
105. Choi YH, Liu F, Park JS, Kim SW. Characterization of a targeted gene carrier, lactose-polyethylene glycol-grafted poly-L-lysine, and its complex with plasmid DNA. Human Gene Ther. 1999;10:2657–61.
106. Park TG, Jeong JH, Kim SW. Current status of polymeric gene delivery systems. Adv. Drug Del. Rev. 2006;58(4):467–86.
107. van der Meel R, Vehmeijer L, Kok R J, Storm G, van Gaal E V, Ligand-targeted particulate Nano-medicines undergoing clinical evaluation: current status, in Intracellular Delivery III: editors, Prokop a, Weissig V, springer, 2015; Chap.7: 163–200.
108. Bodles-Brakhop AM, Heller R, Draghia-Akli R. Electroporation for the delivery of DNA-based vaccines and Immunotherapeutics: current clinical developments. Molecular Ther. 2009;17(4):585–92.
109. Pulkkanen KJ, Herttuala SY. Gene therapy for malignant glioma: current clinical status. Molecular Ther. 2005;12(4):585–98. https://doi.org/10.1016/j.ymthe.2005.07.357.
110. Rainov N G, A Phase III Clinical Evaluation of Herpes Simplex Virus Type 1 Thymidine Kinase and Ganciclovir Gene Therapy as an Adjuvant to Surgical Resection and Radiation in Adults with

Previously Untreated Glioblastoma Multiforme. Human Gene Ther. 2000; 11(17): 2389–401. Pmid: 11096443. doi.org/https://doi.org/10.1089/104303400750038499

111. Young LS, Searle PF, David Onion D, Mautner V. Viral gene therapy strategies: from basic science to clinical application. J Pathol. 2006;208:299–318. https://doi.org/10.1002/path.1896.
112. Yoshida J, Mizuno IM, Wakabayashi T. Interferon-β gene therapy for cancer: basic research to clinical application. Cancer Sci. 2004;95(11):858–65.
113. Weichselbaum RR, Kufe DW, Hellman S, Rasmussen HS, King CR, Fischer PH, Mauceri HJ. Radiation-induced tumour necrosis factor-α expression: clinical application of transcriptional and physical targeting of gene therapy. Lacet Oncol. 2002;3(11):665–7.
114. 1Amer MH. Gene therapy for cancer: present status and future perspective. Mol Cell Ther. 2014;2:27–32. https://doi.org/10.1186/2052-8426-2-27.
115. Jayant RD, Sosa D, Kaushik A, Atluri V, Vashist A, Tomitaka A, Nair M. Current status of non-viral gene therapy for CNS disorders. Expert Opin Drug Deliv. 2016;13(10):1433–45. https://doi.org/10.1080/17425247.2016.1188802.
116. Ginn S L, Amaya A K, Alexander I E, Edelstein M, Abedi M R, Gene therapy clinical trials worldwide to 2017: An update. J Gene Med. 2018; 20:e3015 (1–16).
117. Nam K, Jung S, Nam J-P, Kim S W, Poly(ethylenimine) conjugated bio-reducible dendrimer for efficient gene delivery. J Control Release, 2015; 220:447–455. PMID: 26551343.

2

Cell Penetrating Peptides, Novel Vectors for Gene Therapy

Rebecca E. Taylor [1] and Maliha Zahid [2]

[1]Mechanical Engineering, Biomedical Engineering and Electrical and Computer Engineering, Carnegie Mellon University, Pittsburgh, PA 15213, USA
[2]Department of Developmental Biology, University of Pittsburgh School of Medicine, Pittsburgh, PA 15201, USA

ABSTRACT

Cell penetrating peptides (CPPs), also known as protein transduction domains (PTDs), first identified ~25 years ago, are small, 6–30 amino acid long, synthetic, or naturally occurring peptides, able to carry variety of cargoes across the cellular membranes in an intact, functional form. Since their initial description and characterization, the field of cell penetrating peptides as vectors has exploded. The cargoes they can deliver range from other small peptides, full-length proteins, nucleic acids including RNA and

Citation: Taylor, R.E.; Zahid, M. Cell Penetrating Peptides, Novel Vectors for Gene Therapy. Pharmaceutics 2020, 12, 225. https://doi.org/10.3390/pharmaceutics12030225.

DNA, liposomes, nanoparticles, and viral particles as well as radioisotopes and other fluorescent probes for imaging purposes. In this review, we will focus briefly on their history, classification system, and mechanism of transduction followed by a summary of the existing literature on use of CPPs as gene delivery vectors either in the form of modified viruses, plasmid DNA, small interfering RNA, oligonucleotides, full-length genes, DNA origami or peptide nucleic acids.

KEYWORDS: cell penetrating peptides; protein transduction domains; gene therapy; small interfering RNA

INTRODUCTION

The plasma membrane of a cell is essential to its identity and survival, but at the same time presents a barrier to intracellular delivery of potentially diagnostic or therapeutic cargoes. Therefore, the development of approaches to deliver functional cargoes, be they peptides, proteins, nucleic acids, or nanoparticles across cell membranes, a process termed protein transduction, has wide-reaching research and clinical implications. The ability of Trans-Activator of Transcription (Tat) protein of the Human Immunodeficiency Virus (HIV) to transduce cultured cells and lead to viral gene expression [1,2] without requiring a receptor was the first example of a protein that naturally employs a portion of itself to achieve cell penetration and lead to intracellular delivery of the entire HIV viral particle. The chemical cross-linking of a full-length Tat protein to multiple different proteins such as horseradish peroxidase, ß-galactosidase, RNase A and domain III of Pseudomonas exotoxin A served to demonstrate the ability of Tat protein to ferry other large cargoes across the cell membrane [3]. Similarly, the homeobox Antennapedia (Antp) transcription factor of *Drosophila melanogaster* was demonstrated to enter nerve cells in a receptor independent manner where it could then regulate neural morphogenesis [4]. Mapping of the domains within Tat and Antp responsible for the observed transduction led to the identification of the first two cell penetrating peptides (CPPs): the 11 amino acid cationic domain of HIV-1 Tat protein (YGRKKRRQRRR) [5] and the 16 amino acid sequence from the third helix of the Antennapedia domain (RQIKIWFQNRRMKWKK) termed Antp or penetratin [6]. Subsequently, the ability of the small part of the full-length Tat protein to deliver cargoes, including other full length proteins and even large multimeric protein complexes across cell membranes in culture and in vivo following systemic delivery in mice [7] was documented, further highlighting the delivery potential of these unique peptides. Since then, the number of peptides, both cell-specific and non-specific, reported as having cell penetrating properties has increased exponentially [8]. This is particularly true for a wide spectrum of cationic peptides that primarily rely on their cationic charge to interact with proteoglycans on the cell surface (see below). There has been intense interest in identifying both new cell-specific CPPs, as well as strategies to make Tat and other non-specific CPPs act in a more cell-specific manner by taking advantage of tissue characteristics, mostly in the context of cancer. With the accompanying interest and explosion in the number of studies, it has become impossible to provide a comprehensive, all-encompassing single review of these unique peptides. Therefore, out of necessity, the authors chose to provide only a very broad overview of classification, mechanism of transduction and highlight applications of CPPs as vectors for gene and nucleic acid delivery only. Readers are referred to an earlier review of CPPs by the authors [8] as well as several other excellent reviews [9,10,11].

TYPES OF CPPS

Non-Cell-Specific CPPs

CPPs are broadly categorized into non-cell-specific and cell-specific peptides with great sequence heterogeneity (Table 1). The non-cell-specific CPPs can be sub-classified into three classes: cationic, hydrophobic, and amphipathic. Tat and Antp are cationic peptides, rich in arginine and lysine, with the longer Antp peptide having a more defined 3D structure. In addition to these naturally occurring CPP sequences, synthetic cationic peptides including homopolymers of arginine [12], lysine [13] and/or the cationic, amino acid ornithine [14] were demonstrated to function as effective transduction peptides. Even histidine, which becomes protonated at low pH, can function as a CPP at pH below 6.0 and has been used for delivery into tumor cells with lower pH [15]. Arginine-based homopolymers ranging from 6 to 12 amino acids function as CPPs with 8–10 amino acid length identified as having the highest transduction ability [16]. Similarly, 8-mer homopolymers of lysine transduce a variety of cell types with similar efficiencies as homopolymers of arginine [17]. There is a definite optimum length for these homopolymers with greater than 12 amino acids showing reduced transduction efficiency. Unlike the acute cellular toxicity elicited by long poly-lysine molecules, short lysine homopolymers (6–12 mers) have no demonstrable cytotoxic effects, even at high concentrations [13]. Thus, it appears that too little or too much cationic charge within a short region negatively affects transduction.

Table 1. Classification of Cell Penetrating Peptides.

CPPs-Non-Tissue Specific	**Peptide Sequence**	**Origin**
Cationic		
Tat [**5**]	GRKKRRQRRRPPQ	HIV Tat Protein
Ant [**6**]	RQIKIWFQNRRMKWKK	Antennapedia homeodomain
8-Arginine [**12**]	RRRRRRRR	n/a
8-Lysine [**13**]	KKKKKKKK	n/a
PTD-5 [**17**]	RRQRRTSKLMKR	Phage display
Hydrophobic		
Transportan [**165**]	GWTLNSAGYLLGKINLKA-LAALAKKIL	Galanin and mastoparan
MAP [**166**]	KLALKLALKALKAALKLA	Galanin and mastoparan
TP10 [**167**]	AGYLLGKINLKALAALAKKIL	
Pep-7 [**168**]	SDLWEMMMVSLACQY	CHL8 peptide
Amphipathic		
Azurin p18 [**169**]	LSTAADMQGVVTDGMASG	Azurin
Azurin p28 [**170**]	LSTAADMQGVVTDGMAS-GLDKDYLKPDD	Azurin
hCT18-32 [**171**]	KFHTFPQTAIGVGAP	Calcitonin
Bac 7 [**172**]	RRIRPRPPRLPRPRPRPLPF-PRPG	Bactenecin
CPPs-Tissue Specific	**Peptide Sequence**	**Origin**
CTP [**31,66**]	APWHLSSQYSRT	Phage display
K5-FGF [**173**]	AAVALLPAVLLALLP	Phage display
HAP-1 [**27**]	SFHQFARATLAS	Phage display
293P-1 [**174**]	SNNNVRPIHIWP	Phage display
Vascular Endothelium [**33**]	SIGYPLP	Phage display

Amphipathic CPPs are chimeric peptides generated by attaching the hydrophobic domain of the CPP to a nuclear localizing signal (NLS) such as the SV40 NLS through a covalent bond [18]. Usually, hydrophobic CPPs are derived from signal peptide sequences. Signal peptides that allow proteins to be secreted from cells can also facilitate entry of the proteins back across the membrane into cells. Examples of hydrophobic transduction peptides identified to date include leader sequences for keratinocyte growth factor and fibroblast growth factor [19], but likely most leader sequences of secreted proteins could potentially function as CPPs.

Interestingly, even certain pathogenic bacteria use CPPs for delivery of bacterial effector proteins into different types of mammalian cells. For example, the pathogenic bacteria *Yersinia enterocolitica* encodes for the anti-inflammatory protein YopM with two alpha helices, $\alpha 1H$ and $\alpha 2H$, in its amino terminus that function as CPPs similar to Antp [20,21,22]. Similarly, Shigella and Salmonella encode for immune effector proteins that can also enter cells efficiently to modulate the immune response [22].

Cell-Specific CPPs

The other major sub-classification is cell-specific CPPs, identified through different screening methods including plasmid, microorganism surface, ribosome, or phage display of large peptide libraries. The advantage of this approach is that a priori knowledge of a binding partner is not necessary. Such cell-specific CPPs circumvent the issues associated with the non-cell specific CPPs, namely non-specific cellular uptake leading to off-target side effects and the need to administer high doses of CPPs to achieve adequate levels in target organs or cell types of interest. Such high doses are necessary in order for a small fraction of CPPs to escape the liver, kidney, and the reticuloendothelial system to reach the target organ of interest. Such an approach would be particularly troublesome if the target is the brain or a poorly vascularized tissue. Thus, developing tissue or cell-specific CPPs would be particularly attractive as it would improve the efficacy of the delivered cargo with less off-target effects while reducing the overall dose needed, which would be particularly beneficial when scaling up from small to larger animal models, and ultimately for human clinical trials.

Another approach to circumventing this issue is by delivering non-tissue specific CPP bearing cargo in a pro-drug fashion that can be activated under certain conditions or specific environments. This is only feasible if a specific cell type expresses a unique enzymatic activity such as viral or disease specific proteases [23]. Another approach is local delivery (e.g., intra-tumoral, intra-articular, intra-muscular, intra-ocular, intra-tracheal, intra-dermal, etc.) to limit the transduction activity of non-specific CPPs to specific sites. This would depend on and be limited to a specific application or situation and would be feasible only if the target cell is located in an accessible site with limited diffusion such as topical delivery for dermatological applications, the eye for ophthalmological applications [24,25,26], joints for arthritis/degenerative conditions [27], and directly into tumors [28].

IDENTIFICATION OF TISSUE SPECIFIC CPPS

Cell-specific CPPs were identified predominantly by using peptide phage display libraries to screen for peptides able to target specific cell types. The concept of phage display was first proposed by Smith in 1985 [29]. Following this initial report, combinatorial peptide libraries of various lengths using different types of phages (M13, T7) have been used successfully to identify peptides able to facilitate internalization of intact phage. Alternatively, plasmid, antibodies, microorganism surface or ribosome displays of peptide libraries have been employed as well. Phage display requires exposing the target cell or tissue of interest to a large, randomized phage library in which one of the envelope proteins used by the phage for internalization has been modified to display linear or cyclic peptides of various lengths and randomized amino acid sequences [30]. The internalized phage can then be isolated, expanded and used in subsequent rounds of screening. Usually 3–5 rounds of screening results in enrichment of a small number of peptides identifiable by DNA sequencing of the recovered phage. This approach requires enriching for a specific, small subset of phage from a very large library. As such, false positives are a concern and have to be discerned from phage that is indeed bound and internalized by the target cell type. One approach to circumvent this problem is to carry out the first cycle in cell culture using relevant cell types in order to reduce the likelihood of false positives. Subsequent cycles can then be carried out in vivo using the enriched pool from the in vitro cycle [30,31,32]. Such an approach has led to the identification of peptides targeting vascular endothelium [33], synovial tissue [27], dendritic cells [34], pancreatic islet cells [35] and cardiac myocytes [31], and has identified NRG (Asparginine-Arginine-Glycine) and RGD (Arginine-Glycine-Aspartic acid) motifs that target phage to tumor vasculature in nude tumor-bearing mice [36].

MECHANISMS OF TRANSDUCTION

Despite intense study of CPPs, the specific pathway(s) involved in facilitating transduction remain elusive. CPPs are short in length, making the use of standard techniques for identifying binding targets on the cell surface more challenging. Additionally, their rapid cell entry, occurring within minutes at physiological conditions, makes analysis difficult. It is likely that a non-cell specific CPP such as Tat that crosses the blood brain barrier will not share a cell entry pathway with a cell-specific CPP. Even for a particular CPP the mechanism of transduction likely varies depending on the specific cargo fused to it, with changes in parameters such as local milieu and pH almost certainly playing a part. This is further complicated by recent data suggesting that the local concentration of a CPP may influence the internalization pathway used [37, 38, 39]. It should be noted that elucidation of the mechanism of transduction is not only of theoretical interest as the loading of cargoes must be achieved in a way as to not interfere with either the binding or cell internalization mechanism of CPPs. For example changing the hydrophobicity, but not the cationic charge, of a guanidine-rich homo-polymer significantly changed its transduction abilities and ability to internalize cargoes [40].

Although the exact transduction mechanism of CPPs remains elusive, extensive work from multiple investigators has shed considerable light. There is evidence both for mechanisms that are non-endocytic/energy independent or endocytic/energy-dependent [41]. The broad range of cells that are readily transducible by non-specific CPPs such as Tat suggests a role for ubiquitously shared cellular structures such as surface binding to plasma membrane phospholipids or, in particular proteoglycans through electrostatic interactions, as a first step towards cell entry [42,43]. Cell lines deficient in heparan sulfate have significantly reduced transduction by cationic peptides [13,44,45]. This reduction suggests that electrostatic interactions on the cell surface, separate from CPP-lipid interactions, contribute to protein transduction. It is likely that transduction is a two-step process, with the first step being electrostatic interaction of non-specific CPPs with anionic elements, such as glycosaminoglycans on the cell surface that draw the CPPs into close proximity to the plasma membrane. Subsequently, cationic CPPs bearing small cargoes likely enter cells via direct translocation, with uptake of larger cargoes mediated by micropinocytosis, a more energy-dependent and slower process [46]. Transduction has been shown to occur at 4 °C and after depletion of the adenosine triphosphate (ATP) pool [13], albeit at a reduced level, suggesting that it is not exclusively an energy-dependent process. Research also suggests that increasing hydrophobic characteristics of a CPP, as in the case of Tat, increases its efficiency as a transporter [47].

CELL PENETRATING PEPTIDES AS GENE DELIVERY VECTORS

Although CPPs have been used as vectors for delivery of drugs [48,49,50,51], other peptides of therapeutic potential [52,53,54,55,56,57,58,59,60], proteins [56,61,62,63,64,65], radioisotopes [66,67,68,69], quantum dots [70,71] and photosensitizers [72,73], this review will focus on use of CPPs as vectors for nucleic acid delivery, be they genes, oligonucleotides, peptide nucleic acid conjugates, small interfering RNA (siRNA) or the newest application with DNA origami. Although there is literature on all of these applications, most interest and work done to date has been with siRNA. The necessary factors to consider in these applications is the platform's ability to successfully conjugate nucleic acids to CPPs, escape from enzymatic degradation in serum, escape the reticuloendothelial compartments, successfully cross cell membranes, escape from endocytic degradation and, in cases of gene delivery, achieve nuclear localization. All of these hurdles have been addressed with creative strategies that we will try to summarize here. Even though CPPs are attractive alternatives to viral gene delivery, we will begin by briefly summarizing the literature on use of CPPs to modify viral vectors for the purpose of gene therapy.

Viral Vectors

Adenovirus is commonly employed as a vector for gene delivery due to its high transduction efficiency, although it is dependent on the presence of coxsackievirus and adenovirus receptors (CAR) for internalization. Chemical conjugation of adenovirus to Tat, octa-arginine or proline rich CPPs, led to 1–2 log-fold higher transduction of CAR-negative cells in vitro than unmodified adenovirus [74,75,76]. Electrostatic formation of complexes between negatively charged adenoviral particles and positively charged CPPs, such as Tat, penetratin, poly-arginine and Pep1, led to 100-fold greater transduction of CAR-negative cells compared to unmodified adenovirus, though again only in vitro [77]. In contrast, oncolytic adenovirus-serotype 5 modified to express an SCG3 promoter/ASH1 enhancer driven E1A gene expression, further modified to express Tat in the viral capsid protein, led to delays in tumor growth and prolonged survival of nude mice harboring subcutaneous human neuroblastoma xenografts [78]. In a very interesting application, brain-derived neurotrophic factor was fused with Tat and another CPP, HA2, and packaged into adeno-associated virus. Upon intranasal delivery to mice subjected to chronic mild stress, there was significant amelioration of depressive symptoms and increased hippocampal protein levels of brain-derived neurotrophic factor [79]. In another application, Tat was displayed on surface of MS2 bacteriophages carrying miRNA-122. Not only were there tumor inhibitory effects seen in multiple hepatocellular cancer cell lines, but also decreased tumor growth in nude mice bearing the hepatocellular cancer xenografts after they received an injection of the modified bacteriophage [80]. More recently, Sun and colleagues generated a recombinant PP7 bacteriophage using a one-plasmid double expression system to carry pre-microRNA-23b and displaying a CPP peptide that was able to inhibit hepatoma cell migration in vitro via down-regulation of liver-intestine cadherin expression [81]. Tobacco mosaic virus, a plant virus, decorated with Tat was able to carry enhanced green fluorescent protein (EGFP) silencing siRNA in vitro and to transduce EGFP expressing highly metastatic hepatocellular carcinoma cell lines in vivo in mice, with treatment for 10-days achieving over 80% EGFP-negative rates in tumors [82].

Plasmid DNA

As early as 2003, investigators realized the potential of delivering plasmid DNA using CPPs [83,84]. As Tat is rich in positively charged arginine and lysine residues, electrostatic interactions between Tat and negatively charged plasmid DNA allows for complexes to be formed between Luciferase or EGFP expressing plasmids and Tat. Though there was uptake in cells in vitro, it was via an endocytic pathway. Additionally, in vivo transfection rates were low with the internalization of complexes hindered by serum albumin [83]. To escape from endosomes after internalization, stearylation of CPPs has been used [85,86]. Stearyl-transportan 10 formed stable nano-complexes with plasmid DNA, was non-cytotoxic, non-immunogenic, and mediated transfer/expression of Luciferase reporter plasmid, though with only local uptake after an intra-dermal or intra-muscular injection [85]. Modifying another CPP, S4(13)-PV, with a 5 histidine residue attached to its N-terminus resulted in successful packaging of plasmid DNA and siRNA in vitro with successful gene silencing of Survivin as compared with the unmodified SV(13)-PV [87]. Poly-lysine homopolymers have been used to condense with plasmids bearing angiotensin type II receptor using calcium chloride. This approach produced negligible cytotoxicity in four different human cell lines, successful tumor targeting, and marked attenuation of lung cancer growth in vivo [88].

Another approach used is one of dual targeting. This was achieved by synthesizing biotinylated forms of a tumor targeting peptide (CREKA) and a homo-polymer of arginine by linking them together through an avidin moiety. This complex could successfully condense with plasmid DNA carrying the p53 gene [89], with tumor regression seen in mice, but again the injections were adjacent to the tumor, bringing into question whether this reflected true targeting. Other homopolymers of arginine were synthesized with tumor targeting peptides [90] or a pro-apoptotic peptide (AVPI) [91] to enhance selective targeting of tumor cells. The latter conjugate was condensed with plasmid DNA carrying p53, to act synergistically to induce apoptosis in a multi-drug resistant breast cancer line, and indeed proved efficacious in vivo [91]. However, it is difficult to understand why these dual targeting complexes would specifically target tumors, as both poly-arginine and poly-lysine are non-specific CPPs. Excited by these reports, we constructed a tandem peptide by adding 6-arginine to the N-terminus of an optimized version of our cardiac targeting peptide (CTP-B). Based on our prior bio-distribution studies [66], this fluorescently labeled peptide was injected intravenously into mice and allowed to circulate for 15 min. However, the conjugate peptide actually reduced cardiac uptake compared to the parent CTP-B peptide, while increasing kidney and liver uptake (Figure 1).

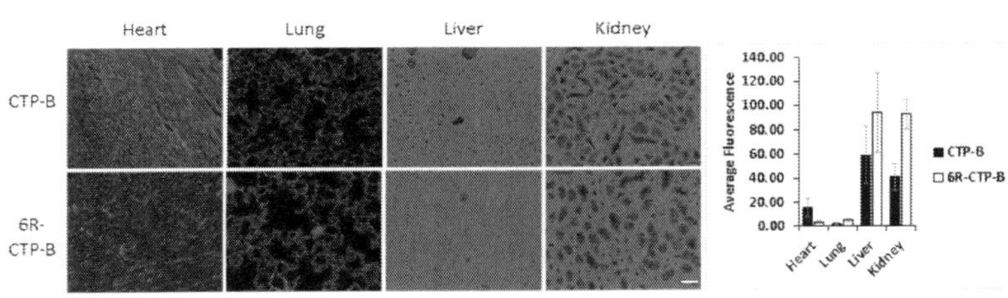

Figure 1. Fluorescent micrographs of heart, lung, liver and kidneys from mice euthanized 15 min after an intravenous injection with fluorescently labeled CTP-B or 6-Arginines-CTP-B. $N = 3$ for each peptide. Scale bar represents 50 μm.

Small Interfering RNA

Although CPPs have been used to deliver siRNA to the heart ex vivo [92,93], or for intradermal delivery [94,95], the vast majority of applications of this approach has been in the context of cancer therapeutics. The hurdles to successful siRNA delivery are similar to delivery of plasmid DNA, namely successful conjugation to CPPs, protection from RNAses in circulation, targeted delivery to organ or cancer of interest, successful uptake, and endosomal escape. Various, creative strategies have been employed to surmount these barriers. Electrostatic charge interaction between cationic CPPs and negatively charged siRNAs have produced self-assembly of CPPs and siRNA into nanoparticles [96,97,98,99]. Escape from endosomal compartments has been achieved by acylation [99], stearylation [100] or histidine modification [101] of the N-terminus of CPPs. Histidine is purported to act as a proton pump within lysosomal compartments leading to swelling, leakiness, and bursting of these compartments releasing the CPP/siRNA conjugate.

In vivo targeting approaches have ranged from dual targeting to dual therapies. Tumor extracellular matrix has a high level of $\alpha v\beta 3/5$ integrin expression that binds cyclic RGD peptide. Tandem peptides combing a CPP with tumor targeting RGD through a disulfide bond complexed with anti-KRAS siRNA formed nanoparticles that significantly delayed tumor growth in a mouse model of pancreatic cancer [102]. In a similar manner, a peptide targeting epidermal growth factor, overexpressed in oral cancer cells, was conjugated to an endosomal disruptive peptide, to successfully deliver siRNA targeting cancerous inhibitor of protein phosphatase 2A (CIP2A), an oncogene [103]. The association with the siRNA was through electrostatic interactions between the peptides and siRNA when mixed in a 60:30:1 ratio with successful silencing of CIP2A in vivo in mice bearing oral cancer tumor xenografts [103]. In another report, a tumor environment responsive nanoparticle was generated, comprised of a polyethylene glycol shell with a tumor pH-responsive polymer core. The core contained the tumor targeting RGD peptide along with homo-polymer of arginine associated through electrostatic interactions with siRNA silencing bromodomain 4. This construct was able to inhibit prostate cancer growth significantly [97].

Differences in tumor microenvironment, like lower pH, increased matrix metalloproteinase 2 expression and increased glutathione in the cytosol, have been employed to deliver siRNA using activatable CPPs to silence Rac1 in order to reduce hepatic metastases in colon cancer [104] and suppress c-MYC gene expression in breast cancer cell lines in vivo [105]. As anti-angiogenic therapy, stearylated poly-arginine was modified with histidine, loaded with anti-VEGF siRNA through electrostatic interaction, and Fausidil, a selective Rho-kinase inhibitor. In vitro and in vivo studies showed strong efficacy for cellular uptake and tumor growth inhibition [101].

Nanoparticles

As liposomes are a subset of nanoparticles (NPs), in this section we will focus on selected literature pertaining to non-lipid-based NPs modified in various ways with CPPs that showed efficacy in vivo. Targeting of NPs has been studied largely in the context of tumor therapies using Tat or homopolymers of arginine of various lengths. As both of these are non-cell specific CPPs, investigators have harnessed tumor environment specific properties, such as lower pH, increased matrix metalloproteinase 2 expression and increased redox potential, to make these non-specific CPPs act in a more tumor-specific manner. In one of the earliest studies, cationic NPs composed of β-cyclodextrin and low-molecular weight polyethylenimine were labeled with both octa-arginine and folic acid to deliver plasmid DNA to folate-receptor positive tumor cells, both in vitro and in vivo [106]. In another application gelatin-silica NPs were modified with different CPPs, including a fusogenic peptide comprised of Tat and influenza hemaglutanin A2, to successfully deliver plasmid DNA with endosomal escape and nuclear targeting properties in vivo [107]. A tumor activatable CPP dual-triggered by lowered pH and matrix metalloproteinase 2 was used to label NPs carrying dual anti-tumor therapies, doxorubicin and siRNA targeting vascular endothelial growth factor. This led to effective shut down of blood vessel formation and to apoptosis within the tumor [108]. In another approach, NPs were modified with photo- and pH-responsive CPPs. The cell penetrating ability of the cationic CPPs was quenched by a pH-sensitive, negatively charged inhibitory peptide that was released in the lower pH microenvironment of the tumor. These NPs, loaded with siRNA, accumulated within tumor cells upon near-infrared light illumination, resulting in increased antitumor efficacy in vivo [109]. Another approach employed was PEGylation of glycosaminoglycan-binding peptides coupled with DNA through electrostatic charge interactions, which formed NPs that targeted bronchial epithelial cell lines with precision cut lung slices in vitro showing that PEGylation rates of >40% were the optimal formulation [110]. Additionally, PEGylated supra-magnetic iron oxide NPs [111,112], Nobel metal NPs [113] or gold particles [114] have also been modified with CPPs to show internalization into tumor or stem cells. These latter studies were, however, all in vitro.

Liposomes

Cationic liposomes (LP) form spontaneous complexes with nucleic acids, which can be sterically stabilized through PEGylation. One study explored the modification of such LPs with CPPs and found that full peptide coverage resulted in less internalization into cells than intermediate coverage, with optimum coverage being cell specific [115]. Additionally, cationic LPs, though effective in vitro, have demonstrated significant cytotoxicity in vivo. This issue can be addressed by using neutral PEGylated LPs modified by CPPs. In one study, neutral PEGylated LPs modified by octa-arginine homo-polymer showed negligible cytotoxicity, enhanced cellular association, and gene silencing capacity in vitro [116]. Harnessing the increased redox milieu of tumors, a Tat-functionalized LP was loaded with paclitaxel, a common cancer chemotherapeutic, with Tat conjugated to a pegylated (PEG) moiety through a cleavable disulfide linker. At the tumor site, PEG was detached by exogenous reducing agent glutathione, resulting in exposure of Tat, with subsequent enhanced tumor uptake and inhibition of proliferation of murine melanoma cell lines, both in vitro and in vivo [117]. In another approach, alanine-alanine-asparagine, a substrate for the endoprotease legumain, was added to the fourth lysine of Tat leading to a branched peptide version. This modified branching Tat peptide was used to label LPs loaded with doxorubicin, leading to decreased non-specific uptake and increased uptake by tumor cells, resulting in increase in anti-tumor activity and decrease in systemic side effects [118]. These data again demonstrate the ability to harness the cell penetrating capabilities of a non-specific CPP such as Tat in a tumor environment specific manner.

DNA Origami

DNA nanostructures, first described by Nadrian Seeman [119], are novel structures formed by the hybridization of multiple single-stranded DNA (ssDNA) oligomers to create a variety of precise structures, with structure, size, and capabilities of DNA nanofabrication growing significantly in the intervening years (Figure 2A,B). In particular, the development of scaffolded DNA origami by Rothemund uses long ssDNA, typically from the M13 bacteriophage, in combination with hundreds of short ssDNAs that act as "staples" to form nanostructures with arbitrary shapes (Figure 2D) [120]. Approaches using only short ssDNA also demonstrated capability of forming two-dimensional [121,122] and three-dimensional structures (Figure 2C,E) [123,124]. Together, these methods treat DNA as a biopolymer, and its information-containing structure is used to drive nanostructure assembly. These DNA origami platforms can be decorated with biomolecules, including peptides, proteins, and functional molecules including fluorophores, aptamers, quantum dots, and gold nanoparticles, allowing them the ability to target cells [125,126]. In one example, the efficiency of intracellular delivery of DNA origami could be increased 22-fold when it was decorated with the iron transport protein transferrin, with efficiency increasing with increasing number of transferrin molecules attached [127]. The versatility of decoration of DNA origami has enabled the co-delivery of chemotherapeutics along with gene therapy. Liu et al. recently demonstrated the delivery of both doxorubicin and the p53 gene using a triangular DNA origami platform decorated with multiple targeting aptamers (Figure 3A) [128].

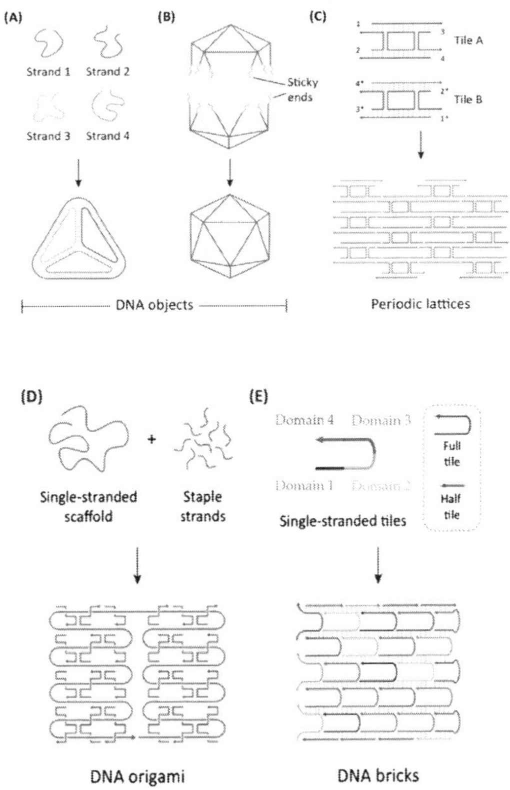

Figure 2. Structural DNA nanotechnology approaches for drug delivery. Common methods for creating DNA nanostructures include (A,B) junction and lattice-based structures like DNA nanocages, (C) periodic lattices, (D) scaffolded DNA origami and (E) scaffold-free assembly of single-stranded tiles (SSTs), otherwise known as DNA "DNA bricks". Panels (A through E) reprinted with permission from Madhanagopal, B. R. et al. [129]. Copyright 2018, Elsevier.

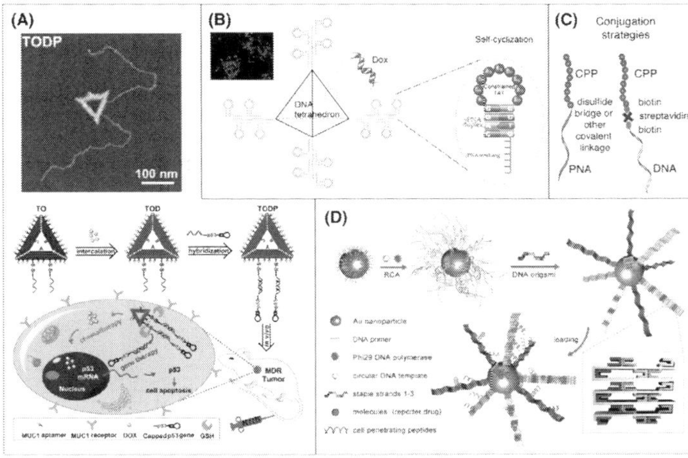

Figure 3. Drug delivery nanostructures can integrate multiple types of functional decoration and for cell-specific and high-efficiency transduction. (A) Decoration with MUC1 aptamers, DOX, and p53 gene cassettes enabled a DNA origami triangular construct to target tumor cells and delivery a combination therapy of chemotherapy agent DOX alongside p53 genes. Reprinted with permission from Liu, J. et al. [128]. Copyright 2018, American Chemical Society. (B) DNA tetrahedron structures were decorated with both DOX and Tat cell targeting peptides for targeted therapy. When Tat peptides were self-cyclized by the hybridization of short flanking γPNAs, the self-cyclized Tat constructs were taken up at 10-fold higher rates. Reprinted with permission from Tan, X. et al. [133]. Copyright 2018, American Chemical Society. (C) While CPP is typically conjugated to peptide nucleic acids (PNA) using disulfide bonds or other covalent methods, CPP is typically ionically bound to PNA using biotin-streptavidin attachment. (D) CPP-decorated DNA origami ribbons and gold nanoparticles were used together to create superstructures with high molecular loading capacity for both cellular imaging and drug delivery. Reprinted with permission from Yan, J. et al. [130]. Copyright 2015, John Wiley and Sons.

To date, only a few DNA nanotechnology studies used CPPs. One study used gold nanoparticles coated with CPP-decorated filamentous DNA origami to create a 3D superstructure with high transduction efficiency [130]. This approach demonstrated the versatility of DNA nanotechnology and its potential for cellular imaging and drug delivery [130]. Another multivalent approach for CPP and structural DNA nanotechnology used both aptamers and CPP to target Ramos cells and then drive the nanopore-like structure into the tumor cells [131]. The combination of aptamers with CPP in this study showed improved cellular uptake and targeting. DNA tetrahedrons have been used as vehicle for delivering siRNAs into cells to silence genes [132], and for the purpose of optimizing delivery, numerous targeting cationic and amphipathic CPPs including Tat, penetratin, and MAP were used. However, likely due to the overexpression of folate receptors in tumor membranes, decoration with folate as a targeting model produced more efficient uptake than platforms decorated with CPPs. In another application, a gammaPNA (γPNA) hairpin was used to create a non-covalently cyclized form of Tat to enhance cell uptake [133]; when DNA tetrahedra were used to increase the multi-valency of this approach, cellular delivery efficiency increased further (Figure 3B).

Like a "Swiss army knife" with many capabilities, DNA nanotechnology has the potential to create multivalent, dynamic and responsive gene delivery platforms capable of unprecedented control over targeting and delivery while integrating multiple distinct approaches using CPPs [134,135]. However, for optimization of this technology, robust and low-cost stabilization methods are critical for protecting DNA origami from low salt denaturation and enzymatic degradation in vivo.

Peptide Nucleic Acids

Peptide nucleic acids (PNAs) are peptide-like polymers with nucleic acid side chains. PNAs were first developed as a synthetic DNA mimic in which the deoxyribose phosphate backbone was replaced with an uncharged and achiral polyamide backbone [136]. With a similar axial rise and identical nucleobase side chains, PNAs are notable for their capacity to bind DNA with extraordinarily high affinity [137,138,139,140,141]. Furthermore, the pseudopeptide backbone of PNA is not a substrate for nucleases or proteases [137], giving them stability and potential to form nano-systems with "trojan horse"-like properties. Without modification, PNAs are not readily taken up by cells in vivo [142], but covalent conjugation to CPPs is straight-forward since both can be made using solid phase peptide synthesis (Figure 2C). In practice however, CPPs and PNAs are often separately synthesized and then conjugated via a disulfide bond [143]. Depending on the PNA nucleobase composition, they are capable of binding to complementary DNA and RNA via both Watson-Crick and Hoogsteen base pairing rules [144]. Important for cell delivery applications like gene therapy, covalent linkage to CPPs does not reduce the biological activity of uncharged PNA [142]. Once inside the cell, PNAs are well-suited for the steric blocking method of gene expression control, since they bind tightly to RNA and DNA while resisting degradation.

The 16-residue CPP penetratin has most commonly been studied in PNA-CPP conjugates [145]. Notably, PNA-penetratin studies showed that the CPP does not interfere with the binding of PNA to target DNA [146], but later studies did find that cellular uptake of PNA-penetratin varies as a function of cell type, temperature and concentration [143]. The first demonstration of targeted delivery of CPP-coupled PNA came in 1998 when penetratin and transportan were covalently coupled to a 21mer PNA [147]. These conjugates showed efficient uptake in Bowes melanoma cells, binding to mRNA and blocking galanin gene expression to modify pain response. Another study demonstrated that PNA coupled to a retro-inverso delivery peptide were rapidly taken up by neurons, and that these antisense conjugates were able to depress the amount of target mRNA in culture [148]. Dragulescu-Andrasi et al. created cell-permeable guanidine-based PNA-oligoarginine analogues [149]. Gamma-position modifications with l-serine [150] and hydrophilic (R)-diethylene glycol "miniPEG" [141] were able to pre-organize these γPNAs into helical structures with higher binding affinity and higher solubility. Modification of PNA may also reduce the required dose of PNA in therapeutic applications, because such modification can decrease the rate of physiological clearance [151]. Endosomal escape remains a challenge for PNA-CPP conjugates post-endocytosis, which further necessitates high dose administration. To increase therapeutic efficacy, CPP-PNAs can be co-administered with chloroquine or Ca^{2+} to facilitate conjugate escape into the cytosol [152].

A strategy combining DNA nanotechnology, PNAs and CPPs to form NPs has shown promising results with integration of modified γPNAs further improving cellular uptake [153]. Specialized NPs using DNA nanotechnology offer unique opportunities for compactly integrating precise ratios and arrangements of distinct functional molecules such as PNA and CPP onto DNA nanostructures. As mentioned in the previous section, one such approach was a DNA tetrahedron-based beacon decorated with Tat flanked by short complementary γPNAs (Figure 3B) [133]. In this study, as the complementary γPNAs hybridized, they formed a hairpin causing the CPP to "self-cyclize" to form into a conformationally constrained and higher activity state. Uptake studies of the beacon system showed a 10-fold increase in uptake of self-cyclized Tat-PNA systems as compared with linear ones.

CRISPR-Cas

CRISPR-Cas (clustered, regularly interspaced, short palindromic repeats-associated system) represents an efficient tool for gene editing and consists of a guide RNA and the Cas9 protein delivered to cells using either plasmid or virus-based vectors. However, using CPPs, direct delivery leads to less cellular toxicity, and fewer off-target mutations [154,155,156] as well as delivery to hard-to-transfect cell lines [157]. Protocols were optimized to use this approach [158].

HURDLES TO CLINICAL APPLICATION

Few clinical trials using CPPs have been published to date, mostly in the context of cancer diagnosis and therapeutics [159,160,161]. Twenty-seven patients with breast cancer received intra-operative activatable fluorescent peptide to improve tumor margin detection and assist with complete tumor resection [162]. Intraoperative imaging of surgical specimens allowed for real-time tumor detection and tumor-free margin resection [162]. Another small trial of 31 patients undergoing coronary artery angioplasty used the cell penetrating lipopeptide protease-activated receptor-1-based pepducin as an anti-platelet agent and this approach was demonstrated to be safe [163]. However, the trial was too small to assess whether this offered superior therapeutic outcomes over standard therapies.

AVI-5126 is a CPP conjugated to a morpholino designed to knock-down the human C-MYC gene. It was the first CPP-conjugated morpholino tested in a safety and efficacy clinical trial to prevent blockage of veins harvested for cardiovascular bypass surgery by inhibiting cell proliferation. The vein was excised and immersed in a solution containing 10 mM AVI-5126 and then re-implanted as a bypass graft. The study, titled "Clinical Study to Assess the Safety and Efficacy of ex-Vivo Vein Graft Exposure to AVI-5126 in Coronary Artery By-Pass Grafting to Reduce Clinical Graft Failure", was terminated early, not due to safety concerns, but due to low likelihood of clinical efficacy. The same CPP and morpholino conjugate were also tested in a restenosis of coronary artery after balloon angioplasty trial, enrolling 30 patients, but it was terminated for unknown reasons with no safety or efficacy data published. It is possible that the advent of drug-eluting stents and their remarkable efficacy has reduced the interest in the clinical development of this conjugate.

Give the small number of clinical trials with CPPs, there clearly are significant issues to overcome before clinical use of CPP-based therapeutics becomes a reality. The first issue is the potential immunogenicity of the CPP containing protein or peptide, especially for chronic or repetitive treatments. It is unlikely that the small CPPs themselves would be immunogenic, but their cargoes can vary in size and the novel antigenic epitopes generated by fusion with the CPP could theoretically pose an issue. The second issue is lack of oral bioavailability of CPPs linked to their cargoes, and hence need for either topical or intravenous administration. One exception may be cyclized arginine rich CPPs with d- or l-naphthylalanine that showed some evidence of oral bioavailability with increased endosomal escape and improved cytosolic delivery [164]. The third issue, as discussed above, is the lack of cell specificity for the cationic and hydrophobic CPPs. The lack of specificity decreases the therapeutic window, increasing the dose used and hence potential for adverse effects. Fourth, since the two major routes of elimination for CPPs are the kidney and liver, analysis of toxicity of a CPP-based therapeutic in these and other tissues is required by the FDA prior to initiation of clinical trials. The testing of CPP-based therapeutics has to be carried out under Good Laboratory Practices (GLP) conditions and pharmacology/toxicity analyses by a GLP certified lab can be prohibitively expensive, with few federal, non-private funding mechanisms available. Lastly, CPP-based therapeutics will likely be expensive to produce and the cost of the therapeutic could be an issue, especially when treating a chronic condition that requires frequent dosing for extended time-periods. As a reference, it is estimated to typically take ~2.6 billion US dollars to bring a therapeutic to market. However, depending on efficacy and targeting, especially for cancers where therapies are time-limited, the cost-benefit ratio could be very favorable. For other conditions, delivering cargoes that last a few weeks or months, like siRNA, could reduce the frequency of administration, and potentially reduce treatment costs.

SUMMARY

The identification of cell penetrating peptides (CPP) or protein transduction domains (PTD) a quarter of a century ago has opened up new avenues to deliver peptides, proteins, nucleic acids, and nanoparticles, including viral particles, more efficiently into cells. Tissue specific CPPs, as well as non-specific CPPs that were engineered to target certain cell types, show great potential for clinical use. Their utility in diagnostic imaging is already being realized in the arena of tumor imaging. Despite the myriad positive results using CPPs in pre-clinical models for diagnosis and/or treatment of disease, the clinical applications of CPPs have been slow to develop due to the many hurdles to implementing new therapeutics.

PATENTS

M.Z. and Paul D. Robbins (University of Minnesota, Minnesota, MN, USA) hold a patent on the use of cardiac targeting peptide as a cardiac vector (Cardiac-specific protein targeting domain, U.S. Patent Serial No. 9249184).

AUTHOR CONTRIBUTIONS

R.E.T. contributed the sections on DNA origami and peptide nucleic acids, as well as the graphical abstract, and provided critical review of the rest of the sections. M.Z. provided the rest of the sections. All authors have read and agreed to the published version of the manuscript.

FUNDING

M.Z. is supported by American Heart Association Scientist Development Award 17SDG33411180, and by a grant awarded under the Pitt Innovation Challenge (PinCh) through the Clinical and Translational Science Institute of the University of Pittsburgh, through National Institutes of Health, UL1TR001857. R.E.T. is supported in part by National Science Foundation grants 1739308 and 1944130 and by the Air Force Office of Scientific Research grant number FA9550-18-1-0199.

ACKNOWLEDGMENTS

The authors would like to thank Raymond A. Frizzell and Bruce Armitage for their insightful reviews of the manuscript.

CONFLICTS OF INTEREST

M.Z. also serves as Chief Scientific Officer and is on the Board of Directors of the startup Vivasc Therapeutics Inc. and holds substantial future equity in it. This company had no role in the design of the study or writing of this review; in the collection, analyses or interpretation of the data presented; in the writing of the manuscript, and in the decision to publish the results.

REFERENCES

1. Frankel, A.D.; Pabo, C.O. Cellular uptake of the tat protein from human immunodeficiency virus. *Cell* 1988, *55*, 1189–1193.
2. Green, M.; Loewenstein, P.M. Autonomous functional domains of chemically synthesized human immunodeficiency virus tat trans-activator protein. *Cell* 1988, *55*, 1179–1188.
3. Fawell, S.; Seery, J.; Daikh, Y.; Moore, C.; Chen, L.L.; Pepinsky, B.; Barsoum, J. Tat-mediated delivery of heterologous proteins into cells. *Proc. Natl. Acad. Sci. USA* 1994, *91*, 664–668.
4. Joliot, A.; Pernelle, C.; Deagostini-Bazin, H.; Prochiantz, A. Antennapedia homeobox peptide regulates neural morphogenesis. *Proc. Natl. Acad. Sci. USA* 1991, *88*, 1864–1868.
5. Green, M.; Ishino, M.; Loewenstein, P.M. Mutational analysis of HIV-1 Tat minimal domain peptides: Identification of trans-dominant mutants that suppress HIV-LTR-driven gene expression. *Cell* 1989, *58*, 215–223.
6. DeRossi, D.; Joliot, A.H.; Chassaing, G.; Prochiantz, A. The third helix of the Antennapedia homeodomain translocates through biological membranes. *J. Biol. Chem.* 1994, *269*, 10444–10450.
7. Schwarze, S.R. In Vivo Protein Transduction: Delivery of a Biologically Active Protein into the Mouse. *Science* 1999, *285*, 1569–1572.
8. Zahid, M.; Robbins, P.D. Cell-Type Specific Penetrating Peptides: Therapeutic Promises and Challenges. *Molecules* 2015, *20*, 13055–13070.
9. Feni, L.; Neundorf, I.; Sunna, A.; Care, A.; Bergquist, P.L. The Current Role of Cell-Penetrating Peptides in Cancer Therapy. *Adv. Exp. Med. Biol.* 2017, *1030*, 279–295.
10. Guidotti, G.; Brambilla, L.; Rossi, D. Cell-Penetrating Peptides: From Basic Research to Clinics. *Trends Pharmacol. Sci.* 2017, *38*, 406–424.
11. Kalafatovic, D.; Giralt, E. Cell-Penetrating Peptides: Design Strategies beyond Primary Structure and Amphipathicity. *Molecules* 2017, *22*, 1929.
12. Wender, P.A.; Mitchell, D.J.; Pattabiraman, K.; Pelkey, E.T.; Steinman, L.; Rothbard, J.B. The design, synthesis, and evaluation of molecules that enable or enhance cellular uptake: Peptoid molecular transporters. *Proc. Natl. Acad. Sci. USA* 2000, *97*, 13003–13008.

13. Mai, J.C.; Shen, H.; Watkins, S.C.; Cheng, T.; Robbins, P.D. Efficiency of Protein Transduction Is Cell Type-dependent and Is Enhanced by Dextran Sulfate. *J. Biol. Chem.* 2002, *277*, 30208–30218.
14. Madani, F.; Lindberg, S.; Langel, Ü.; Futaki, S.; Gräslund, A. Mechanisms of Cellular Uptake of Cell-Penetrating Peptides. *J. Biophys.* 2011, *2011*, 1–10.
15. Fei, L.; Yap, L.-P.; Conti, P.S.; Shen, W.-C.; Zaro, J.L. Tumor targeting of a cell penetrating peptide by fusing with a pH-sensitive histidine-glutamate co-oligopeptide. *Biomaterials* 2014, *35*, 4082–4087.
16. Tunnemann, G.; Ter-Avetisyan, G.; Martin, R.M.; Stockl, M.; Herrmann, A.; Cardoso, M.C. Live-cell analysis of cell penetration ability and toxicity of oligo-arginines. *J. Pept. Sci.* 2008, *14*, 469–476.
17. Mi, Z.; Mai, J.; Lu, X.; Robbins, P.D. Characterization of a Class of Cationic Peptides Able to Facilitate Efficient Protein Transduction in Vitro and in Vivo. *Mol. Ther.* 2000, *2*, 339–347.
18. Xu, Y.; Liang, W.; Qiu, Y.; Cespi, M.; Palmieri, G.F.; Mason, A.J.; Lam, J.K. Incorporation of a Nuclear Localization Signal in pH Responsive LAH4-L1 Peptide Enhances Transfection and Nuclear Uptake of Plasmid DNA. *Mol. Pharm.* 2016, *13*, 3141–3152.
19. Nakayama, F.; Yasuda, T.; Umeda, S.; Asada, M.; Imamura, T.; Meineke, V.; Akashi, M. Fibroblast growth factor-12 (FGF12) translocation into intestinal epithelial cells is dependent on a novel cell-penetrating peptide domain: Involvement of internalization in the in vivo role of exogenous FGF12. *J. Biol. Chem.* 2011, *286*, 25823–25834.
20. Höfling, S.; Scharnert, J.; Cromme, C.; Bertrand, J.; Pap, T.; Schmidt, M.A.; Rüter, C. Manipulation of pro-inflammatory cytokine production by the bacterial cell-penetrating effector protein YopM is independent of its interaction with host cell kinases RSK1 and PRK2. *Virulence* 2014, *5*, 761–771.
21. Gomarasca, M.; Martins, T.F.C.; Greune, L.; Hardwidge, P.R.; Schmidt, M.A.; Rüter, C. Bacterium-Derived Cell-Penetrating Peptides Deliver Gentamicin to Kill Intracellular Pathogens. *Antimicrob. Agents Chemother.* 2017, *61*, e02545-16.
22. Norkowski, S.; Körner, B.; Greune, L.; Stolle, A.-S.; Lubos, M.-L.; Hardwidge, P.R.; Schmidt, M.A.; Rüter, C. Bacterial LPX motif-harboring virulence factors constitute a species-spanning family of cell-penetrating effectors. *Cell. Mol. Life Sci.* 2017, *75*, 2273–2289.

23. Braun, G.B.; Sugahara, K.N.; Yu, O.M.; Kotamraju, V.R.; Mölder, T.; Lowy, A.M.; Ruoslahti, E.; Teesalu, T. Urokinase-controlled tumor penetrating peptide. *J. Control. Release* 2016, *232*, 188–195.

24. De Cogan, F.; Hill, L.J.; Lynch, A.; Morgan-Warren, P.J.; Lechner, J.; Berwick, M.R.; Peacock, A.; Chen, M.; Scott, R.; Xu, H.; et al. Topical Delivery of Anti-VEGF Drugs to the Ocular Posterior Segment Using Cell-Penetrating Peptides. *Investig. Opthalmol. Vis. Sci.* 2017, *58*, 2578.

25. Tai, L.; Liu, C.; Jiang, K.; Chen, X.; Wei, G.; Lu, W.; Pan, W. Noninvasive delivery of oligonucleotide by penetratin-modified polyplexes to inhibit protein expression of intraocular tumor. *Nanomed. Nanotechnol. Biol. Med.* 2017, *13*, 2091–2100.

26. Ildefonso, C.J.; Jaime, H.; Brown, E.E.; Iwata, R.L.; Ahmed, C.M.; Massengill, M.T.; Biswal, M.R.; Boye, S.E.; Hauswirth, W.; Ash, J.D.; et al. Targeting the Nrf2 Signaling Pathway in the Retina WIth a Gene-Delivered Secretable and Cell-Penetrating Peptide. *Investig. Opthalmol. Vis. Sci.* 2016, *57*, 372–386.

27. Mi, Z.; Lu, X.; Mai, J.C.; Ng, B.G.; Wang, G.; Lechman, E.R.; Watkins, S.C.; Rabinowich, H.; Robbins, P.D. Identification of a synovial fibroblast-specific protein transduction domain for delivery of apoptotic agents to hyperplastic synovium. *Mol. Ther.* 2003, *8*, 295–305.

28. Mai, J.C.; Mi, Z.; Kim, S.H.; Ng, B.; Robbins, P.D. A proapoptotic peptide for the treatment of solid tumors. *Cancer Res.* 2001, *61*, 7709–7712.

29. Smith, G. Filamentous fusion phage: Novel expression vectors that display cloned antigens on the virion surface. *Science* 1985, *228*, 1315–1317.

30. Zahid, M.; Robbins, P.D. Identification and Characterization of Tissue-Specific Protein Transduction Domains Using Peptide Phage Display. *Adv. Struct. Saf. Stud.* 2010, *683*, 277–289.

31. Zahid, M.; Phillips, B.E.; Albers, S.M.; Giannoukakis, N.; Watkins, S.C.; Robbins, P.D. Identification of a Cardiac Specific Protein Transduction Domain by In Vivo Biopanning Using a M13 Phage Peptide Display Library in Mice. *PLoS ONE* 2010, *5*, e12252.

32. Hyvönen, M.; Laakkonen, P. Identification and Characterization of Homing Peptides Using In Vivo Peptide Phage Display. *Adv. Struct.*

Saf. Stud. 2015, *1324*, 205–222.

33. Nicklin, S.A.; White, S.; Watkins, S.J.; Hawkins, R.E.; Baker, A.H. Selective targeting of gene transfer to vascular endothelial cells by use of peptides isolated by phage display. *Circulation* 2000, *102*, 231–237.
34. Chamarthy, S.P.; Jia, L.; Kovacs, J.R.; Anderson, K.R.; Shen, H.; Firestine, S.M.; Meng, W.S. Gene delivery to dendritic cells facilitated by a tumor necrosis factor alpha-competing peptide. *Mol. Immunol.* 2004, *41*, 741–749.
35. Rehman, K.K.; Bai, X.-Y.; Miao, D.; Goltzman, D.; Karaplis, A.C. Protection of Islets byin SituPeptide-mediated Transduction of the Ikappa Kinase Inhibitor Nemo-binding Domain Peptide. *J. Biol. Chem.* 2003, *278*, 9862–9868.
36. Arap, W.; Pasqualini, R.; Ruoslahti, E. Cancer Treatment by Targeted Drug Delivery to Tumor Vasculature in a Mouse Model. *Science* 1998, *279*, 377–380.
37. Magzoub, M.; Eriksson, L.E.G.; Gräslund, A. Conformational states of the cell-penetrating peptide penetratin when interacting with phospholipid vesicles: Effects of surface charge and peptide concentration. *Biochim. Biophys. Acta* 2002, *1563*, 53–63.
38. Jiao, C.-Y.; Delaroche, D.; Burlina, F.; Alves, I.; Chassaing, G.; Sagan, S. Translocation and Endocytosis for Cell-penetrating Peptide Internalization. *J. Biol. Chem.* 2009, *284*, 33957–33965.
39. Gump,J.M.;June,R.K.;Dowdy,S.F.Revisedroleofglycosaminoglycans in TAT protein transduction domain-mediated cellular transduction. *J. Biol. Chem.* 2009, *285*, 1500–1507.
40. Backlund, C.; Takeuchi, T.; Futaki, S.; Tew, G.N. Relating structure and internalization for ROMP-based protein mimics. *Biochim. Biophys. Acta* 2016, *1858*, 1443–1450.
41. Illien, F.; Rodriguez, N.; Amoura, M.; Joliot, A.; Pallerla, M.; Cribier, S.; Burlina, F.; Sagan, S. Quantitative fluorescence spectroscopy and flow cytometry analyses of cell-penetrating peptides internalization pathways: Optimization, pitfalls, comparison with mass spectrometry quantification. *Sci. Rep.* 2016, *6*, 36938.
42. Åmand, H.L.; Rydberg, H.A.; Fornander, L.H.; Lincoln, P.; Nordén, B.; Esbjörner, E.K. Cell surface binding and uptake of arginine- and lysine-rich penetratin peptides in absence and presence of proteoglycans. *Biochim. Biophys. Acta (BBA) Biomembr.* 2012, *1818*,

2669–2678.

43. Takechi, Y.; Tanaka, H.; Kitayama, H.; Yoshii, H.; Tanaka, M.; Saito, H. Comparative study on the interaction of cell-penetrating polycationic polymers with lipid membranes. *Chem. Phys. Lipids* 2012, *165*, 51–58.
44. Rusnati, M.; Coltrini, D.; Oreste, P.; Zoppetti, G.; Albini, A.; Noonan, D.; d'Adda di Fagagna, F.; Giacca, M.; Presta, M. Interaction of HIV-1 Tat protein with heparin. Role of the backbone structure, sulfation, and size. *J. Biol. Chem.* 1997, *272*, 11313–11320.
45. Tyagi, M.; Rusnati, M.; Presta, M.; Giacca, M. Internalization of HIV-1 Tat Requires Cell Surface Heparan Sulfate Proteoglycans. *J. Biol. Chem.* 2000, *276*, 3254–3261.
46. Ma, D.-X.; Shi, N.-Q.; Qi, X. Distinct transduction modes of arginine-rich cell-penetrating peptides for cargo delivery into tumor cells. *Int. J. Pharm.* 2011, *419*, 200–208.
47. Li, Y.; Rosal, R.V.; Brandt-Rauf, P.W.; Fine, R.L. Correlation between hydrophobic properties and efficiency of carrier-mediated membrane transduction and apoptosis of a p53 C-terminal peptide. *Biochem. Biophys. Res. Commun.* 2002, *298*, 439–449.
48. Nguyen, L.T.; Yang, X.-Z.; Du, X.; Wang, J.-W.; Zhang, R.; Zhao, J.; Wang, F.-J.; Dong, Y.; Li, P.-F. Enhancing tumor-specific intracellular delivering efficiency of cell-penetrating peptide by fusion with a peptide targeting to EGFR. *Amino Acids* 2015, *47*, 997–1006.
49. Yan, C.; Gu, J.; Hou, D.; Jing, H.; Wang, J.; Guo, Y.; Katsumi, H.; Sakane, T.; Yamamoto, A. Improved tumor targetability of Tat-conjugated PAMAM dendrimers as a novel nanosized anti-tumor drug carrier. *Drug Dev. Ind. Pharm.* 2014, *41*, 617–622.
50. De La Torre, B.G.; Hornillos, V.; Luque-Ortega, J.R.; Abengozar, M.A.; Amat-Guerri, F.; Acuña, A.U.; Rivas, L.; Andreu, D. A BODIPY-embedding miltefosine analog linked to cell-penetrating Tat(48-60) peptide favors intracellular delivery and visualization of the antiparasitic drug. *Amino Acids* 2014, *46*, 1047–1058.
51. Zhang, P.; Cheetham, A.G.; Lin, Y.-A.; Cui, H. Self-Assembled Tat Nanofibers as Effective Drug Carrier and Transporter. *ACS Nano* 2013, *7*, 5965–5977.
52. Fu, B.; Long, W.; Zhang, Y.; Zhang, A.; Miao, F.; Shen, Y.; Pan, N.; Gan, G.; Nie, F.; He, Y.; et al. Enhanced antitumor effects of the BRBP1 compound peptide BRBP1-TAT-KLA on human brain metastatic breast

cancer. *Sci. Rep.* 2015, *5*, 8029.

53. Davoudi, Z.; Akbarzadeh, A.; Rahmatiyamchi, M.; Movassaghpour, A.; Alipour, M.; Nejati-Koshki, K.; Sadeghi, Z.; Dariushnejad, H.; Zarghami, N. Molecular target therapy of AKT and NF-kB signaling pathways and multidrug resistance by specific cell penetrating inhibitor peptides in HL-60 cells. *Asian Pac. J. Cancer Prev.* 2014, *15*, 4353–4358.
54. Xie, X.; Kerrigan, J.E.; Minko, T.; Garbuzenko, O.; Lee, K.-C.; Scarborough, A.; Abali, E.E.; Budak-Alpdogan, T.; Johnson-Farley, N.; Banerjee, D.; et al. Antitumor and modeling studies of a penetratin-peptide that targets E2F-1 in small cell lung cancer. *Cancer Biol. Ther.* 2013, *14*, 742–751.
55. Liu, R.; Xi, L.; Luo, D.; Ma, X.; Yang, W.; Xi, Y.; Wang, H.; Qian, M.; Fan, L.; Xia, X.; et al. Enhanced targeted anticancer effects and inhibition of tumor metastasis by the TMTP1 compound peptide TMTP1-TAT-NBD. *J. Control. Release* 2012, *161*, 893–902.
56. Hotchkiss, R.S.; McConnell, K.W.; Bullok, K.; Davis, C.G.; Chang, K.C.; Schwulst, S.J.; Dunne, J.C.; Dietz, G.P.; Bähr, M.; McDunn, J.; et al. TAT-BH4 and TAT-Bcl-xL peptides protect against sepsis-induced lymphocyte apoptosis in vivo. *J. Immunol.* 2006, *176*, 5471–5477.
57. Tan, M.; Lan, K.-H.; Yao, J.; Lu, C.-H.; Sun, M.; Neal, C.L.; Lu, J.; Yu, D. Selective Inhibition of ErbB2-Overexpressing Breast CancerIn vivoby a Novel TAT-Based ErbB2-Targeting Signal Transducers and Activators of Transcription 3–Blocking Peptide. *Cancer Res.* 2006, *66*, 3764–3772.
58. Bidwell, G. Application of thermally responsive polypeptides directed against c-Myc transcriptional function for cancer therapy. *Mol. Cancer Ther.* 2005, *4*, 1076–1085.
59. Perea, S.E.; Puchades, Y.; Mendoza, O.; Vispo, N.S.; Torrens, I.; Santos, A.; Acevedo, B.; Falcón, V.; Reyes, O.; Silva, R.; et al. Antitumor Effect of a Novel Proapoptotic Peptide that Impairs the Phosphorylation by the Protein Kinase 2 (Casein Kinase 2). *Cancer Res.* 2004, *64*, 7127–7129.
60. Wang, R.; Wang, H.Y. Enhancement of antitumor immunity by prolonging antigen presentation on dendritic cells. *Nat. Biotechnol.* 2002, *20*, 149–154.
61. Zhang, T.; Wu, K.; Ding, C.; Sun, K.; Guan, Z.; Wang, X.; Hsieh, J.-T.;

He, D.; Fan, J. Inhibiting bladder tumor growth with a cell penetrating R11 peptide derived from the p53 C-terminus. *Oncotarget* 2015, *6*, 37782–37791.

62. Chu, X.; Wu, B.; Fan, H.; Hou, J.; Hao, J.; Hu, J.; Wang, B.; Liu, G.; Li, C.; Meng, S. PTD-fused p53 as a potential antiviral agent directly suppresses HBV transcription and expression. *Antivir. Res.* 2016, *127*, 41–49.
63. Wu, Y.; Xie, G.; Xu, Y.; Ma, L.; Tong, C.; Fan, D.; Du, F.; Yu, H. PEP-1-MsrA ameliorates inflammation and reduces atherosclerosis in apolipoprotein E deficient mice. *J. Transl. Med.* 2015, *13*, 316.
64. Dietz, G.P.; Valbuena, P.C.; Dietz, B.; Meuer, K.; Mueller, P.; Weishaupt, J.H.; Bahr, M. Application of a blood-brain-barrier-penetrating form of GDNF in a mouse model for Parkinson's disease. *Brain Res.* 2006, *1082*, 61–66.
65. Lee, J.; Jung, E.; Park, J.; Park, D. Transdermal delivery of interferon-gamma (IFN-gamma) mediated by penetratin, a cell-permeable peptide. *Biotechnol. Appl. Biochem.* 2005, *42*, 169–173.
66. Zahid, M.; Feldman, K.S.; Garcia-Borrero, G.; Feinstein, T.N.; Pogodzinski, N.; Xu, X.; Yurko, R.; Czachowski, M.; Wu, Y.; Mason, N.S.; et al. Cardiac Targeting Peptide, a Novel Cardiac Vector: Studies in Bio-Distribution, Imaging Application, and Mechanism of Transduction. *Biomolecules* 2018, *8*, 147.
67. Stangl, S.; Tei, L.; De Rose, F.; Reder, S.; Martinelli, J.; Sievert, W.; Shevtsov, M.; Öllinger, R.; Rad, R.; Schwaiger, M.; et al. Preclinical Evaluation of the Hsp70 Peptide Tracer TPP-PEG24-DFO[89Zr] for Tumor-Specific PET/CT Imaging. *Cancer Res.* 2018, *78*, 6268–6281.
68. Zhao, M.; Yang, W.; Zhang, M.; Li, G.; Wang, S.; Wang, Z.; Ma, X.; Kang, F.; Wang, J. Evaluation of 68Ga-labeled iNGR peptide with tumor-penetrating motif for microPET imaging of CD13-positive tumor xenografts. *Tumor Biol.* 2016, *37*, 12123–12131.
69. Dong, P.; Cai, H.; Chen, L.; Li, Y.; Yuan, C.; Wu, X.; Shen, G.; Zhou, H.; Zhang, W.; Li, L. Biodistribution and evaluation of131I-labeled neuropilin-binding peptide for targeted tumor imaging. *Contrast Media Mol. Imaging* 2016, *11*, 467–474.
70. Falanga, A.; Vitiello, M.T.; Cantisani, M.; Tarallo, R.; Guarnieri, D.; Mignogna, E.; Netti, P.A.; Pedone, C.; Galdiero, M.; Galdiero, S. A peptide derived from herpes simplex virus type 1 glycoprotein H:

Membrane translocation and applications to the delivery of quantum dots. *Nanomed. Nanotechnol. Biol. Med.* 2011, *7*, 925–934.

71. Brunetti, J.; Riolo, G.; Gentile, M.; Bernini, A.; Paccagnini, E.; Falciani, C.; Lozzi, L.; Scali, S.; Depau, L.; Pini, A.; et al. Near-infrared quantum dots labelled with a tumor selective tetrabranched peptide for in vivo imaging. *J. Nanobiotechnol.* 2018, *16*, 21.
72. Avula, U.M.; Yoon, H.K.; Lee, C.H.; Kaur, K.; Ramirez, R.J.; Takemoto, Y.; Ennis, S.R.; Morady, F.; Herron, T.; Berenfeld, O.; et al. Cell-selective arrhythmia ablation for photomodulation of heart rhythm. *Sci. Transl. Med.* 2015, *7*, 311ra172.
73. Bourré, L.; Giuntini, F.; Eggleston, I.; Mosse, C.A.; MacRobert, A.J.; Wilson, M. Effective photoinactivation of Gram-positive and Gram-negative bacterial strains using an HIV-1 Tat peptide–porphyrin conjugate. *Photochem. Photobiol. Sci.* 2010, *9*, 1613–1620.
74. Eto, Y.; Yoshioka, Y.; Asavatanabodee, R.; Kida, S.; Maeda, M.; Mukai, Y.; Mizuguchi, H.; Kawasaki, K.; Okada, N.; Nakagawa, S. Transduction of adenovirus vectors modified with cell-penetrating peptides. *Peptides* 2009, *30*, 1548–1552.
75. Yu, D.; Jin, C.; Ramachandran, M.; Xu, J.; Nilsson, B.; Korsgren, O.; Le Blanc, K.; Uhrbom, L.; Forsberg-Nilsson, K.; Westermark, B.; et al. Adenovirus Serotype 5 Vectors with Tat-PTD Modified Hexon and Serotype 35 Fiber Show Greatly Enhanced Transduction Capacity of Primary Cell Cultures. *PLoS ONE* 2013, *8*, e54952.
76. Kida, S.; Eto, Y.; Yoshioka, Y.; Nakagawa, S.; Kawasaki, K.; Maeda, M. Evaluation of synthetic cell-penetrating peptides, Pro-rich peptide and octaargine derivatives, as adenovirus vector carrier. *Protein Pept. Lett.* 2010, *17*, 164–167.
77. Nigatu, A.S.; Vupputuri, S.; Flynn, N.; Neely, B.J.; Ramsey, J.D. Evaluation of Cell-Penetrating Peptide/Adenovirus Particles for Transduction of CAR-Negative Cells. *J. Pharm. Sci.* 2013, *102*, 1981–1993.
78. Jin, C.; Yu, D.; Cancer, M.; Nilsson, B.; Leja, J.; Essand, M. Tat-PTD-modified oncolytic adenovirus driven by the SCG3 promoter and ASH1 enhancer for neuroblastoma therapy. *Hum. Gene Ther.* 2013, *24*, 766–775.
79. Ma, X.-C.; Liu, P.; Zhang, X.-L.; Jiang, W.-H.; Jia, M.; Wang, C.-X.; Dong, Y.-Y.; Dang, Y.-H.; Gao, C.-G. Intranasal Delivery of

Recombinant AAV Containing BDNF Fused with HA2TAT: A Potential Promising Therapy Strategy for Major Depressive Disorder. *Sci. Rep.* 2016, *6*, 22404.

80. Wang, G.; Jia, T.; Xu, X.; Chang, L.; Zhang, R.; Fu, Y.; Li, Y.; Yang, X.; Zhang, K.; Lin, G.; et al. Novel miR-122 delivery system based on MS2 virus like particle surface displaying cell-penetrating peptide TAT for hepatocellular carcinoma. *Oncotarget* 2016, *7*, 59402–59416.
81. Sun, Y.; Sun, Y.; Zhao, R. Establishment of MicroRNA delivery system by PP7 bacteriophage-like particles carrying cell-penetrating peptide. *J. Biosci. Bioeng.* 2017, *124*, 242–249.
82. Tian, Y.; Zhou, M.; Shi, H.; Gao, S.; Xie, G.; Zhu, M.; Wu, M.; Chen, J.; Niu, Z.-W. Integration of Cell-Penetrating Peptides with Rod-like Bionanoparticles: Virus-Inspired Gene-Silencing Technology. *Nano Lett.* 2018, *18*, 5453–5460.
83. Ignatovich, I.A.; Dizhe, E.B.; Pavlotskaya, A.V.; Akifiev, B.N.; Burov, S.V.; Orlov, S.; Perevozchikov, A. Complexes of Plasmid DNA with Basic Domain 47–57 of the HIV-1 Tat Protein Are Transferred to Mammalian Cells by Endocytosis-mediated Pathways. *J. Biol. Chem.* 2003, *278*, 42625–42636.
84. Rudolph, C.; Plank, C.; Lausier, J.; Schillinger, U.; Müller, R.H.; Rosenecker, J. Oligomers of the Arginine-rich Motif of the HIV-1 TAT Protein Are Capable of Transferring Plasmid DNA into Cells. *J. Biol. Chem.* 2003, *278*, 11411–11418.
85. Lehto, T.; Simonson, O.E.; Mäger, I.; Ezzat, K.; Sork, H.; Copolovici, D.; Viola, J.R.; Zaghloul, E.; Lundin, P.; Moreno, P.; et al. A Peptide-based Vector for Efficient Gene Transfer in Vitro and in Vivo. *Mol. Ther.* 2011, *19*, 1457–1467.
86. Zhang, W.; Song, J.; Liang, R.; Zheng, X.; Chen, J.; Li, G.; Zhang, B.; Wang, K.; Yan, X.; Wang, R. Stearylated antimicrobial peptide [D]-K6L9 with cell penetrating property for efficient gene transfer. *Peptides* 2013, *46*, 33–39.
87. Cardoso, A.M.; Trabulo, S.; Cardoso, A.L.; Maia, S.; Gomes, P.A.C.; Jurado, A.S.; De Lima, M.C.P. Comparison of the Efficiency of Complexes Based on S413-PV Cell-Penetrating Peptides in Plasmid DNA and siRNA Delivery. *Mol. Pharm.* 2013, *10*, 2653–2666.
88. Alhakamy, N.A.; Ishiguro, S.; Uppalapati, D.; Berkland, C.; Tamura, M. AT2R Gene Delivered by Condensed Polylysine Complexes

Attenuates Lewis Lung Carcinoma after Intravenous Injection or Intratracheal Spray. *Mol. Cancer Ther.* 2015, *15*, 209–218.

89. Qu, W.; Chen, W.-H.; Kuang, Y.; Zeng, X.; Cheng, S.-X.; Zhou, X.; Zhuo, R.-X.; Zhang, X.-Z. Avidin–Biotin Interaction Mediated Peptide Assemblies as Efficient Gene Delivery Vectors for Cancer Therapy. *Mol. Pharm.* 2012, *10*, 261–269.
90. Gong, C.; Pan, D.; Qiu, F.; Sun, P.; Zhang, Y.-H. Selective DNA Delivery to Tumor Cells Using an Oligoarginine-LTVSPWY Peptide. *PLoS ONE* 2014, *9*, e110632.
91. Wang, H.; Wang, H.; Liang, J.; Jiang, Y.; Guo, Q.; Peng, H.; Xu, Q.; Huang, Y. Cell-Penetrating Apoptotic Peptide/p53 DNA Nanocomplex as Adjuvant Therapy for Drug-Resistant Breast Cancer. *Mol. Pharm.* 2014, *11*, 3352–3360.
92. Nam, H.Y.; Kim, J.; Kim, S.W.; Bull, D.A. Cell Targeting Peptide Conjugation to siRNA Polyplexes for Effective Gene Silencing in Cardiomyocytes. *Mol. Pharm.* 2012, *9*, 1302–1309.
93. Li, H.; Zheng, X.; Koren, V.; Vashist, Y.K.; Tsui, T.Y. Highly efficient delivery of siRNA to a heart transplant model by a novel cell penetrating peptide-dsRNA binding domain. *Int. J. Pharm.* 2014, *469*, 206–213.
94. Ruan, W.; Zhai, Y.; Yu, K.; Wu, C.; Xu, Y. Coated microneedles mediated intradermal delivery of octaarginine/BRAF siRNA nanocomplexes for anti-melanoma treatment. *Int. J. Pharm.* 2018, *553*, 298–309.
95. Uchida, T.; Kanazawa, T.; Kawai, M.; Takashima, Y.; Okada, H. Therapeutic Effects on Atopic Dermatitis by Anti-RelA Short Interfering RNA Combined with Functional Peptides Tat and AT1002. *J. Pharmacol. Exp. Ther.* 2011, *338*, 443–450.
96. Freire, J.M.; De Figueiredo, I.R.; Valle, J.; Veiga, A.S.; Andreu, D.; Enguita, F.J.; Castanho, M.A.R.B. siRNA-cell-penetrating peptides complexes as a combinatorial therapy against chronic myeloid leukemia using BV173 cell line as model. *J. Control. Release* 2017, *245*, 127–136.
97. Suh, J.S.; Lee, H.J.; Nam, H.; Jo, B.S.; Lee, D.W.; Kim, J.-H.; Lee, J.Y.; Chung, C.P.; Lee, G.; Park, Y.J. Control of cancer stem cell like population by intracellular target identification followed by the treatment with peptide-siRNA complex. *Biochem. Biophys. Res. Commun.* 2017, *491*, 827–833.
98. Tuttolomondo, M.; Casella, C.; Hansen, P.L.; Polo, E.; Herda, L.M.;

Dawson, K.A.; Ditzel, H.J.; Mollenhauer, J. Human DMBT1-Derived Cell-Penetrating Peptides for Intracellular siRNA Delivery. *Mol. Ther. Nucleic Acids* 2017, *8*, 264–276.

99. Morais, C.; Cardoso, A.M.; Cunha, P.; Aguiar, L.; Vale, N.; Lage, E.V.; Pinheiro, M.; Nunes, C.; Gomes, P.A.C.; Reis, S.; et al. Acylation of the S413-PV cell-penetrating peptide as a means of enhancing its capacity to mediate nucleic acid delivery: Relevance of peptide/lipid interactions. *Biochim. Biophys. Acta (BBA) Biomembr.* 2018, *1860*, 2619–2634.
100. Tai, Z.; Wang, X.; Tian, J.; Gao, Y.; Zhang, L.; Yao, C.; Wu, X.; Zhang, W.; Zhu, Q.; Gao, S. Biodegradable Stearylated Peptide with Internal Disulfide Bonds for Efficient Delivery of siRNA In Vitro and In Vivo. *Biomacromolecules* 2015, *16*, 1119–1130.
101. Liu, Y.; Wu, X.; Gao, Y.; Zhang, J.; Zhang, D.; Gu, S.; Zhu, G.; Liu, G.-L.; Li, X.-Y. Aptamer-functionalized peptide H3CR5C as a novel nanovehicle for codelivery of fasudil and miRNA-195 targeting hepatocellular carcinoma. *Int. J. Nanomed.* 2016, *11*, 3891–3905.
102. Lo, J.H.; Hao, L.; Muzumdar, M.; Raghavan, S.; Kwon, E.J.; Pulver, E.M.; Hsu, F.; Aguirre, A.J.; Wolpin, B.M.; Fuchs, C.S.; et al. iRGD-guided Tumor-penetrating Nanocomplexes for Therapeutic siRNA Delivery to Pancreatic Cancer. *Mol. Cancer Ther.* 2018, *17*, 2377–2388.
103. Alexander-Bryant, A.A.; Zhang, H.; Attaway, C.C.; Pugh, W.; Eggart, L.; Sansevere, R.M.; Andino, L.M.; Dinh, L.; Cantini, L.P.; Jakymiw, A. Dual peptide-mediated targeted delivery of bioactive siRNAs to oral cancer cells in vivo. *Oral Oncol.* 2017, *72*, 123–131.
104. Bao, Y.; Guo, H.; Lu, Y.; Feng, W.; Sun, X.; Tang, C.; Wang, X.; Shen, M. Blocking hepatic metastases of colon cancer cells using an shRNA against Rac1 delivered by activatable cell-penetrating peptide. *Oncotarget* 2016, *7*, 77183–77195.
105. Yang, Y.; Xia, X.; Dong, W.; Wang, H.; Li, L.; Ma, P.; Sheng, W.; Xu, X.; Liu, Y. Acid Sensitive Polymeric Micelles Combining Folate and Bioreducible Conjugate for Specific Intracellular siRNA Delivery. *Macromol. Biosci.* 2016, *16*, 759–773.
106. Jiang, Q.-Y.; Lai, L.-H.; Shen, J.; Wang, Q.-Q.; Xu, F.-J.; Tang, G. Gene delivery to tumor cells by cationic polymeric nanovectors coupled to folic acid and the cell-penetrating peptide octaarginine. *Biomaterials* 2011, *32*, 7253–7262.

107. Ye, S.; Tian, M.-M.; Wang, T.-X.; Ren, L.; Wang, N.; Shen, L.; Shang, T. Synergistic effects of cell-penetrating peptide Tat and fusogenic peptide HA2-enhanced cellular internalization and gene transduction of organosilica nanoparticles. *Nanomed. Nanotechnol. Biol. Med.* 2012, *8*, 833–841.

108. Huang, S.; Shao, K.; Liu, Y.; Kuang, Y.; Li, J.; An, S.; Guo, Y.; Ma, H.; Jiang, C. Tumor-Targeting and Microenvironment-Responsive Smart Nanoparticles for Combination Therapy of Antiangiogenesis and Apoptosis. *ACS Nano* 2013, *7*, 2860–2871.

109. Yang, Y.; Xie, X.; Yang, Y.; Li, Z.; Yu, F.; Gong, W.; Li, Y.; Zhang, H.; Wang, Z.; Mei, X. Polymer Nanoparticles Modified with Photo- and pH-Dual-Responsive Polypeptides for Enhanced and Targeted Cancer Therapy. *Mol. Pharm.* 2016, *13*, 1508–1519.

110. Osman, G.; Rodriguez, J.; Chan, S.Y.; Chisholm, J.; Duncan, G.; Kim, N.; Tatler, A.L.; Shakesheff, K.M.; Hanes, J.; Suk, J.S.; et al. PEGylated enhanced cell penetrating peptide nanoparticles for lung gene therapy. *J. Control. Release* 2018, *285*, 35–45.

111. Dowaidar, M.; Abdelhamid, H.N.; Hällbrink, M.; Freimann, K.; Kurrikoff, K.; Zou, X.; Langel, Ü. Magnetic Nanoparticle Assisted Self-assembly of Cell Penetrating Peptides-Oligonucleotides Complexes for Gene Delivery. *Sci. Rep.* 2017, *7*, 9159.

112. Ben Djemaa, S.; David, S.; Aubert, K.H.; Falanga, A.; Galdiero, S.; Allard-Vannier, E.; Chourpa, I.; Munnier, E. Formulation and in vitro evaluation of a siRNA delivery nanosystem decorated with gH625 peptide for triple negative breast cancer theranosis. *Eur. J. Pharm. Biopharm.* 2018, *131*, 99–108.

113. Peng, L.-H.; Niu, J.; Zhang, C.-Z.; Yu, W.; Wu, J.-H.; Shan, Y.-H.; Wang, X.-R.; Shen, Y.; Mao, Z.; Liang, W.; et al. TAT conjugated cationic noble metal nanoparticles for gene delivery to epidermal stem cells. *Biomaterials* 2014, *35*, 5605–5618.

114. Park, H.; Tsutsumi, H.; Mihara, H. Cell penetration and cell-selective drug delivery using α-helix peptides conjugated with gold nanoparticles. *Biomaterials* 2013, *34*, 4872–4879.

115. Wonder, E.; Simón-Gracia, L.; Scodeller, P.; Majzoub, R.N.; Kotamraju, V.R.; Ewert, K.; Teesalu, T.; Safinya, C. Competition of charge-mediated and specific binding by peptide-tagged cationic liposome–DNA nanoparticles in vitro and in vivo. *Biomaterials* 2018, *166*, 52–63.

116. Fisher, R.K.; Mattern-Schain, S.I.; Best, M.D.; Kirkpatrick, S.S.; Freeman, M.B.; Grandas, O.H.; Mountain, D.J. Improving the efficacy of liposome-mediated vascular gene therapy via lipid surface modifications. *J. Surg. Res.* 2017, *219*, 136–144.
117. Fu, H.; Shi, K.; Hu, G.; Yang, Y.; Kuang, Q.; Lu, L.; Zhang, L.; Chen, W.; Dong, M.; Chen, Y.; et al. Tumor-Targeted Paclitaxel Delivery and Enhanced Penetration Using TAT-Decorated Liposomes Comprising Redox-Responsive Poly(Ethylene Glycol). *J. Pharm. Sci.* 2015, *104*, 1160–1173.
118. Liu, Z.; Xiong, M.; Gong, J.; Zhang, Y.; Bai, N.; Luo, Y.; Li, L.; Wei, Y.; Liu, Y.; Tan, X.; et al. Legumain protease-activated TAT-liposome cargo for targeting tumours and their microenvironment. *Nat. Commun.* 2014, *5*, 4280.
119. Seeman, N.C. Nucleic acid junctions and lattices. *J. Theor. Biol.* 1982, *99*, 237–247.
120. Rothemund, P.W.K. Folding DNA to create nanoscale shapes and patterns. *Nature* 2006, *440*, 297–302.
121. Yin, P.; Hariadi, R.; Sahu, S.; Choi, H.; Park, S.H.; LaBean, T.; Reif, J.H. Programming DNA Tube Circumferences. *Science* 2008, *321*, 824–826.
122. Rothemund, P.W.K.; Ekani-Nkodo, A.; Papadakis, N.; Kumar, A.; Fygenson, D.K.; Winfree, E. Design and Characterization of Programmable DNA Nanotubes. *J. Am. Chem. Soc.* 2004, *126*, 16344–16352.
123. Ke, Y.; Ong, L.L.; Sun, W.; Song, J.; Dong, M.; Shih, W.M.; Yin, P. DNA brick crystals with prescribed depths. *Nat. Chem.* 2014, *6*, 994–1002.
124. Ke, Y.; Ong, L.L.; Shih, W.M.; Yin, P. Three-Dimensional Structures Self-Assembled from DNA Bricks. *Science* 2012, *338*, 1177–1183.
125. Sacca, B.; Niemeyer, C.M. Functionalization of DNA nanostructures with proteins. *Chem. Soc. Rev.* 2011, *40*, 5910.
126. Sacca, B.; Meyer, R.; Erkelenz, M.; Kiko, K.; Arndt, A.; Schroeder, H.; Rabe, K.S.; Niemeyer, C.M. Orthogonal protein decoration of DNA origami. *Angew. Chem. Int. Ed. Engl.* 2010, *49*, 9378–9383.
127. Schaffert, D.H.; Okholm, A.H.; Sørensen, R.S.; Nielsen, J.S.; Tørring, T.; Rosen, C.B.; Kodal, A.L.B.; Mortensen, M.R.; Gothelf, K.V.; Kjems, J. Intracellular Delivery of a Planar DNA Origami Structure

by the Transferrin-Receptor Internalization Pathway. *Small* 2016, *12*, 2634–2640.

128. Liu, J.; Song, L.; Liu, S.; Jiang, Q.; Liu, Q.; Li, N.; Wang, Z.-G.; Ding, B. A DNA-Based Nanocarrier for Efficient Gene Delivery and Combined Cancer Therapy. *Nano Lett.* 2018, *18*, 3328–3334.
129. Madhanagopal, B.R.; Zhang, S.; Demirel, E.; Wady, H.; Chandrasekaran, A.R. DNA Nanocarriers: Programmed to Deliver. *Trends Biochem. Sci.* 2018, *43*, 997–1013.
130. Yan, J.; Hu, C.; Wang, P.; Zhao, B.; Ouyang, X.; Zhou, J.; Liu, R.; He, D.; Fan, C.; Song, S. Growth and origami folding of DNA on nanoparticles for high-efficiency molecular transport in cellular imaging and drug delivery. *Angew. Chem. Int. Ed. Engl.* 2015, *54*, 2431–2435.
131. Guo, X.-L.; Yuan, D.-D.; Song, T.; Li, X.-M. DNA nanopore functionalized with aptamer and cell-penetrating peptide for tumor cell recognition. *Anal. Bioanal. Chem.* 2017, *4*, 713–797.
132. Lee, H.; Lytton-Jean, A.; Chen, Y.; Love, K.T.; Park, A.I.; Karagiannis, E.; Sehgal, A.; Querbes, W.; Zurenko, C.S.; Jayaraman, M.; et al. Molecularly self-assembled nucleic acid nanoparticles for targeted in vivo siRNA delivery. *Nat. Nanotechnol.* 2012, *7*, 389–393.
133. Tan, X.; Bruchez, M.P.; Armitage, B. Closing the Loop: Constraining TAT Peptide by γPNA Hairpin for Enhanced Cellular Delivery of Biomolecules. *Bioconjug. Chem.* 2018, *29*, 2892–2898.
134. Liu, Y.; Kumar, S.; Taylor, R. Mix-and-match nanobiosensor design: Logical and spatial programming of biosensors using self-assembled DNA nanostructures. *Wiley Interdiscip. Rev. Nanomed. Nanobiotechnol.* 2018, *10*, e1518.
135. Ijäs, H.; Nummelin, S.; Shen, B.; Kostiainen, M.A.; Linko, V. Dynamic DNA Origami Devices: From Strand-Displacement Reactions to External-Stimuli Responsive Systems. *Int. J. Mol. Sci.* 2018, *19*, 2114.
136. Nielsen, P.; Egholm, M.; Berg, R.; Buchardt, O. Sequence-selective recognition of DNA by strand displacement with a thymine-substituted polyamide. *Science* 1991, *254*, 1497–1500.
137. Demidov, V.V.; Potaman, V.N.; Frank-Kamenetskil, M.; Egholm, M.; Buchard, O.; Sönnichsen, S.H.; Nlelsen, P.E. Stability of peptide nucleic acids in human serum and cellular extracts. *Biochem. Pharmacol.* 1994, *48*, 1310–1313.
138. Takiya, T.; Seto, Y.; Yasuda, H.; Suzuki, T.; Kawai, K. An empirical

approach for thermal stability (Tm) prediction of PNA/DNA duplexes. *Nucleic Acids Symp. Ser.* 2004, *48*, 131–132.

139. Uhlmann, E.; Peyman, A.; Breipohl, G.; Will, D.W. PNA. Synthetic Polyamide Nucleic Acids with Unusual Binding Properties. *Angew. Chem. Int. Ed. Engl.* 1998, *37*, 2796–2823.

140. Eriksson, M.; Nielsen, P.E. PNA-nucleic acid complexes. Structure, stability and dynamics. *Q. Rev. Biophys.* 1996, *29*, 369–394.

141. Sahu, B.; Sacui, I.; Rapireddy, S.; Zanotti, K.J.; Bahal, R.; Armitage, B.A.; Ly, D.H. Synthesis and characterization of conformationally preorganized, (R)-diethylene glycol-containing gamma-peptide nucleic acids with superior hybridization properties and water solubility. *J. Org. Chem.* 2011, *76*, 5614–5627.

142. Nielsen, P.E. Addressing the challenges of cellular delivery and bioavailability of peptide nucleic acids (PNA). *Q. Rev. Biophys.* 2005, *38*, 345–350.

143. Koppelhus, U.; Awasthi, S.K.; Zachar, V.; Holst, H.U.; Ebbesen, P.; Nielsen, P.E. Cell-Dependent Differential Cellular Uptake of PNA, Peptides, and PNA-Peptide Conjugates. *Antisense Nucleic Acid Drug Dev.* 2002, *12*, 51–63.

144. Egholm, M.; Buchardt, O.; Christensen, L.; Behrens, C.; Freier, S.M.; Driver, D.A.; Berg, R.H.; Kim, S.K.; Norden, B.; Nielsen, P.E. PNA hybridizes to complementary oligonucleotides obeying the Watson–Crick hydrogen-bonding rules. *Nature* 1993, *365*, 566–568.

145. Shiraishi, T.; Nielsen, P.E. Cellular Delivery of Peptide Nucleic Acids (PNAs). *Adv. Struct. Saf. Stud.* 2013, *1050*, 193–205.

146. Quijano, E.; Bahal, R.; Ricciardi, A.; Saltzman, W.M.; Glazer, P.M. Therapeutic Peptide Nucleic Acids: Principles, Limitations, and Opportunities. *Yale J. Biol. Med.* 2017, *90*, 583–598.

147. Pooga, M.; Soomets, U.; Hällbrink, M.; Valkna, A.; Saar, K.; Rezaei, K.; Kahl, U.; Hao, J.-X.; Xu, X.-J.; Wiesenfeld-Hallin, Z.; et al. Cell penetrating PNA constructs regulate galanin receptor levels and modify pain transmission in vivo. *Nat. Biotechnol.* 1998, *16*, 857–861.

148. Aldrian-Herrada, G.; Desarmenien, M.G.; Orcel, H.; Boissin-Agasse, L.; Mery, J.; Brugidou, J.; Rabie, A. A peptide nucleic acid (PNA) is more rapidly internalized in cultured neurons when coupled to a retro-inverso delivery peptide. The antisense activity depresses the target mRNA and protein in magnocellular oxytocin neurons. *Nucleic Acids*

Res. 1998, *26*, 4910–4916.

149. Dragulescu-Andrasi, A.; Rapireddy, S.; He, G.; Bhattacharya, B.; Hyldig-Nielsen, J.J.; Zon, G.; Ly, D.H. Cell-permeable peptide nucleic acid designed to bind to the 5'-untranslated region of E-cadherin transcript induces potent and sequence-specific antisense effects. *J. Am. Chem. Soc.* 2006, *128*, 16104–16112.

150. Dragulescu-Andrasi, A.; Rapireddy, S.; Frezza, B.M.; Gayathri, C.; Gil, R.R.; Ly, D.H. A Simple γ-Backbone Modification Preorganizes Peptide Nucleic Acid into a Helical Structure. *J. Am. Chem. Soc.* 2006, *128*, 10258–10267.

151. Rogers, F.A.; Lin, S.S.; Hegan, D.C.; Krause, D.S.; Glazer, P.M. Targeted Gene Modification of Hematopoietic Progenitor Cells in Mice Following Systemic Administration of a PNA-peptide Conjugate. *Mol. Ther.* 2011, *20*, 109–118.

152. Shiraishi, T.; Pankratova, S.; Nielsen, P.E. Calcium Ions Effectively Enhance the Effect of Antisense Peptide Nucleic Acids Conjugated to Cationic Tat and Oligoarginine Peptides. *Chem. Biol.* 2005, *12*, 923–929.

153. Babar, I.A.; Cheng, C.J.; Booth, C.J.; Liang, X.; Weidhaas, J.; Saltzman, W.M.; Slack, F.J. Nanoparticle-based therapy in an in vivo microRNA-155 (miR-155)-dependent mouse model of lymphoma. *Proc. Natl. Acad. Sci. USA* 2012, *109*, E1695–E1704.

154. Wang, H.-X.; Song, Z.; Lao, Y.-H.; Xu, X.; Gong, J.; Cheng, D.; Chakraborty, S.; Park, J.S.; Li, M.; Huang, D.; et al. Nonviral gene editing via CRISPR/Cas9 delivery by membrane-disruptive and endosomolytic helical polypeptide. *Proc. Natl. Acad. Sci. USA* 2018, *115*, 4903–4908.

155. Lostalé-Seijo, I.; Louzao, I.; Juanes, M.; Montenegro, J. Peptide/Cas9 nanostructures for ribonucleoprotein cell membrane transport and gene edition. *Chem. Sci.* 2017, *8*, 7923–7931.

156. Ramakrishna, S.; Dad, A.-B.K.; Beloor, J.; Gopalappa, R.; Lee, S.-K.; Kim, H.H. Gene disruption by cell-penetrating peptide-mediated delivery of Cas9 protein and guide RNA. *Genome Res.* 2014, *24*, 1020–1027.

157. Del'Guidice, T.; Lepetit-Stoffaes, J.P.; Bordeleau, L.J.; Roberge, J.; Theberge, V.; Lauvaux, C.; Barbeau, X.; Trottier, J.; Dave, V.; Roy, D.C.; et al. Membrane permeabilizing amphiphilic peptide delivers recombinant transcription factor and CRISPR-Cas9/Cpf1

ribonucleoproteins in hard-to-modify cells. *PLoS ONE* 2018, *13*, e0195558.

158. Suresh, B.; Ramakrishna, S.; Kim, H.H. Cell-Penetrating Peptide-Mediated Delivery of Cas9 Protein and Guide RNA for Genome Editing. *Adv. Struct. Saf. Stud.* 2016, *1507*, 81–94.

159. Coriat, R.; Faivre, S.; Mir, O.; Dreyer, C.; Ropert, S.; Bouattour, M.; Desjardins, R.; Goldwasser, F.; Raymond, E. Pharmacokinetics and safety of DTS-108, a human oligopeptide bound to SN-38 with an esterase-sensitive cross-linker in patients with advanced malignancies: A Phase I study. *Int. J. Nanomed.* 2016, *11*, 6207–6216.

160. Lulla, R.R.; Goldman, S.; Yamada, T.; Beattie, C.W.; Bressler, L.; Pacini, M.; Pollack, I.F.; Fisher, P.G.; Packer, R.J.; Dunkel, I.J.; et al. Phase I trial of p28 (NSC745104), a non-HDM2-mediated peptide inhibitor of p53 ubiquitination in pediatric patients with recurrent or progressive central nervous system tumors: A Pediatric Brain Tumor Consortium Study. *Neuro Oncol.* 2016, *18*, 1319–1325.

161. Warso, M.A.; Richards, J.M.; Mehta, D.; Christov, K.; Schaeffer, C.; Bressler, L.R.; Yamada, T.; Majumdar, D.; Kennedy, S.; Beattie, C.W.; et al. A first-in-class, first-in-human, phase I trial of p28, a non-HDM2-mediated peptide inhibitor of p53 ubiquitination in patients with advanced solid tumours. *Br. J. Cancer* 2013, *108*, 1061–1070.

162. Unkart, J.T.; Wapnir, I.L.; E González, J.; Harootunian, A.; Wallace, A.M.; Chen, S.L. Intraoperative Tumor Detection Using a Ratiometric Activatable Fluorescent Peptide: A First-in-Human Phase 1 Study. *Ann. Surg. Oncol.* 2017, *24*, 3167–3173.

163. Gurbel, P.A.; Bliden, K.P.; Turner, S.E.; Tantry, U.S.; Gesheff, M.G.; Barr, T.P.; Covic, L.; Kuliopulos, A. Cell-Penetrating Pepducin Therapy Targeting PAR1 in Subjects with Coronary Artery Disease. *Arter. Thromb. Vasc. Biol.* 2016, *36*, 189–197.

164. Qian, Z.; Martyna, A.; Hard, R.L.; Wang, J.; Appiah-Kubi, G.; Coss, C.; Phelps, M.; Rossman, J.; Pei, D. Discovery and Mechanism of Highly Efficient Cyclic Cell-Penetrating Peptides. *Biochemistry* 2016, *55*, 2601–2612.

165. Pooga, M.; Hallbrink, M.; Zorko, M.; Langle, U. Cell penetration by transportan. *FASEB J.* 1998, *12*, 67–77.

166. Robbins, P.F.; A Kantor, J.; Salgaller, M.; Hand, P.H.; Fernsten, P.D.; Schlom, J. Transduction and expression of the human carcinoembryonic

antigen gene in a murine colon carcinoma cell line. *Cancer Res.* 1991, *51*, 3657–3662.

167. Jones, S.; Farquhar, M.; Martin, A.; Howl, J. Intracellular translocation of the decapeptide carboxyl terminal of Gi3 alpha induces the dual phosphorylation of p42/p44 MAP kinases. *Biochim. Biophys. Acta* 2005, *1745*, 207–214.
168. Gao, C.; Mao, S.; Ditzel, H.J.; Farnaes, L.; Wirsching, P.; Lerner, R.A.; Janda, K.D. A cell-penetrating peptide from a novel pVII-pIX phage-displayed random peptide library. *Bioorg. Med. Chem.* 2002, *10*, 4057–4065.
169. Lim, J.; Kim, J.; Duong, T.; Lee, G.; Yoon, J.; Kim, H.; Ruley, H.E.; El-Rifai, W.; Jo, D. Antitumor activity of cell-permeable p18(INK4c) with enhanced membrane and tissue penetration. *Mol. Ther.* 2012, *20*, 1540–1549.
170. Lim, J.; Duong, T.; Do, N.; Do, P.; Kim, J.; Kim, H.; El-Rifai, W.; Ruley, H.E.; Jo, D. Antitumor activity of cell-permeable RUNX3 protein in gastric cancer cells. *Clin. Cancer Res.* 2013, *19*, 680–690.
171. Findlay, D.M.; Houssami, S.; Lin, H.Y.; Myers, D.E.; Brady, C.L.; Darcy, P.K.; Ikeda, K.; Martin, T.J.; Sexton, P.M. Truncation of the porcine calcitonin receptor cytoplasmic tail inhibits internalization and signal transduction but increases receptor affinity. *Mol. Endocrinol.* 1994, *8*, 1691–1700.
172. Sadler, K.; Eom, K.D.; Yang, J.-L.; Dimitrova, Y.; Tam, J.P. Translocating Proline-Rich Peptides from the Antimicrobial Peptide Bactenecin 7. *Biochemistry* 2002, *41*, 14150–14157.
173. McConnell, S.J.; Thon, V.J.; Spinella, D.G. Isolation of fibroblast growth factor receptor binding sequences using evolved phage display libraries. *Comb. Chem. High Throughput Screen.* 1999, *2*, 155–163.
174. Zong, X.; Jiang, D.-Y.; Li, G.-J.; Cai, J.-L. Construction of keratinocyte growth factor phage active peptides for the promotion of epidermal cell proliferation. *Zhonghua Yi Xue Za Zhi* 2013, 1058–1062.

3

Advances in Viral Vector-Based TRAIL Gene Therapy for Cancer

Lyse A. Norian [1], Britnie R. James [2] and Thomas S. Griffith [1,2]

[1]Department of Urology, University of Iowa Carver College of Medicine, Iowa City, IA 52242, USA
[2]Interdisciplinary Graduate Program in Immunology, University of Iowa Carver College of Medicine, Iowa City, IA 52242, USA

ABSTRACT

Numerous biologic approaches are being investigated as anti-cancer therapies in an attempt to induce tumor regression while circumventing the toxic side effects associated with standard chemo- or radiotherapies. Among these, tumor necrosis factor–related apoptosis-inducing ligand (TRAIL) has shown particular promise in pre-clinical and early clinical trials, due to its preferential ability to induce apoptotic cell death in cancer cells and its minimal toxicity. One limitation of TRAIL use is the fact that many

Citation: Norian, L.A.; James, B.R.; Griffith, T.S. Advances in Viral Vector-Based TRAIL Gene Therapy for Cancer. Cancers 2011, 3, 603-620. https://doi.org/10.3390/cancers3010603.

tumor types display an inherent resistance to TRAIL-induced apoptosis. To circumvent this problem, researchers have explored a number of strategies to optimize TRAIL delivery and to improve its efficacy via co-administration with other anti-cancer agents. In this review, we will focus on TRAIL-based gene therapy approaches for the treatment of malignancies. We will discuss the main viral vectors that are being used for TRAIL gene therapy and the strategies that are currently being attempted to improve the efficacy of TRAIL as an anti-cancer therapeutic.

Keywords: TRAIL; gene therapy; viral vectors; cancer; apoptosis

INTRODUCTION: TRAIL AS AN ANTI-CANCER BIOLOGIC AGENT

Vast amounts of time and money have been spent developing new, more efficacious treatment options for cancer patients. Despite the many significant medical breakthroughs that have been made in recent years, advanced cancers remain difficult to treat, and standard therapies such as chemotherapy and radiotherapy are often associated with substantial toxicities that can limit their use in the clinic. For these reasons, biologic therapies have been increasingly explored as alternative approaches due to their potential to specifically target tumor cells for eradication, while leaving untransformed cells intact. By restricting cell death to only malignant cells, many of the side effects associated with standard therapies can be potentially avoided.

TNF-related apoptosis-inducing ligand (TRAIL) is a biologic therapy that has shown great promise in pre-clinical studies, due to its ability to preferentially induce apoptotic cell death in transformed cells. Apoptosis is a tightly regulated cellular process, and differs from necrosis in that it does not trigger inflammatory host responses. Apoptosis proceeds by two main pathways, referred to as intrinsic, which is controlled by interactions of pro-apoptotic and anti-apoptotic members of the B-cell leukemia/lymphoma 2 (Bcl-2) protein family [1], and extrinsic, which is induced via ligation of apoptosis-inducing "death receptors" on the cell surface [2]. Malignant cells often carry mutations in the proteins that control intrinsic and extrinsic apoptotic signaling pathways, resulting in a decreased susceptibility to cell death. Consequently, many tumor types have shown a resistance to TRAIL-induced cell death, thereby reducing its efficacy as a single agent [2-4]. To circumvent this deficiency, many groups have explored the use of TRAIL-based combinatorial approaches to cancer treatment, with the goal of overriding specific checkpoints that limit the clinical application of TRAIL "standalone" therapies.

Therapeutic TRAIL use has been investigated in many different formats. Initial approaches employed systemic or local administration of recombinant TRAIL protein to ligate the TRAIL receptors on cancer cells and activate the extrinsic cell death machinery [5]. As the cell surface receptors for TRAIL were identified, antibody-based therapies were used to specifically bind the death-inducing TRAIL receptors on cancer cells and trigger apoptosis [5-7]. A number of recent reviews have covered these approaches [4,8], so we will instead focus on gene therapy-based methodologies that stimulate the TRAIL/TRAIL receptor cell death pathway in tumor cells. Within this context, we will summarize the TRAIL receptor family, the intracellular pathways that culminate in apoptosis following ligation of those receptors, the approaches that have been undertaken to enhance tumor cell susceptibility to TRAIL-mediated cell death, and the viral vectors that are being used to induce TRAIL gene expression.

STRUCTURE AND FUNCTION OF TRAIL

Prior to the discovery of TRAIL, two other members of the TNF family—the prototypic TNF and the closely related Fas Ligand (FasL)—had been identified as potent inducers of tumor cell death. Unfortunately, significant toxicity (*i.e.*, induction of normal cell/tissue death) was observed after systemic administration of TNF and FasL (or anti-Fas mAb), which put the utility of these molecules as anti-cancer therapeutics into question. In 1995, Wiley *et al.* reported the cloning of the full length TRAIL cDNA [9]. A few months later, Pitti *et al.* published a report describing the same protein, which they called Apo-2 ligand [10]. These two reports defined TRAIL as a Type II transmembrane protein that can be released as a soluble form following cleavage of the C-terminal extracellular domain. Evaluation of the extracellular domain of TRAIL revealed a moderate degree of homology with other TNF family members, including FasL (28% amino acid identity), TNF (23%), lymphotoxin-α (23%), and lymphotoxin-β (22%). Although the amino acid homology between TRAIL and these proteins was modest, the three-dimensional crystal structure of TRAIL was found to be quite similar to that of TNF and CD40 ligand [11]. TRAIL monomers are made up of two anti-parallel β-pleated sheets that form a β sandwich core; the monomers then interact in a head-to-tail fashion to form a bell-shaped trimer [11]. This is an important hallmark of numerous TNF family members, in that trimerization leads to much greater biologic activity than is observed for either the monomeric or dimeric forms [9]. In humans, TRAIL mRNA is present in a variety of tissues including the spleen, thymus, prostate, ovary, small intestine, colon and placenta [9]. Within the hematopoietic compartment, TRAIL is expressed on activated T lymphocytes, B cells, NK cells, monocytes, dendritic cells, and neutrophils [3,12-16].

Early *in vitro* studies on TRAIL function revealed that it preferentially induced apoptosis in transformed cells, while leaving normal cells and tissues intact [9], and multiple *in vivo* studies have confirmed that the primary targets of TRAIL-induced apoptosis are malignant cells [17-21]. Further investigation into TRAIL function in tumor-free model systems has demonstrated that it can, in fact, induce apoptosis in specific populations of untransformed cells such as activated lymphocytes and influenza-infected epithelial cells of the lung [22-24]. So, depending on the pathophysiological condition, the apoptosis-inducing ability of TRAIL may not be as selective for transformed cells as was originally believed. Despite the fact that noncancerous cells can also be killed by TRAIL under certain circumstances, the development of TRAIL as an anti-cancer agent has continued and the vast majority of work with TRAIL has focused on its tumoricidal activity.

THE TRAIL RECEPTOR FAMILY

TRAIL (either soluble or membrane-bound) can bind to one of several receptors. In humans, four membrane-bound TRAIL receptors have been identified: TRAIL receptor-1 [Death Receptor 4 (DR4)] [25], TRAIL receptor-2 [Death Receptor 5 (DR5)] [26,27], TRAIL receptor-3 [Decoy Receptor 1 (DcR1)/TRAIL receptor without an intracellular domain (TRID)] [26,28,29], and TRAIL receptor-4 [Decoy Receptor 2 (DcR2)/ TRAIL receptor with a truncated death domain (TRUNDD)] [29-32]. The cytoplasmic tails of TRAIL-R1 and -R2 contain functional death domains that transduce apoptotic signals after receptor trimerization [32]. In contrast, neither TRAIL-R3 nor -R4 possess a functional intracellular death domain, and consequently have been referred to as “decoy receptors” because they bind TRAIL with similar affinity as TRAIL-R1 and -R2 but cannot initiate apoptotic signaling. TRAIL-R3 and -R4 can therefore be considered natural competitive inhibitors that, when expressed at the right ratio with TRAIL-R1 and/or -R2, diminish the net apoptotic activity of TRAIL protein [28]. TRAIL can also bind to the soluble TNF receptor superfamily member osteoprotegerin [33], which functions primarily as a decoy receptor for Receptor Activator of NF-κB Ligand (RANKL) [34]. In mice, TRAIL is thought to bind to three receptors: TRAIL-R/DR5, mDcTRAIL-R1, and mDcTRAIL-R2. Of these murine TRAIL receptors, only DR5 has been characterized at the functional level [35,36].

SIGNALING CASCADES THAT REGULATE TRAIL-INDUCED APOPTOSIS

Upon ligation of TRAIL trimers to TRAIL-R1 or -R2 in humans, the extrinsic apoptotic pathway is activated via a series of intracellular changes that culminate in the apoptotic death of the TRAIL receptor-bearing cell (Figure 1). One of the most proximal events of the TRAIL-induced signaling cascade is formation of a multimeric protein structure called the death-inducing signaling complex (DISC). The functional DISC is comprised of several proteins, including the ligated death receptors, Fas-associated death domain protein (FADD), and procaspases 8 and 10 [37,38]. Following DISC formation, the downstream signaling cascade is initiated via autocatalytic cleavage of procaspase 8 into its active form [38]. Active caspase 8 then amplifies the apoptotic signal by cleaving and activating the effector caspases 3, 6, and 7. Caspases 3, 6, and 7 actively promote cell death by cleaving multiple protein targets that are responsible for maintaining cellular integrity, resulting in the cellular hallmarks of apoptosis: plasma membrane blebbing, inter-nucleosomal DNA cleavage, and nuclear shrinking. Active caspase 8 can also cleave the pro-apoptotic Bcl-2 protein Bid, thereby simultaneously triggering the intrinsic apoptotic pathway [39,40]. Transduction of pro-apoptotic signaling from the DISC is tightly regulated, and cellular FADD-like IL-1β-converting enzyme inhibitory protein (cFLIP) is a critical inhibitor at this proximal point in the apoptosis signal transduction pathway [41].

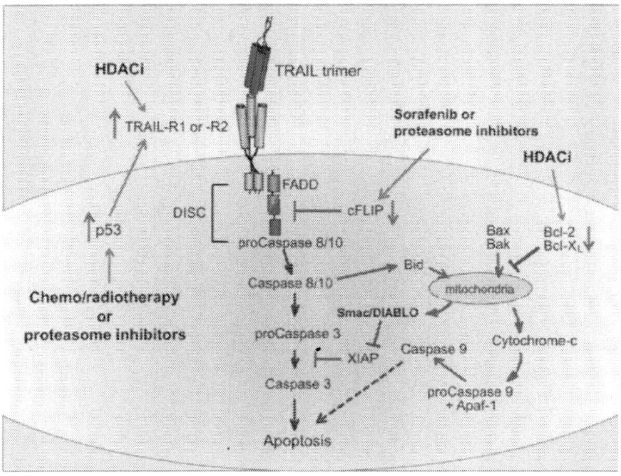

Figure 1. Therapeutic intervention points in the signaling pathways that mediate TRAIL-induced apoptosis. Intracellular pathways that culminate in apoptosis following TRAIL-R1 and -R2 ligation are illustrated. Blue arrows indicate apoptotic signaling pathways. Brown arrows indicate molecular intervention points for therapies that augment TRAIL-induced apoptosis. Red arrows indicate up- or down-regulation of specific molecular targets.

Cells expressing functional TRAIL receptors can be classified as either Type I or Type II, depending on their differential requirements for involvement of the intrinsic pathway in triggering apoptosis. Type I cells undergo apoptosis in response to extrinsic signals that lead to caspase 8 cleavage. Type II cells do not undergo apoptosis upon activation of the extrinsic pathway alone; they also require activation of the intrinsic apoptotic pathway and inactivation of intracellular proteins that inhibit caspase signaling cascades, such as X-linked Inhibitor of Apoptosis Protein (XIAP) [42]. Initiation of the intrinsic apoptotic pathway leads to a loss of mitochondrial membrane potential. This allows cytochrome c to escape from the mitochondria into the cytosol, and also releases Second Mitochondria-derived Activator of Caspases/Direct Inhibitor of Apoptosis Protein Binding Protein with Low Isoelectric Point (Smac/DIABLO), which blocks the function of caspase inhibitors such as XIAP (Figure 1) [43,44]. Cytosolic cytochrome c promotes apoptosis by contributing to the formation of the apoptosome, which also includes ATP and Apoptotic Peptidase-Activating Factor 1 (APAF-1), and results in activation of caspase 9 [45,46]. Thus, co-induction of the intrinsic apoptotic pathway results in greater overall caspase activity, and a stronger pro-apoptotic signal that robustly promotes cell death.

OVERCOMING TUMOR CELL RESISTANCE TO TRAIL-INDUCED APOPTOSIS

The focus on TRAIL as an anti-cancer agent originated from early studies that showed its ability to induce apoptosis was largely restricted to malignant cells [9]. In one report, 32 of 39 human tumor cell lines were sensitive to TRAIL-induced apoptosis *in vitro*, whereas untransformed cells were resistant [19]. This same study demonstrated that repeated intravenous injections of human TRAIL caused no observable toxicity in non-human primates [19]. Since that time, numerous reports have demonstrated the preferential ability of TRAIL to induce apoptosis in tumor cells, both *in vitro* and *in vivo*, while producing minimal toxic side effects in hosts [9,19-21,47-49]. Despite these encouraging results, other reports have shown that many tumor types are resistant to TRAIL (reviewed in [2,4,5]). To circumvent this limitation, a wide array of combinatorial therapies based on TRAIL administration has been examined, and several of these are outlined below.

Chemotherapy and Radiotherapy

Because chemo- and radiotherapy are the standard of care for many malignancies, their ability to synergize with TRAIL administration has been investigated. These therapeutic approaches tend to induce the tumor suppressor p53, which normally accumulates in cells following DNA damage [50]. Increased p53 expression can lead to increased TRAIL-R1 and –R2 expression [51,52], providing a mechanism for heightened TRAIL sensitivity (Figure 1) [53,54]. Therefore, combining TRAIL administration with standard chemo- or radiotherapy may prove more efficacious than administering TRAIL alone, as suggested by promising results in pre-clinical models of glioma, renal cell carcinoma, breast, prostate, and bladder cancer [55-59].

Proteasome Inhibitors

A variety of drugs can increase tumor cell sensitivity to TRAIL-mediated apoptosis, usually by bypassing the intracellular checkpoints that render cells resistant to TRAIL. For example, the proteasome inhibitor bortezomib, approved for the treatment of multiple myeloma, promotes apoptosis by preventing ubiquitin-mediated degradation of pro-apoptotic proteins, increasing p53 and TRAIL-R2 expression, and decreasing cFLIP expression (Figure 1) [60-62]. A recent report examining the combined effects of bortezomib and TRAIL on 15 different squamous cell carcinoma lines found that bortezomib also enhanced TRAIL efficacy through increased recruitment of caspase 8 and FADD into the DISC, and augmentation of the intrinsic apoptotic pathway [63]. For these reasons, the combination of TRAIL and bortezomib has shown promise as a therapeutic for TRAIL-resistant breast, colon and kidney tumors [64].

Sorafenib

Sorafenib is a protein tyrosine kinase inhibitor approved for use in the treatment of renal cell carcinoma, and it has also received attention for its ability to sensitize tumor cells to TRAIL-mediated apoptosis by shifting the intracellular ratio of pro-apoptotic and anti-apoptotic molecules [65]. This can occur through several mechanisms, including sorafenib-triggered proteolytic degradation of cFLIP (Figure 1) [66] and decreased expression of the anti-apoptotic Bcl-2 family member Mcl-1 [67].

Histone Deacetylase Inhibitors

Histone deacetylase inhibitors (HDACi) are a new class of antitumor drugs that epigenetically alter gene expression through increased histone acetylation [68], which in turn leads to the transcription of genes, such as tumor suppressors, that are normally repressed during tumor outgrowth. HDACi are being investigated as stand-alone agents for a variety of cancers, but they are also being explored for their ability to sensitize tumor cells to TRAIL-mediated killing. HDACi upregulate TRAIL-R2 expression on tumor cells, thereby leading to a greater susceptibility to TRAIL-induced apoptosis (Figure 1) [69-73]. In addition, HDAC inhibition can modulate TRAIL-induced apoptosis in tumor cells by other molecular mechanisms, including increased caspase activation and decreased expression of anti-apoptotic molecules such as Bcl-2 [71,74,75].

VIRAL VECTOR-MEDIATED GENE THERAPY

Depending on the disease, gene therapy has become a viable therapeutic alternative for patients over the last decade because this technology can be used to restore proper gene expression in host cells, to express proteins with the intent of stimulating immunity (*i.e.* immunization), and/or to introduce genes that encode cytotoxic proteins. There has also been a great deal of work performed in the gene therapy arena focused on improving the delivery of therapeutic genes by modifying viral vectors to increase their tropism for target cells, or by restricting transgene expression via incorporation of cell/tissue-specific promoters. Regardless of its intended use, one of the most important aspects of viral vector-based gene therapy is that this type of treatment should be capable of treating the disease in question while remaining benign to the host.

Viruses act as natural vehicles for delivering genes as they readily transfer their genetic material to infected host cells [76,77]. For use as therapeutic agents, viral vectors are frequently genetically modified to circumvent productive infection and toxicity, while still allowing for expression of virally-derived proteins in host cells. A number of viral vectors have been described, from both RNA and DNA genome origins, and each has unique advantages in terms of the types of cells that can be infected and the ability to express transgenes either transiently or permanently. Among the most frequently studied viral vectors are adenovirus (Ad) and adeno-associated virus (AAV) [76-78]. Ad- and AAV-based vectors have been used in a number of pre-clinical studies examining the potential of viral vector-mediated TRAIL administration via gene therapy. Here we will discuss the properties that make both Ad and AAV desirable candidates for TRAIL-based gene therapy.

Adenovirus

Discovered in 1953, Ad is a non-enveloped virus with a double stranded 26-45 kb DNA genome [79]. There are 51 identified human Ad serotypes, with types 2 (Ad2) and 5 (Ad5) being the best characterized [76,77]. Ad2 and Ad5 have high tropism for numerous cell types, do not cause extensive disease in humans, and were the first established vectors for gene therapy [80,81]. Because Ad do not integrate into the host genome, the transgenes are transiently expressed [76]. The transient gene expression exhibited with Ad vectors favors gene therapy situations where permanent transgene expression is not required.

Ad vectors can be engineered to remain replication-sufficient, conditionally replication-sufficient, or replication-deficient in the host. First-generation Ad vectors contained a deletion of the early gene 1 (*E1A*), which is activated upon entry into the host cell and vital for transcription and function of the other early genes [82]. Deletion of *E1A* also renders Ad replication-deficient and allows space within the genome for transgene insertion. *E3* deletion has no apparent effect on viral infectivity [83], and in most first-generation Ad5 vectors, both *E1A* and *E3* were deleted to create a replication-deficient vector that could accommodate larger transgene insertions [83]. Many second-generation Ad5 vectors have also had the *E2* region deleted to alleviate host inflammatory responses and potential toxicity [84].

For entry into a cell by receptor-mediated endocytosis, Group C adenoviruses (e.g. Ad2 and Ad5) require interaction between the viral fiber capsid protein with the coxsackievirus and adenovirus receptor (CAR), and the viral penton base binding to α_v integrins [85-87]. Thus, the success of these Ad-based therapies is primarily dictated by CAR recognition [88]. CAR is ubiquitously expressed in most benign epithelial tissues, yet marked variations in CAR levels have been demonstrated using different cancer cell lines of the same tissue origin [89]. Cellular resistance to Ad infection can, therefore, further complicate (and potentially limit) the utility of Ad-mediated gene delivery.

Modifications to the Ad vector can aid in overcoming cellular resistance. A modification to the viral fiber, via insertion of an Arg-Gly-Asp (RGD) integrin-binding motif, allows for CAR-independent binding interactions [86,89]. Fiber-modified Ad vectors encoding TRAIL have been used to induce apoptosis of human pancreatic cancer *in vitro* and *in vivo* [90] and colon cancer *in vitro* [91]. Importantly, the modified Ad-TRAIL-RGD vector also elicited cytotoxic effects in adenovirus-resistant tumors and cancer cells with low to negative CAR expression [92]. Similarly, Ad5 vectors that co-express Group B-derived Ad35 capsid fibers can recognize CD46 instead of CAR, which increases the tropism of the vector for tumor cells [93,94]. An Ad5/35 vector encoding TRAIL exhibited enhanced tumor cell killing in a model of human glioblastoma *in vitro* and *in vivo* [94]. Alternatively, tumor cell treatment with HDACi can result in increased CAR and α_v integrin expression, which enhances adenoviral infection and transgene expression [72].

Site-specific cell targeting for directed transgene expression is another modification that can be made in Ad vectors. For example, targeting Ad-mediated transgene delivery to tumor cells has been achieved through expression of the transgene under the human telomerase reverse transcriptase (hTERT) promoter [95]. Telomerase activity can be used as a marker for tumor cell activity, so expression of TRAIL under the hTERT promoter allows for site-specific expression of TRAIL [95]. Tissue specificity can also be achieved using tissue specific promoters, such as prostate-specific enhancing sequences (PSES). The addition of this promoter was found to limit TRAIL expression to prostate-specific antigen (PSA)- and prostate-specific membrane antigen (PSMA)-expressing cells [96]. Thus, the refinement of Ad vectors for TRAIL-based gene therapy continues, with the goal of increasing TRAIL transgene expression specifically in transformed tumor cells, thereby increasing the potential therapeutic benefits of this approach while simultaneously minimizing toxicity.

We were the first to describe the *in vitro* and *in vivo* tumoricidal activity of Ad-TRAIL [97,98]. These initial studies were designed to only examine Ad-TRAIL-induced tumor cell death, and a number of other investigators have evaluated similar TRAIL-encoding Ad vectors in a variety of tumor models [21,91,99]. Our more recent studies have examined the impact of Ad-TRAIL-induced tumor cell death on the subsequent induction of systemic anti-tumor immunity. Using a murine model of RCC, we demonstrated that combinatorial therapy with Ad-TRAIL and CpG-containing oligonucleotides increased tumor regression and prolonged animal survival [100]. Data from this report also showed that Ad-TRAIL/CpG therapy led to the generation of immunological memory, since mice that went on to clear the primary tumor after treatment were also able to resist a second tumor challenge. These preclinical results suggesting the therapeutic potential of Ad-TRAIL gave us the necessary "proof-of-concept" data to justify initiation of a phase I clinical trial in men with prostate cancer. Our preliminary results from that trial showed that intra-prostatic injection of Ad-TRAIL was well-tolerated in patients, and produced no adverse side effects [5]. In addition, there was evidence of apoptotic death (DNA fragmentation via TUNEL staining), which suggested that TRAIL expressed from the transferred transgene was functional. Unfortunately, this phase I trial was not designed to test therapeutic efficacy. Thus, it will be interesting to continue our evaluation of Ad-TRAIL clinically.

Adeno-Associated Virus

In addition to Ad vectors for gene therapy, AAV vectors have also been engineered and tested as gene transfer vectors. AAV is a small, nonenveloped, single stranded DNA virus [101]. Unlike Ad, only six AAV serotypes have been identified in primates, with AAV-2 being the most extensively studied and used in gene therapy. AAV has not been conclusively shown to cause human disease, and elicit minimal cellular immune responses in the host, with little inflammation and cellular activation. AAV infection is essentially non-productive without the presence of a helper virus, usually Ad. Consequently, the generation of AAV vectors is more difficult than other recombinant viral vectors because AAV vectors require co-transfection with helper protein plasmids, and low viral titers are frequently obtained [78,102].

However, AAV vectors are desirable for, and have been successfully used for, gene therapy due to their other unique features. For example, AAV has high tropism for many proliferating and quiescent cell types, and in human cells AAV is able to site-specifically integrate into chromosome 19. Despite chromosome integration, AAV transgene expression is not life-long in the host; however, expression does have the potential to persist for several years *in vivo* [78,101,102]. AAV can also be found extrachromosomally as an episomal AAV genome, which decreases the risk of random integration (another feature that makes it desirable as a gene therapy vector). Due to the small size of AAV, only transgene inserts smaller than 5 kb can be inserted. Interestingly, AAV has the ability to heterodimerize, which allows genes to be "split" between two vectors, resulting in expression of an intact transgene only after post-transcriptional modifications occur [78]. Modifications can be made to the AAV vector to increase tropism, enhance transduction and to evade the immune response [103]. Also, mosaic vectors consisting of components from multiple AAV can be used to allow for tissue-specific targeting and to broaden cell/tissue tropism. Together, these unique features of AAV have made it a successful vector in pre-clinical research for gene therapy.

The positive features of AAV have made it a popular vector option for TRAIL gene administration. In a model of colorectal cancer, AAV-encoded expression of TRAIL resulted in rapid transgene expression, followed by apoptosis of several tumor cell lines and either complete eradication of tumors or blunted progression of existing tumors [104]. Similar results were seen for an AAV vector encoding soluble TRAIL, using A549 established tumors in nude mice [105]. Mosaic AAV vectors contain a mixture of capsid proteins from multiple AAV capable of binding different cellular receptors allowing for a broader cellular tropism [106]. Mosaic vector, AAV2/5, consists of both serotypes 2, which binds hepran sulfate proteoglycans, and 5, which binds sialic acid. Shi e*t al.* utilized AAV2/5 encoding TRAIL and showed increased tropism for the A549 cell line *in vitro* and *in vivo*, while histology and primary cell culture showed no detectable toxicity to normal tissues [107]. Similar to Ad, tissue- and tumor-specific targeting of the AAV vector can be established with insertion of target sequences [95,108]. Tumor cell-specific AAV vectors have been engineered with TRAIL expression under the control of the telomerase reverse transcriptase (hTERT) promoter [95]. Treatment of human hepatocellular carcinoma cells with AAV-hTERT-TRAIL demonstrated cancer-specific cytoxicity [109,110]. *In* *vivo* experiments exhibited hepatocellular specific tumor cell death and tumor regression when treated with AAV-hTERT-TRAIL [109-111]. Tissue-targeting AAV vectors have also been engineered for site-specific TRAIL expression. Ma *et al.* achieved hepatocellular-specific TRAIL expression by fusing the human insulin signal peptide to TRAIL (AAV-ISN-T) within the vector [112]. Oral administration of AAV-ISN-T to mice bearing subcutaneous SMMC-7721 (human liver cancer) tumors produced long-lasting expression of soluble TRAIL in the liver, which resulted in tumor regression [112].

Other Vector Options

With low cytoxicity in normal tissues and site-specific therapeutic effects, treatment with Ad- or AAV-vector encoded TRAIL in pre-clinical studies has provided strong evidence for future gene therapy of cancer patients. While TRAIL-encoding Ad and AAV have been the most popular vectors so far, other TRAIL-encoding vectors have been investigated. Lentiviral-mediated TRAIL expression was shown to be successful for specific induction of apoptosis in human lung cancer cells *in vitro*, but was not as efficacious *in vivo* [113]. A genetically engineered synthetic biomimetic vector has also been used to co-administer plasmid DNA specific for TRAIL. Transfection of ZR-75-1 breast cancer cells with the biomimetic TRAIL vector demonstrated increased tumor cell killing [114], indicating that this approach warrants additional investigation.

CONCLUSIONS

In summary, the use of TRAIL-based gene therapy for cancer patients holds a great deal of promise. Known limitations of TRAIL, such as resistance to TRAIL-mediated killing, and cancer cell resistance to viral vector infection are being overcome in a variety of ways. Pre-clinical studies in mice have demonstrated the safety and efficacy of viral-based gene therapy using TRAIL, and phase I clinical trial results have demonstrated the safety of this approach. These advances pave the way for future clinical applications of TRAIL gene therapy as a treatment option for a variety of malignancies.

ACKNOWLEDGEMENTS

This work was supported by the National Institutes of Health Grant CA109446.

REFERENCES

1. Wei, M.C.; Zong, W.X.; Cheng, E.H.; Lindsten, T.; Panoutsakopoulou, V.; Ross, A.J.; Roth, K.A.; MacGregor, G.R.; Thompson, C.B.; Korsmeyer, S.J. Proapoptotic Bax and Bak: A requisite gateway to mitochondrial dysfunction and death. *Science* 2001, *292*, 727–730.
2. Testa, U. Trail/trail-r in hematologic malignancies. *J. Cell. Biochem.* 2010, *110*, 21–34.
3. Zamai, L.; Ahmad, M.; Bennett, I.M.; Azzoni, L.; Alnemri, E.S.; Perussia, B. Natural killer (NK) cell-mediated cytotoxicity: Differential use of TRAIL and Fas ligand by immature and mature primary human NK cells. *J. Exp. Med.* 1998, *188*, 2375–2380.
4. Mellier, G.; Huang, S.; Shenoy, K.; Pervaiz, S. Trailing death in cancer. *Mol. Aspects Med.* 2010, *31*, 93–112.
5. Holoch, P.A.; Griffith, T.S. TNF-related apoptosis-inducing ligand (TRAIL): A new path to anti-cancer therapies. *Eur. J. Pharmacol.* 2009, *625*, 63–72.
6. Rosevear, H.M.; Lightfoot, A.J.; Griffith, T.S. Conatumumab, a fully human mAb against death receptor 5 for the treatment of cancer. *Curr. Opin. Investig. Drugs* 2010, *11*, 688–698.
7. Griffith, T.S.; Rauch, C.T.; Smolak, P.J.; Waugh, J.Y.; Boiani, N.; Lynch, D.H.; Smith, C.A.; Goodwin, R.G.; Kubin, M.Z. Functional analysis of TRAIL receptors using monoclonal antibodies. *J. Immunol.* 1999, *162*, 2597–2605.
8. Wiezorek, J.; Holland, P.; Graves, J. Death receptor agonists as a targeted therapy for cancer. *Clin. Cancer Res.* 2010, *16*, 1701–1708.
9. Wiley, S.R.; Schooley, K.; Smolak, P.J.; Din, W.S.; Huang, C.P.; Nicholl, J.K.; Sutherland, G.R.; Smith, T.D.; Rauch, C.; Smith, C.A.; *et al.* Identification and characterization of a new member of the TNF family that induces apoptosis. *Immunity* 1995, *3*, 673–682.
10. Pitti, R.M.; Marsters, S.A.; Ruppert, S.; Donahue, C.J.; Moore, A.; Ashkenazi, A. Induction of apoptosis by Apo-2 ligand, a new member of the tumor necrosis factor cytokine family. *J. Biol. Chem.* 1996, *271*, 12687–12690.
11. Cha, S.S.; Kim, M.S.; Choi, Y.H.; Sung, B.J.; Shin, N.K.; Shin, H.C.; Sung, Y.C.; Oh, B.H. 2.8 A resolution crystal structure of human TRAIL, a cytokine with selective antitumor activity. *Immunity* 1999, *11*, 253–261.

12. Griffith, T.S.; Wiley, S.R.; Kubin, M.Z.; Sedger, L.M.; Maliszewski, C.R.; Fanger, N.A. Monocyte-mediated tumoricidal activity via the tumor necrosis factor-related cytokine, TRAIL. *J. Exp. Med.* 1999, *189*, 1343–1354.
13. Fanger, N.A.; Maliszewski, C.R.; Schooley, K.; Griffith, T.S. Human dendritic cells mediate cellular apoptosis via tumor necrosis factor-related apoptosis-inducing ligand (TRAIL). *J. Exp. Med.* 1999, *190*, 1155–1164.
14. Kemp, T.J.; Moore, J.M.; Griffith, T.S. Human B cells express functional TRAIL/Apo-2 ligand after CpG-containing oligodeoxynucleotide stimulation. *J. Immunol.* 2004, *173*, 892–899.
15. Kemp, T.J.; Ludwig, A.T.; Earel, J.K.; Moore, J.M.; Vanoosten, R.L.; Moses, B.; Leidal, K.; Nauseef, W.M.; Griffith, T.S. Neutrophil stimulation with Mycobacterium bovis bacillus calmette-guerin (BCG) results in the release of functional soluble TRAIL/Apo-2L. *Blood* 2005, *106*, 3474–3482.
16. Simons, M.P.; Leidal, K.G.; Nauseef, W.M.; Griffith, T.S. TNF-related apoptosis-inducing ligand (TRAIL) is expressed throughout myeloid development, resulting in a broad distribution among neutrophil granules. *J. Leukoc. Biol.* 2008, *83*, 621–629.
17. Yang, F.; Shi, P.; Xi, X.; Yi, S.; Li, H.; Sun, Q.; Sun, M. Recombinant adenoviruses expressing TRAIL demonstrate antitumor effects on non-small cell lung cancer (NSCLC). *Med. Oncol.* 2006, *23*, 191–204.
18. Walczak, H.; Miller, R.E.; Ariail, K.; Gliniak, B.; Griffith, T.S.; Kubin, M.; Chin, W.; Jones, J.; Woodward, A.; Le, T.; Smith, C.; Smolak, P.; Goodwin, R.G.; Rauch, C.T.; Schuh, J.C.; Lynch, D.H. Tumoricidal activity of tumor necrosis factor-related apoptosis-inducing ligand *in vivo*. *Nat. Med.* 1999, *5*, 157–163.
19. Ashkenazi, A.; Pai, R.C.; Fong, S.; Leung, S.; Lawrence, D.A.; Marsters, S.A.; Blackie, C.; Chang, L.; McMurtrey, A.E.; Hebert, A.; DeForge, L.; Koumenis, I.L.; Lewis, D.; Harris, L.; Bussiere, J.; Koeppen, H.; Shahrokh, Z.; Schwall, R.H. Safety and antitumor activity of recombinant soluble Apo2 ligand. *J. Clin. Invest.* 1999, *104*, 155–162.
20. Lin, T.; Zhang, L.; Davis, J.; Gu, J.; Nishizaki, M.; Ji, L.; Roth, J.A.; Xiong, M.; Fang, B. Combination of TRAIL gene therapy and chemotherapy enhances antitumor and antimetastasis effects in chemosensitive and chemoresistant breast cancers. *Mol. Ther.* 2003, *8*,

441–448.
21. Huang, X.; Lin, T.; Gu, J.; Zhang, L.; Roth, J.A.; Stephens, L.C.; Yu, Y.; Liu, J.; Fang, B. Combined TRAIL and Bax gene therapy prolonged survival in mice with ovarian cancer xenograft. *Gene Ther.* 2002, *9*, 1379–1386.
22. Gurung, P.; Kucaba, T.A.; Schoenberger, S.P.; Ferguson, T.A.; Griffith, T.S. TRAIL-expressing CD8+ T cells mediate tolerance following soluble peptide-induced peripheral T cell deletion. *J. Leukoc. Biol.* 2010, *88*, 1217–1225.
23. Brincks, E.L.; Katewa, A.; Kucaba, T.A.; Griffith, T.S.; Legge, K.L. CD8 T cells utilize TRAIL to control influenza virus infection. *J. Immunol.* 2008, *181*, 4918–4925.
24. Janssen, E.M.; Droin, N.M.; Lemmens, E.E.; Pinkoski, M.J.; Bensinger, S.J.; Ehst, B.D.; Griffith, T.S.; Green, D.R.; Schoenberger, S.P. CD4+ T-cell help controls CD8+ T-cell memory via TRAIL-mediated activation-induced cell death. *Nature* 2005, *434*, 88–93.
25. Pan, G.; O'Rourke, K.; Chinnaiyan, A.M.; Gentz, R.; Ebner, R.; Ni, J.; Dixit, V.M. The receptor for the cytotoxic ligand TRAIL. *Science* 1997, *276*, 111–113.
26. Pan, G.; Ni, J.; Wei, Y.F.; Yu, G.; Gentz, R.; Dixit, V.M. An antagonist decoy receptor and a death domain-containing receptor for TRAIL. *Science* 1997, *277*, 815–818.
27. Walczak, H.; Degli-Esposti, M.A.; Johnson, R.S.; Smolak, P.J.; Waugh, J.Y.; Boiani, N.; Timour, M.S.; Gerhart, M.J.; Schooley, K.A.; Smith, C.A.; Goodwin, R.G.; Rauch, C.T. TRAIL-R2: A novel apoptosis-mediating receptor for TRAIL. *EMBO J.* 1997, *16*, 5386–5397.
28. Sheridan, J.P.; Marsters, S.A.; Pitti, R.M.; Gurney, A.; Skubatch, M.; Baldwin, D.; Ramakrishnan, L.; Gray, C.L.; Baker, K.; Wood, W.I.; Goddard, A.D.; Godowski, P.; Ashkenazi, A. Control of TRAIL-induced apoptosis by a family of signaling and decoy receptors. *Science* 1997, *277*, 818–821.
29. Degli-Esposti, M.A.; Dougall, W.C.; Smolak, P.J.; Waugh, J.Y.; Smith, C.A.; Goodwin, R.G. The novel receptor TRAIL-R4 induces NF-kappaB and protects against TRAIL-mediated apoptosis, yet retains an incomplete death domain. *Immunity* 1997, *7*, 813–820.
30. Marsters, S.A.; Sheridan, J.P.; Pitti, R.M.; Huang, A.; Skubatch, M.; Baldwin, D.; Yuan, J.; Gurney, A.; Goddard, A.D.; Godowski, P.;

Ashkenazi, A. A novel receptor for Apo2L/TRAIL contains a truncated death domain. *Curr. Biol.* 1997, *7*, 1003–1006.

31. Pan, G.; Ni, J.; Yu, G.; Wei, Y.F.; Dixit, V.M. TRUNDD, a new member of the TRAIL receptor family that antagonizes TRAIL signalling. *FEBS Lett.* 1998, *424*, 41–45.
32. Hymowitz, S.G.; Christinger, H.W.; Fuh, G.; Ultsch, M.; O'Connell, M.; Kelley, R.F.; Ashkenazi, A.; de Vos, A.M. Triggering cell death: The crystal structure of Apo2L/TRAIL in a complex with Death Receptor 5. *Mol. Cell* 1999, *4*, 563–571.
33. Emery, J.G.; McDonnell, P.; Burke, M.B.; Deen, K.C.; Lyn, S.; Silverman, C.; Dul, E.; Appelbaum, E.R.; Eichman, C.; DiPrinzio, R.; Dodds, R.A.; James, I.E.; Rosenberg, M.; Lee, J.C.; Young, P.R. Osteoprotegerin is a receptor for the cytotoxic ligand TRAIL. *J. Biol. Chem.* 1998, *273*, 14363–14367.
34. Lacey, D.L.; Timms, E.; Tan, H.L.; Kelley, M.J.; Dunstan, C.R.; Burgess, T.; Elliott, R.; Colombero, A.; Elliott, G.; Scully, S.; Hsu, H.; Sullivan, J.; Hawkins, N.; Davy, E.; Capparelli, C.; Eli, A.; Qian, Y.X.; Kaufman, S.; Sarosi, I.; Shalhoub, V.; Senaldi, G.; Guo, J.; Delaney, J.; Boyle, W.J. Osteoprotegerin ligand is a cytokine that regulates osteoclast differentiation and activation. *Cell* 1998, *93*, 165–176.
35. Wu, G.S.; Burns, T.F.; Zhan, Y.; Alnemri, E.S.; El-Deiry, W.S. Molecular cloning and functional analysis of the mouse homologue of the Killer/DR5 tumor necrosis factor-related apoptosis-inducing ligand (TRAIL) death receptor. *Cancer Res.* 1999, *59*, 2770–2775.
36. Schneider, P.; Olson, D.; Tardivel, A.; Browning, B.; Lugovskoy, A.; Gong, D.; Dobles, M.; Hertig, S.; Hofmann, K.; Van Vlijmen, H.; Hsu, Y.M.; Burkly, L.C.; Tschopp, J.; Zheng, T.S. Identification of a new murine tumor necrosis factor receptor locus that contains two novel murine receptors for tumor necrosis factor-related apoptosis-inducing ligand (TRAIL). *J. Biol. Chem.* 2003, *278*, 5444–5454.
37. Sprick, M.R.; Weigand, M.A.; Rieser, E.; Rauch, C.T.; Juo, P.; Blenis, J.; Krammer, P.H.; Walczak, H. FADD/MORT1 and caspase-8 are recruited to TRAIL receptors 1 and 2 and are essential for apoptosis mediated by TRAIL receptor 2. *Immunity* 2000, *12*, 599–609.
38. Kischkel, F.C.; Lawrence, D.A.; Chuntharapai, A.; Schow, P.; Kim, K.J.;

Ashkenazi, A. Apo2L/TRAIL-dependent recruitment of endogenous FADD and caspase-8 to death receptors 4 and 5. *Immunity* 2000, *12*, 611–620.

39. Li, H.; Zhu, H.; Xu, C.J.; Yuan, J. Cleavage of Bid by caspase 8 mediates the mitochondrial damage in the Fas pathway of apoptosis. *Cell* 1998, *94*, 491–501.
40. Griffith, T.S.; Chin, W.A.; Jackson, G.C.; Lynch, D.H.; Kubin, M.Z. Intracellular regulation of TRAIL-induced apoptosis in human melanoma cells. *J. Immunol.* 1998, *161*, 2833–2840.
41. Irmler, M.; Thome, M.; Hahne, M.; Schneider, P.; Hofmann, K.; Steiner, V.; Bodmer, J.L.; Schroter, M.; Burns, K.; Mattmann, C.; Rimoldi, D.; French, L.E.; Tschopp, J. Inhibition of death receptor signals by cellular FLIP. *Nature* 1997, *388*, 190–195.
42. Barnhart, B.C.; Alappat, E.C.; Peter, M.E. The CD95 type I/type II model. *Semin. Immunol.* 2003, *15*, 185–193.
43. Du, C.; Fang, M.; Li, Y.; Li, L.; Wang, X. Smac, a mitochondrial protein that promotes cytochrome c-dependent caspase activation by eliminating IAP inhibition. *Cell* 2000, *102*, 33–42.
44. Verhagen, A.M.; Ekert, P.G.; Pakusch, M.; Silke, J.; Connolly, L.M.; Reid, G.E.; Moritz, R.L.; Simpson, R.J.; Vaux, D.L. Identification of DIABLO, a mammalian protein that promotes apoptosis by binding to and antagonizing IAP proteins. *Cell* 2000, *102*, 43–53.
45. Schulze-Osthoff, K.; Ferrari, D.; Los, M.; Wesselborg, S.; Peter, M.E. Apoptosis signaling by death receptors. *Eur. J. Biochem.* 1998, *254*, 439–459.
46. Li, P.; Nijhawan, D.; Budihardjo, I.; Srinivasula, S.M.; Ahmad, M.; Alnemri, E.S.; Wang, X. Cytochrome c and dATP-dependent formation of APAF-1/caspase-9 complex initiates an apoptotic protease cascade. *Cell* 1997, *91*, 479–489.
47. Kelley, S.K.; Harris, L.A.; Xie, D.; Deforge, L.; Totpal, K.; Bussiere, J.; Fox, J.A. Preclinical studies to predict the disposition of Apo2L/tumor necrosis factor-related apoptosis-inducing ligand in humans: Characterization of *in vivo* efficacy, pharmacokinetics, and safety. *J. Pharmacol. Exp. Ther.* 2001, *299*, 31–38.
48. Pollack, I.F.; Erff, M.; Ashkenazi, A. Direct stimulation of apoptotic signaling by soluble Apo2L/tumor necrosis factor-related apoptosis-

inducing ligand leads to selective killing of glioma cells. *Clin. Cancer Res.* 2001, *7*, 1362–1369.

49. Gazitt, Y. TRAIL is a potent inducer of apoptosis in myeloma cells derived from multiple myeloma patients and is not cytotoxic to hematopoietic stem cells. *Leukemia* 1999, *13*, 1817–1824.
50. Kemp, C.J.; Sun, S.; Gurley, K.E. p53 induction and apoptosis in response to radio- and chemotherapy *in vivo* is tumor-type-dependent. *Cancer Res.* 2001, *61*, 327–332.
51. Liu, X.; Yue, P.; Khuri, F.R.; Sun, S.Y. p53 upregulates Death Receptor 4 expression through an intronic p53 binding site. *Cancer Res.* 2004, *64*, 5078–5083.
52. Takimoto, R.; El-Deiry, W.S. Wild-type p53 transactivates the Killer/DR5 gene through an intronic sequence-specific DNA-binding site. *Oncogene* 2000, *19*, 1735–1743.
53. Gliniak, B.; Le, T. Tumor necrosis factor-related apoptosis-inducing ligand's antitumor activity *in vivo* is enhanced by the chemotherapeutic agent CPT-11. *Cancer Res.* 1999, *59*, 6153–6158.
54. Chinnaiyan, A.M.; Prasad, U.; Shankar, S.; Hamstra, D.A.; Shanaiah, M.; Chenevert, T.L.; Ross, B.D.; Rehemtulla, A. Combined effect of tumor necrosis factor-related apoptosis-inducing ligand and ionizing radiation in breast cancer therapy. *Proc. Natl. Acad. Sci. USA* 2000, *97*, 1754–1759.
55. Hingtgen, S.; Ren, X.; Terwilliger, E.; Classon, M.; Weissleder, R.; Shah, K. Targeting multiple pathways in gliomas with stem cell and viral delivered s-TRAILl and temozolomide. *Mol. Cancer Ther.* 2008, *7*, 3575–3585.
56. Keane, M.M.; Ettenberg, S.A.; Nau, M.M.; Russell, E.K.; Lipkowitz, S. Chemotherapy augments TRAIL-induced apoptosis in breast cell lines. *Cancer Res.* 1999, *59*, 734–741.
57. Nimmanapalli, R.; Perkins, C.L.; Orlando, M.; O'Bryan, E.; Nguyen, D.; Bhalla, K.N. Pretreatment with paclitaxel enhances Apo-2 ligand/tumor necrosis factor-related apoptosis-inducing ligand-induced apoptosis of prostate cancer cells by inducing Death Receptors 4 and 5 protein levels. *Cancer Res.* 2001, *61*, 759–763.
58. Mizutani, Y.; Yoshida, O.; Miki, T.; Bonavida, B. Synergistic cytotoxicity and apoptosis by Apo-2 ligand and adriamycin against bladder cancer cells. *Clin. Cancer Res.* 1999, *5*, 2605–2612.

59. Matsubara, H.; Mizutani, Y.; Hongo, F.; Nakanishi, H.; Kimura, Y.; Ushijima, S.; Kawauchi, A.; Tamura, T.; Sakata, T.; Miki, T. Gene therapy with TRAIL against renal cell carcinoma. *Mol. Cancer Ther.* 2006, *5*, 2165–2171.
60. Bonvini, P.; Zorzi, E.; Basso, G.; Rosolen, A. Bortezomib-mediated 26s proteasome inhibition causes cell-cycle arrest and induces apoptosis in CD-30+ anaplastic large cell lymphoma. *Leukemia* 2007, *21*, 838–842.
61. Williams, S.A.; McConkey, D.J. The proteasome inhibitor bortezomib stabilizes a novel active form of p53 in human LnCAP-Pro5 prostate cancer cells. *Cancer Res.* 2003, *63*, 7338–7344.
62. Sayers, T.J.; Brooks, A.D.; Koh, C.Y.; Ma, W.; Seki, N.; Raziuddin, A.; Blazar, B.R.; Zhang, X.; Elliott, P.J.; Murphy, W.J. The proteasome inhibitor PS-341 sensitizes neoplastic cells to TRAIL-mediated apoptosis by reducing levels of c-FLIP. *Blood* 2003, *102*, 303–310.
63. Seki, N.; Toh, U.; Sayers, T.J.; Fujii, T.; Miyagi, M.; Akagi, Y.; Kusukawa, J.; Kage, M.; Shirouzu, K.; Yamana, H. Bortezomib sensitizes human esophageal squamous cell carcinoma cells to TRAIL-mediated apoptosis via activation of both extrinsic and intrinsic apoptosis pathways. *Mol. Cancer Ther.* 2010, *9*, 1842–1851.
64. Brooks, A.D.; Ramirez, T.; Toh, U.; Onksen, J.; Elliott, P.J.; Murphy, W.J.; Sayers, T.J. The proteasome inhibitor bortezomib (Velcade) sensitizes some human tumor cells to Apo2L/TRAIL-mediated apoptosis. *Ann. N. Y. Acad. Sci.* 2005, *1059*, 160–167.
65. Rosato, R.R.; Almenara, J.A.; Coe, S.; Grant, S. The multikinase inhibitor sorafenib potentiates TRAIL lethality in human leukemia cells in association with Mcl-1 and $cFLIP_L$ down-regulation. *Cancer Res.* 2007, *67*, 9490–9500.
66. Llobet, D.; Eritja, N.; Yeramian, A.; Pallares, J.; Sorolla, A.; Domingo, M.; Santacana, M.; Gonzalez-Tallada, F.J.; Matias-Guiu, X.; Dolcet, X. The multikinase inhibitor sorafenib induces apoptosis and sensitises endometrial cancer cells to TRAIL by different mechanisms. *Eur. J. Cancer* 2010, *46*, 836–850.
67. Kim, S.H.; Ricci, M.S.; El-Deiry, W.S. Mcl-1: A gateway to TRAIL sensitization. *Cancer Res.* 2008, *68*, 2062–2064.
68. Johnstone, R.W. Histone-deacetylase inhibitors: Novel drugs for the treatment of cancer. *Nat. Rev. Drug Discov.* 2002, *1*, 287–299.
69. Norian, L.A.; Kucaba, T.A.; Earel, J.K.; Knutson, T.; Vanoosten, R.L.;

Griffith, T.S. Synergistic induction of apoptosis in primary B-CLL cells after treatment with recombinant tumor necrosis factor-related apoptosis-inducing ligand and histone deacetylase inhibitors. *J. Oncol.* 2009, *2009*, 408038.

70. Nakata, S.; Yoshida, T.; Horinaka, M.; Shiraishi, T.; Wakada, M.; Sakai, T. Histone deacetylase inhibitors upregulate Death Receptor 5/TRAIL-R2 and sensitize apoptosis induced by TRAIL/Apo2-L in human malignant tumor cells. *Oncogene* 2004, *23*, 6261–6271.
71. Fulda, S.; Debatin, K.M. Resveratrol-mediated sensitisation to TRAIL-induced apoptosis depends on death receptor and mitochondrial signalling. *Eur. J. Cancer* 2005, *41*, 786–798.
72. VanOosten, R.L.; Moore, J.M.; Karacay, B.; Griffith, T.S. Histone deacetylase inhibitors modulate renal cell carcinoma sensitivity to TRAIL/Apo-2L-induced apoptosis by enhancing TRAIL-R2 expression. *Cancer Biol. Ther.* 2005, *4*, 1104–1112.
73. Vanoosten, R.L.; Moore, J.M.; Ludwig, A.T.; Griffith, T.S. Depsipeptide (FR901228) enhances the cytotoxic activity of TRAIL by redistributing TRAIL receptor to membrane lipid rafts. *Mol. Ther.* 2005, *11*, 542–552.
74. Fulda, S.; Debatin, K.M. HDAC inhibitors: Double edge sword for TRAIL cancer therapy? *Cancer Biol. Ther.* 2005, *4*, 1113–1115.
75. Fulda, S. Modulation of TRAIL-induced apoptosis by HDAC inhibitors. *Curr. Cancer Drug Targets* 2008, *8*, 132–140.
76. Walther, W.; Stein, U. Viral vectors for gene transfer: A review of their use in the treatment of human diseases. *Drugs* 2000, *60*, 249–271.
77. Young, L.S.; Searle, P.F.; Onion, D.; Mautner, V. Viral gene therapy strategies: From basic science to clinical application. *J. Pathol.* 2006, *208*, 299–318.
78. Lai, C.M.; Lai, Y.K.; Rakoczy, P.E. Adenovirus and adeno-associated virus vectors. *DNA Cell. Biol.* 2002, *21*, 895–913.
79. Ginsberg, H.S.; Pereira, H.G.; Valentine, R.C.; Wilcox, W.C. A proposed terminology for the adenovirus antigens and virion morphological subunits. *Virology* 1966, *28*, 782–783.
80. Ballay, A.; Levrero, M.; Buendia, M.A.; Tiollais, P.; Perricaudet, M. *In vitro* and *in vivo* synthesis of the hepatitis B virus surface antigen and of the receptor for polymerized human serum albumin from recombinant human adenoviruses. *EMBO J.* 1985, *4*, 3861–3865.
81. Karlsson, S.; Van Doren, K.; Schweiger, S.G.; Nienhuis, A.W.;

Gluzman, Y. Stable gene transfer and tissue-specific expression of a human globin gene using adenoviral vectors. *EMBO J.* 1986, *5*, 2377–2385.

82. Jones, N.; Shenk, T. An adenovirus type 5 early gene function regulates expression of other early viral genes. *Proc. Natl. Acad. Sci. USA* 1979, *76*, 3665–3669.
83. Kelly, T.J., Jr.; Lewis, A.M., Jr. Use of nondefective adenovirus-simian virus 40 hybrids for mapping the simian virus 40 genome. *J. Virol.* 1973, *12*, 643–652.
84. Yang, Y.; Nunes, F.A.; Berencsi, K.; Gonczol, E.; Engelhardt, J.F.; Wilson, J.M. Inactivation of E2a in recombinant adenoviruses improves the prospect for gene therapy in cystic fibrosis. *Nat. Genet.* 1994, *7*, 362–369.
85. Bergelson, J.M.; Cunningham, J.A.; Droguett, G.; Kurt-Jones, E.A.; Krithivas, A.; Hong, J.S.; Horwitz, M.S.; Crowell, R.L.; Finberg, R.W. Isolation of a common receptor for coxsackie B viruses and adenoviruses 2 and 5. *Science* 1997, *275*, 1320–1323.
86. Bergelson, J.M.; Krithivas, A.; Celi, L.; Droguett, G.; Horwitz, M.S.; Wickham, T.; Crowell, R.L.; Finberg, R.W. The murine CAR homolog is a receptor for coxsackie B viruses and adenoviruses. *J. Virol.* 1998, *72*, 415–419.
87. Wickham, T.J.; Mathias, P.; Cheresh, D.A.; Nemerow, G.R. Integrins alpha v beta 3 and alpha v beta 5 promote adenovirus internalization but not virus attachment. *Cell* 1993, *73*, 309–319.
88. Hemmi, S.; Geertsen, R.; Mezzacasa, A.; Peter, I.; Dummer, R. The presence of human coxsackievirus and adenovirus receptor is associated with efficient adenovirus-mediated transgene expression in human melanoma cell cultures. *Hum. Gene Ther.* 1998, *9*, 2363–2373.
89. Pearson, A.S.; Koch, P.E.; Atkinson, N.; Xiong, M.; Finberg, R.W.; Roth, J.A.; Fang, B. Factors limiting adenovirus-mediated gene transfer into human lung and pancreatic cancer cell lines. *Clin. Cancer Res.* 1999, *5*, 4208–4213.
90. Jacob, D.; Davis, J.; Zhu, H.; Zhang, L.; Teraishi, F.; Wu, S.; Marini, F.C., 3rd.; Fang, B. Suppressing orthotopic pancreatic tumor growth with a fiber-modified adenovector expressing the TRAIL gene from the human telomerase reverse transcriptase promoter. *Clin. Cancer Res.* 2004, *10*, 3535–3541.

91. Jacob, D.; Bahra, M.; Schumacher, G.; Neuhaus, P.; Fang, B. Gene therapy in colon cancer cells with a fiber-modified adenovector expressing the TRAIL gene driven by the hTERT promoter. *Anticancer Res.* 2004, *24*, 3075–3079.

92. Jacob, D.; Schumacher, G.; Bahra, M.; Davis, J.; Zhu, H.B.; Zhang, L.D.; Teraishi, F.; Neuhaus, P.; Fang, B.L. Fiber-modified adenoviral vector expressing the tumor necrosis factor-related apoptosis-inducing ligand gene from the human telomerase reverse transcriptase promoter induces apoptosis in human hepatocellular carcinoma cells. *World J. Gastroenterol.* 2005, *11*, 2552–2556.

93. Mizuguchi, H.; Hayakawa, T. Adenovirus vectors containing chimeric type 5 and type 35 fiber proteins exhibit altered and expanded tropism and increase the size limit of foreign genes. *Gene* 2002, *285*, 69–77.

94. Wohlfahrt, M.E.; Beard, B.C.; Lieber, A.; Kiem, H.P. A capsid-modified, conditionally replicating oncolytic adenovirus vector expressing TRAIL leads to enhanced cancer cell killing in human glioblastoma models. *Cancer Res.* 2007, *67*, 8783–8790.

95. Wang, Y.G.; Wang, J.H.; Zhang, Y.H.; Gu, Q.; Liu, X.Y. Antitumor effect of a novel adeno-associated virus vector targeting to telomerase activity in tumor cells. *Acta Biochim. Biophys. Sin. (Shanghai)* 2004, *36*, 492–500.

96. Jimenez, J.A.; Li, X.; Zhang, Y.P.; Bae, K.H.; Mohammadi, Y.; Pandya, P.; Kao, C.; Gardner, T.A. Antitumor activity of Ad-IU2, a prostate-specific replication-competent adenovirus encoding the apoptosis inducer, TRAIL. *Cancer Gene Ther.* 2010, *17*, 180–191.

97. Griffith, T.S.; Anderson, R.D.; Davidson, B.L.; Williams, R.D.; Ratliff, T.L. Adenoviral-mediated transfer of the TNF-related apoptosis-inducing ligand/Apo-2 ligand gene induces tumor cell apoptosis. *J. Immunol.* 2000, *165*, 2886–2894.

98. Griffith, T.S.; Broghammer, E.L. Suppression of tumor growth following intralesional therapy with TRAIL recombinant adenovirus. *Mol. Ther.* 2001, *4*, 257–266.

99. Armeanu, S.; Lauer, U.M.; Smirnow, I.; Schenk, M.; Weiss, T.S.; Gregor, M.; Bitzer, M. Adenoviral gene transfer of tumor necrosis factor-related apoptosis-inducing ligand overcomes an impaired response of hepatoma cells but causes severe apoptosis in primary human hepatocytes. *Cancer Res.* 2003, *63*, 2369–2372.

100. VanOosten, R.L.; Griffith, T.S. Activation of tumor-specific CD8+ T cells after intratumoral Ad5-TRAIL/CpG oligodeoxynucleotide combination therapy. *Cancer Res.* 2007, *67*, 11980–11990.
101. Berns, K.I.; Giraud, C. Biology of adeno-associated virus. *Curr. Top. Microbiol. Immunol.* 1996, *218*, 1–23.
102. Lu, Y. Recombinant adeno-associated virus as delivery vector for gene therapy--a review. *Stem Cells Dev.* 2004, *13*, 133–145.
103. Kwon, I.; Schaffer, D.V. Designer gene delivery vectors: Molecular engineering and evolution of adeno-associated viral vectors for enhanced gene transfer. *Pharm. Res.* 2008, *25*, 489–499.
104. Mohr, A.; Henderson, G.; Dudus, L.; Herr, I.; Kuerschner, T.; Debatin, K.M.; Weiher, H.; Fisher, K.J.; Zwacka, R.M. AAV-encoded expression of TRAIL in experimental human colorectal cancer leads to tumor regression. *Gene Ther.* 2004, *11*, 534–543.
105. Yoo, J.; Choi, S.; Hwang, K.S.; Cho, W.K.; Jung, C.R.; Kwon, S.T.; Im, D.S. Adeno-associated virus-mediated gene transfer of a secreted form of TRAIL inhibits tumor growth and occurrence in an experimental tumor model. *J. Gene Med.* 2006, *8*, 163–174.
106. Wu, Z.; Asokan, A.; Samulski, R.J. Adeno-associated virus serotypes: Vector toolkit for human gene therapy. *Mol. Ther.* 2006, *14*, 316–327.
107. Shi, J.; Zheng, D.; Liu, Y.; Sham, M.H.; Tam, P.; Farzaneh, F.; Xu, R. Overexpression of soluble TRAIL induces apoptosis in human lung adenocarcinoma and inhibits growth of tumor xenografts in nude mice. *Cancer Res.* 2005, *65*, 1687–1692.
108. Bremer, E.; de Bruyn, M.; Samplonius, D.F.; Bijma, T.; ten Cate, B.; de Leij, L.F.; Helfrich, W. Targeted delivery of a designed sTRAIL mutant results in superior apoptotic activity towards EGFR-positive tumor cells. *J. Mol. Med.* 2008, *86*, 909–924.
109. Wang, Y.; Huang, F.; Cai, H.; Zhong, S.; Liu, X.; Tan, W.S. Potent antitumor effect of TRAIL mediated by a novel adeno-associated viral vector targeting to telomerase activity for human hepatocellular carcinoma. *J. Gene Med.* 2008, *10*, 518–526.
110. Zhang, Y.; Qu, Z.H.; Cui, M.; Guo, C.; Zhang, X.M.; Ma, C.H.; Sun, W.S. Combined endostatin and TRAIL gene transfer suppresses human hepatocellular carcinoma growth and angiogenesis in nude mice. *Cancer Biol. Ther.* 2009, *8*, 466–473.

111. Wang, Y.; Ma, L.; Wang, S.; Bao, Y.; Ni, C.; Guan, N.; Zhao, J.; Fan, X. Assessment of CAR- or CD46-dependent adenoviral vector-mediated TRAIL gene therapy in clinical adenocarcinoma lung cancer cells. *Oncology* 2009, *77*, 366–377.
112. Ma, H.; Liu, Y.; Liu, S.; Xu, R.; Zheng, D. Oral adeno-associated virus-sTRAIL gene therapy suppresses human hepatocellular carcinoma growth in mice. *Hepatology* 2005, *42*, 1355–1363.
113. Wenger, T.; Mattern, J.; Haas, T.L.; Sprick, M.R.; Walczak, H.; Debatin, K.M.; Buchler, M.W.; Herr, I. Apoptosis mediated by lentiviral TRAIL transfer involves transduction-dependent and - independent effects. *Cancer Gene Ther.* 2007, *14*, 316–326.
114. Mangipudi, S.S.; Canine, B.F.; Wang, Y.; Hatefi, A. Development of a genetically engineered biomimetic vector for targeted gene transfer to breast cancer cells. *Mol. Pharm.* 2009, *6*, 1100–1109.

4

Advances in Lipid Nanoparticles for siRNA Delivery

Yuen Yi C. Tam, Sam Chen and Pieter R. Cullis

Department of Biochemistry and Molecular Biology, University of British Columbia, 2350 Health Sciences Mall, Vancouver, B.C. V6T 1Z3, Canada

ABSTRACT

Technological advances in both siRNA (small interfering RNA) and whole genome sequencing have demonstrated great potential in translating genetic information into siRNA-based drugs to halt the synthesis of most disease-causing proteins. Despite its powerful promises as a drug, siRNA requires a sophisticated delivery vehicle because of its rapid degradation in the circulation, inefficient accumulation in target tissues and inability to cross cell membranes to access the cytoplasm where it functions. Lipid nanoparticle (LNP) containing ionizable amino lipids is the leading delivery technology for siRNA, with five products in clinical trials and more in the

Citation: Tam, Y.Y.C.; Chen, S.; Cullis, P.R. Advances in Lipid Nanoparticles for siRNA Delivery. Pharmaceutics 2013, 5, 498-507. https://doi.org/10.3390/pharmaceutics5030498.

pipeline. Here, we focus on the technological advances behind these potent systems for siRNA-mediated gene silencing.

Keywords: lipid nanoparticle; siRNA; ionizable amino lipid; pKa; PEG lipid; targeting

INTRODUCTION

RNA interference (RNAi), since its discovery in the 1990s, has rapidly become the most powerful tool for studying functional genomics and validating gene targets *in vitro* and *in vivo*, as well as developing gene-specific medicines [1,2]. The small interfering RNA (siRNA) is a double-stranded oligonucleotide composed of typically 19–25 base pairs. It binds to the RNA-induced silencing complex (RISC) and guides the cleavage of the targeted mRNA sequence. Advances in design algorithms and nucleotide chemistry for siRNA, combined with the completion of the human genome and advances in whole genome sequencing technologies, have presented exciting opportunities to rapidly translate genomic information (such as whole genome sequencing of tumour tissue to identify causal genes) into siRNA-based therapeutics to block the production of disease-causing proteins [2,3,4]. siRNAs can be designed to target most genes in the genome, from one gene at a time to several genes simultaneously, vastly expanding the number of "druggable" targets.

The major limitation of siRNA use in the clinic is the development of safe and effective systemic delivery systems. "Naked" siRNA is highly unstable and rapidly degraded by nucleases in biological fluids, does not accumulate in target tissues and cannot readily cross target cell membranes to access its cytoplasmic site of action [5,6]. Although there are significant advancements in siRNA design and chemical modifications that both increase stability and reduce immunogenicity [7,8,9,10], there is still a need for efficient cellular uptake and target site accumulation. Appropriate delivery vehicles are therefore essential for realizing the potential of siRNA technology. Delivery strategies include both polymer- and lipid-based systems [11,12]. Among all delivery platforms, lipid nanoparticles (LNPs) containing ionizable amino lipids represent the leading system for intravenously (*i.v.*) administered RNAi-based therapeutics. It has been shown that LNP siRNA systems are capable of hepatocyte gene silencing *in vivo* at doses as low as 0.005 mg/kg in animal models following *i.v.* injection, which is currently the worldwide gold standard for siRNA-based therapeutics [13,14]. Five LNP siRNA formulations are currently in various stages of clinical development. Alnylam Pharmaceuticals has three products in their pipeline, which target transthyretin (TTR) for the treatment of TTR-mediated amyloidosis (ALN-TTR02), vascular endothelial growth factor (VEGF) and kinesin spindle protein (KSP) for hepatocellular carcinoma (ALN-VSP), and proprotein convertase subtilisin/kexin type 9 (PCSK9) for hypercholesterolemia (ALN-PCS). In addition, Tekmira Pharmaceuticals has developed TKM-PLK1 that targets polo-like kinase 1 (PLK1) for solid tumors and TKM-Ebola to treat Ebola virus infection. All five products show very safe clinical profiles and promising activity. They are described in detail in other reviews [15,16,17,18,19] and their respective company websites (http://www.alnylam.com/Programs-and-Pipeline/index.php and http://www.tekmirapharm.com/Programs/Products.asp). This review is focused on the technological bases behind these advanced LNP systems.

LNP FORMULATION: COMPOSITION, STRUCTURE AND SIZE

The advanced LNP siRNA systems are lipid-based particles with diameters less than 100 nm as measured by dynamic light scattering and cryo-transmission electron microscopy (cryoTEM). They are composed of an ionizable amino lipid (e.g., heptatriaconta-6,9,28,31-tetraen-19-yl 4-(dimethylamino)butanoate, DLin-MC3-DMA [14]), a phosphatidylcholine (1,2-distearoyl-sn-glycero-3-phosphocholine, DSPC), cholesterol and a coat lipid (polyethylene glycol-dimyristolglycerol, PEG-DMG) at molar ratio of 50:10:38.5:1.5 (Figure 1). CryoTEM revealed that LNP systems formed by mixing an ethanol stream containing the lipid mixture with an aqueous stream containing the siRNA, by either a T-tube [20] or microfluidic-based mixing [21,22], have an electron-dense core instead of the less dense aqueous core characteristic of vesicular structures [21,23]. Indeed, the LNP siRNA systems have an interior lipid core containing siRNA complexed with ionizable cationic lipid, as shown by the absence of ^{31}P NMR signal from free phosphorothioate in the siRNA and the complete protection of siRNA degradation by external RNases [21]. Computer simulation of the self-assembly of lipids suggests a nanostructured core in which siRNA is located in internal inverted micelles complexed with ionizable amino lipid (Figure 2). Furthermore, DSPC interacts with the siRNA phosphate through its choline group, cholesterol is dispersed about evenly between the core and surface, and the polyethylene glycol-lipid conjugate (PEG-lipid) is distributed predominantly on the surface. It is postulated that the rapid mixing of lipid components with the siRNA allows for the formation of nucleating structures composed of siRNA and ionizable amino lipid. As the polarity of the mixing environment increases, coating of the nucleating structures by the remaining lipids occurs until they reach their solubility limits, thus forming the final LNP siRNA structure [21]. The size of LNP siRNA systems made by the microfluidic technique is dictated by the PEG-lipid content [22]. As the amount of PEG-lipid increases from 0.5 mol% to 5 mol%, LNP diameter decreases from approximately 100 nm to 25 nm. siRNA encapsulation efficiencies are unaffected and remain greater than 95% [21,22].

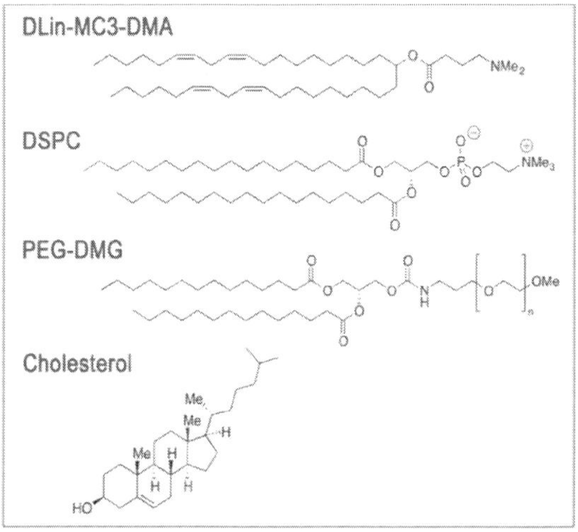

Figure 1. Lipid nanoparticle (LNP) siRNA systems are typically composed of ionizable amino lipids, phosphatidylcholine lipids, cholesterol and polyethylene glycol-lipid conjugate (PEG-lipids). Structure of heptatriaconta-6,9,28,31-tetraen-19-yl 4-(dimethylamino)butanoate (DLin-MC3-DMA), 1,2-distearoyl-*sn*-glycero-3-phosphocholine (DSPC), polyethylene glycol-dimyristolglycerol (PEG-DMG) and cholesterol is shown.

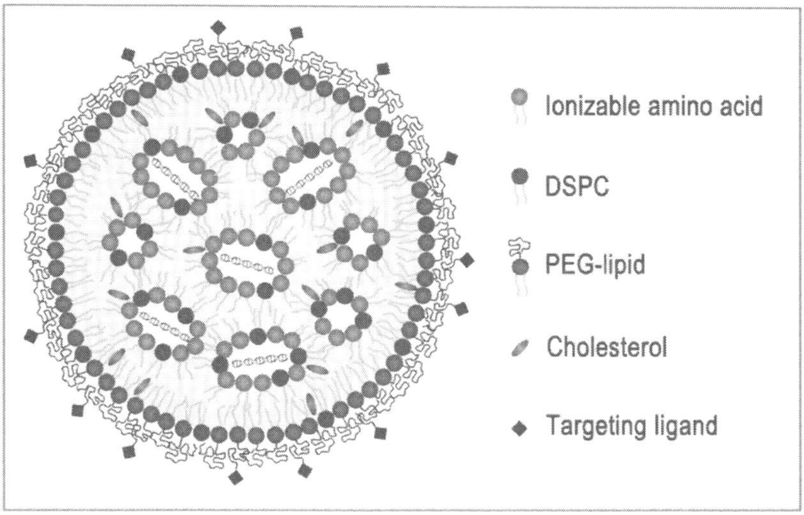

Figure 2. A schematic of LNP siRNA showing a nanostructured core.

ENCAPSULATION AND INTRACELLULAR RELEASE: DUAL FUNCTION OF IONIZABLE AMINO LIPIDS

A major advancement in the design of ionizable cationic lipids was the modulation of their apparent acid dissociation constant (pK_a). pK_a values of 7 and lower have been shown to be critical for encapsulation of nucleic acids and *in vivo* activity [24,25]. In environments where the pH is below the pK_a of the ionizable lipid (e.g., pH 4.0), the amino group is protonated and interacts with the negatively charged nucleic acids, thereby promoting the self-assembly of the formulation components into nanoparticles encapsulating the siRNA. In physiological environments where the pH is above the pK_a of the ionizable lipid (e.g., pH 7.4), the surface of the LNP has an almost neutral charge, which improves circulation and reduces toxicity. Subsequently in the acidic environment of endosomes, the amino group of the ionizable lipid is positively charged and associates with the anionic endosomal lipids. This interaction enables the destabilization of the endosomal membranes and promotes the release of siRNA into the cytosol [26,27].

The first ionizable amino lipid that was used for nucleic acid encapsulation was 1,2-dioleoyl-3-dimethyaminopropane (DODAP), which has a pKa of

6.6–7 and one double bond in each of its acyl chains [24,25,28]. Subsequent work focused on the impact of the number of double bonds in the acyl chain demonstrated that ionizable lipids containing fully saturated acyl chains showed no silencing of luciferase activity *in vitro*, whereas ionizable lipids containing two or three double bonds per acyl chain showed enhanced silencing activity [29]. Encapsulation efficiencies of siRNA seemed to be compromised using ionizable lipids that contain three double bonds per acyl chain. Therefore, the linoleyl lipid has become the acyl chain of choice in subsequent ionizable lipid development.

A structure-activity relationship guided synthesis and screening of a large number of ionizable lipids with various types of linkers connecting the amino group and the acyl chains. These studies have identified DLin-KC2-DMA (2,2-dilinoleyl-4-(2-dimethylaminoethyl)-[1,3]-dioxolane) [13], and DLin-MC3-DMA [14], which are 100-fold and 1000-fold more potent, respectively, in silencing of a hepatic gene (Factor VII) in comparison to the previous generation lipid DLin-DMA (1,2-dilinoleyloxy-*N,N*-dimethyl-3-aminopropane) [29]. The ED_{50} (median effective dose) for LNP containing DLin-MC3-DMA to silence Factor VII in mice and TTR in non-human primates was 0.005 mg/kg and 0.03 mg/kg, respectively [14]. One of the key findings from these studies was the optimum lipid pK_a value of 6.2–6.5 as a dominating factor in determining hepatic gene-silencing activity *in vivo*. DLin-MC3-DMA, having a pKa of 6.44, is currently the most active ionizable lipid being used in clinical trials.

In hopes of further promoting biocompatibility, a novel generation of ionizable amino lipids was recently synthesized with biodegradable ester functionalities in the acyl chains [30]. These novel lipids were well tolerated by the animals and rapidly eliminated from plasma and tissues. Importantly, they exhibited excellent potencies ($ED_{50} < 0.01$ mg/kg in mice, similar to that of DLin-MC3-DMA) in both rodent and non-human primate models. Although the relevance of these findings to human treatment remains to be proven, these preclinical results for biodegradable LNP siRNA formulations show great promise for clinical applications.

STERIC BARRIER: STABILITY AND CIRCULATION LIFETIME

Unprotected lipid-based delivery systems are rapidly cleared by the mononuclear phagocyte system [31,32]. In order to overcome this, hydrophilic polyethylene glycol has been widely used to coat lipid-based delivery systems, which prevents aggregation during the formulation process as well as provide "stealth" like characteristics post *i.v.* administration. The optimized PEG-lipid characteristics in current LNP siRNA systems were largely derived from earlier work on PEGylated liposomes [33,34,35] and stabilized plasmid or antisense lipid particles (SPLP or SALP). It was found that SPLP or SALP with PEG-ceramide that have C_{20} anchors (PEG-CerC_{20}) exhibited poor transfection efficiency in cultured cells compared to its shorter anchor counterparts such as PEG-CerC_{14} or PEG-CerC_8 [36,37,38]. The length of the ceramide lipid anchor dictated how long the PEG-lipid remained associated with the LNP. PEG-lipids with longer acyl anchors remain associated with the LNP longer, thus preventing LNP interaction with cells and subsequently with the endosomal membranes. They also provide the LNP with a more durable coat, which extends circulation lifetime [39,40]. Consistent findings were observed when PEG-ceramides were replaced with PEG-succinoyl-diacylglycerols (PEG-s-DAG) that have various acyl anchor lengths [41]. Due to the ease of synthesis and purification, PEG-s-DAG replaced PEG-Cer in later SPLP formulations and long acyl chain PEG-s-DAG such as PEG-s-distearoylglycerol (PEG-s-DSG) were used to increase circulation lifetime in the hope of exploiting the enhanced permeability and retention effect at tumor sites.

Unfortunately, despite being efficacious and non-toxic after a single bolus injection, PEGylated LNP with long acyl anchors (PEG-DSPE (1,2-distearoyl-*sn*-glycero-3-phosphoethanolamine), PEG-s-DSG, or PEG-CerC_{20}) were rapidly cleared upon repeated administration as a result of a robust immune response to the PEG component [42,43]. The use of rapidly dissociated shorter anchor PEG-lipids such as PEG-s-DMG or PEG-CerC_{14} however, mitigated this immunogenic response. Coincidentally, it was also observed that LNP containing PEG-s-DAG progressively lost the PEG moiety due to its succinate linker leading to particle aggregation and a reduced shelf-life. In response to this, the succinate linker was replaced with a carbamate linker to confer improved chemical stability without affecting efficacy [44].

CELLULAR UPTAKE: ENDOGENOUS AND EXOGENOUS TARGETING LIGANDS

Neutral liposomes have been shown to bind to proteins in serum, exchange components with lipoproteins and acquire factors that can potentially target them to specific cell types [45,46]. In particular, they interact with apolipoprotein (Apo) E and A-I [47,48]. ApoE, but not ApoA-I or ApoA-IV, was further found to enhance uptake of neutral liposomes in HepG2 cells and primary hepatocytes [49]. The role of ApoE in LNP uptake into hepatocytes was confirmed *in vivo* using ApoE-deficient mice [50]. The authors demonstrated that LNP were cleared more slowly from the circulation and were taken up by hepatocytes at least 20-fold less in ApoE-deficient mice than in wild-type animals. Similarly, LNP siRNA systems containing ionizable lipids require ApoE for activity [51]. Silencing of Factor VII was compromised in mice lacking ApoE or low-density lipoprotein (LDL) receptor, suggesting that ApoE acts as an endogenous ligand for LNP siRNA systems that facilitate uptake into hepatocytes via the LDL receptor.

When endogenous ligands are not available, exogenous ligands can be used to enhance uptake of LNP siRNA systems in target cells. Exogenous ligands such as antibodies, antibody fragments and peptides have been widely used in the field of liposome technology [52,53,54,55]. However, they are expensive and difficult to manufacture, as well as potentially immunogenic. In contrast, small molecule ligands conjugated to the distal end of PEG-lipids are simple to synthesize and can be formulated into LNP in a straightforward manner (Figure 2) [51,56,57]. *N*-Acetylgalactosamine (GalNAc), which binds with high affinity to the asialoglycoprotein receptor (ASGPR) found on hepatocytes, have been shown to rescue gene-silencing activity of LNP siRNA systems in ApoE deficient mice [51]. Other exogenous small molecule ligands have shown utility in non-hepatic cells. Anisamide, which interacts with sigma receptors, increases delivery and activity of siRNA containing nanoparticles in lung tumors and metastases [56,58,59]. Furthermore, strophanthidin, a cardiac glycoside that binds to the ubiquitously expressed cell surface receptor Na^+/K^+ ATPase, has been shown to enhance delivery of LNP siRNA to various cell types originating from the ovary, breast, pancreas, lung and prostate [57]. While the delivery of LNP using a promiscuous ligand may be non-specific, target specificity can be conferred using siRNA against disease-causing genes that are expressed in specific tissues.

CONCLUDING REMARKS

Using LNP as the delivery platform for siRNA has undoubtedly increased the potential of siRNA as therapeutics. This is evidenced by five LNP siRNA formulations for various liver-related diseases in clinical trials and more in the pipeline. The increased understanding of individual formulation components and their composition, LNP structure as well as formulation methods have led to the development of increasingly potent LNP siRNA systems. The success of LNP in hepatic applications is due, at least in part, to the liver's ideal physiology, notably being highly perfused with fenestrated endothelium. Other contributing factors include the optimum pK_a of the ionizable lipid, dissociable PEG-lipid, and association of LNP with endogenous ligand ApoE.

A number of challenges remain to be overcome for LNP siRNA to be used as therapeutics in a broad range of diseases. The existing LNP siRNA systems for liver applications have to be modified to extend their utility for non-hepatic tissues such as distal tumors. Novel small molecule targeting ligands will be required to facilitate LNP uptake into these non-hepatic tissues, especially when the siRNA is not tissue-specific. In order to reach tumor cores or tumors with poor vascularization, small LNP may be useful [60,61]. LNP siRNA systems as small as 25 nm can be made using microfluidic micromixing technology [22]; however, the relatively small siRNA payload may compromise activity and alternative methods to increase potency will have to be explored. Furthermore, LNP composition will likely require modifications for other routes of administration such as intraperitoneal, subcutaneous, intranasal or topical.

ACKNOWLEDGMENTS

The authors would like to acknowledge the Canadian Institutes of Health Research (CIHR) grants FRN 111627 and 86587, Alnylam Pharmaceuticals and Tekmira Pharmaceuticals for supporting this work. Karen Lam is gratefully acknowledged for editing the manuscript and illustrating the schematic. Josh Zaifman is gratefully thanked for providing the lipid structures.

CONFLICTS OF INTEREST

The authors declare no conflict of interest.

REFERENCES

1. Novina, C.D.; Sharp, P.A. The rnai revolution. *Nature* 2004, *430*, 161–164.
2. Vaishnaw, A.K.; Gollob, J.; Gamba-Vitalo, C.; Hutabarat, R.; Sah, D.; Meyers, R.; de Fougerolles, T.; Maraganore, J. A status report on rnai therapeutics. *Silence* 2010, *1*, 14.
3. McManus, M.T.; Sharp, P.A. Gene silencing in mammals by small interfering rnas. *Nat. Rev. Genet.* 2002, *3*, 737–747.
4. Miele, E.; Spinelli, G.P.; di Fabrizio, E.; Ferretti, E.; Tomao, S.; Gulino, A. Nanoparticle-based delivery of small interfering RNA: Challenges for cancer therapy. *Int. J. Nanomed.* 2012, *7*, 3637–3657. [Google Scholar]
5. De Fougerolles, A.; Vornlocher, H.P.; Maraganore, J.; Lieberman, J. Interfering with disease: A progress report on siRNA-based therapeutics. *Nat. Rev. Drug Discov.* 2007, *6*, 443–453.
6. Whitehead, K.A.; Langer, R.; Anderson, D.G. Knocking down barriers: Advances in sirna delivery. *Nat. Rev. Drug Discov.* 2009, *8*, 129–138.
7. Manoharan, M.; Akinc, A.; Pandey, R.K.; Qin, J.; Hadwiger, P.; John, M.; Mills, K.; Charisse, K.; Maier, M.A.; Nechev, L.; *et al.* Unique gene-silencing and structural properties of 2›-fluoro-modified sirnas. *Angew. Chem. Int. Ed. Engl.* 2011, *50*, 2284–2288.
8. Manoharan, M. RNA interference and chemically modified small interfering RNAs. *Curr. Opin. Chem. Biol.* 2004, *8*, 570–579.
9. Engels, J.W. Gene silencing by chemically modified sirnas. *New Biotechnol.* 2013, *30*, 302–307.
10. Soutschek, J.; Akinc, A.; Bramlage, B.; Charisse, K.; Constien, R.; Donoghue, M.; Elbashir, S.; Geick, A.; Hadwiger, P.; Harborth, J.; *et al.* Therapeutic silencing of an endogenous gene by systemic administration of modified sirnas. *Nature* 2004, *432*, 173–178.
11. Allen, T.M.; Cullis, P.R. Liposomal drug delivery systems: From concept to clinical applications. *Adv. Drug Deliv. Rev.* 2012, *65*, 36–48.
12. Daka, A.; Peer, D. RNAi-based nanomedicines for targeted personalized therapy. *Adv. Drug Deliv. Rev.* 2012, *64*, 1508–1521.
13. Semple, S.C.; Akinc, A.; Chen, J.; Sandhu, A.P.; Mui, B.L.; Cho, C.K.; Sah, D.W.; Stebbing, D.; Crosley, E.J.; Yaworski, E.; *et al.* Rational design of cationic lipids for siRNA delivery. *Nat. Biotechnol.* 2010, *28*,

172–176.

14. Jayaraman, M.; Ansell, S.M.; Mui, B.L.; Tam, Y.K.; Chen, J.; Du, X.; Butler, D.; Eltepu, L.; Matsuda, S.; Narayanannair, J.K.; *et al.* Maximizing the potency of sirna lipid nanoparticles for hepatic gene silencing *in vivo*. *Angew. Chem. Int. Ed. Engl.* 2012, *51*, 8529–8533.
15. Huang, L.; Liu, Y. *In vivo* delivery of rnai with lipid-based nanoparticles. *Annu. Rev. Biomed. Eng.* 2011, *13*, 507–530.
16. Burnett, J.C.; Rossi, J.J.; Tiemann, K. Current progress of siRNA/shRNA therapeutics in clinical trials. *Biotechnol. J.* 2011, *6*, 1130–1146.
17. Barros, S.A.; Gollob, J.A. Safety profile of RNAi nanomedicines. *Adv. Drug Deliv. Rev.* 2012, *64*, 1730–1737.
18. Tabernero, J.; Shapiro, G.I.; LoRusso, P.M.; Cervantes, A.; Schwartz, G.K.; Weiss, G.J.; Paz-Ares, L.; Cho, D.C.; Infante, J.R.; Alsina, M.; *et al.* First-in-humans trial of an RNA interference therapeutic targeting vegf and ksp in cancer patients with liver involvement. *Cancer Discov.* 2013, *3*, 406–417.
19. Alabi, C.; Vegas, A.; Anderson, D. Attacking the genome: Emerging sirna nanocarriers from concept to clinic. *Curr. Opin. Pharmacol.* 2012, *12*, 427–433.
20. Crawford, R.; Dogdas, B.; Keough, E.; Haas, R.M.; Wepukhulu, W.; Krotzer, S.; Burke, P.A.; Sepp-Lorenzino, L.; Bagchi, A.; Howell, B.J. Analysis of lipid nanoparticles by cryo-em for characterizing siRNA delivery vehicles. *Int. J. Pharm.* 2011, *403*, 237–244.
21. Leung, A.K.; Hafez, I.M.; Baoukina, S.; Belliveau, N.M.; Zhigaltsev, I.V.; Afshinmanesh, E.; Tieleman, D.P.; Hansen, C.L.; Hope, M.J.; Cullis, P.R. Lipid nanoparticles containing sirna synthesized by microfluidic mixing exhibit an electron-dense nanostructured core. *J. Phys. Chem. C Nanomater. Interfaces* 2012, *116*, 18440–18450.
22. Belliveau, N.M.; Huft, J.; Lin, P.J.C.; Chen, S.; Leung, A.K.K.; Leaver, T.J.; Wild, A.W.; Lee, J.B.; Taylor, R.J.; Tam, Y.K.; *et al.* Microfluidic synthesis of highly potent limit-size lipid nanoparticles for *in vivo* delivery of sirna. *Mol. Ther. Nucleic Acids* 2012, *1*, e37.
23. Zhigaltsev, I.V.; Maurer, N.; Edwards, K.; Karlsson, G.; Cullis, P.R. Formation of drug-arylsulfonate complexes inside liposomes: A novel approach to improve drug retention. *J. Control. Release* 2006, *110*, 378–386.

24. Semple, S.C.; Klimuk, S.K.; Harasym, T.O.; Dos Santos, N.; Ansell, S.M.; Wong, K.F.; Maurer, N.; Stark, H.; Cullis, P.R.; Hope, M.J.; *et al.* Efficient encapsulation of antisense oligonucleotides in lipid vesicles using ionizable aminolipids: Formation of novel small multilamellar vesicle structures. *Biochim. Biophys. Acta* 2001, *1510*, 152–166.
25. Maurer, N.; Wong, K.F.; Stark, H.; Louie, L.; McIntosh, D.; Wong, T.; Scherrer, P.; Semple, S.C.; Cullis, P.R. Spontaneous entrapment of polynucleotides upon electrostatic interaction with ethanol-destabilized cationic liposomes. *Biophys. J.* 2001, *80*, 2310–2326.
26. Hafez, I.M.; Maurer, N.; Cullis, P.R. On the mechanism whereby cationic lipids promote intracellular delivery of polynucleic acids. *Gene Ther.* 2001, *8*, 1188–1196.
27. Xu, Y.; Szoka, F.C., Jr. Mechanism of DNA release from cationic liposome/DNA complexes used in cell transfection. *Biochemistry* 1996, *35*, 5616–5623.
28. Bailey, A.L.; Cullis, P.R. Modulation of membrane fusion by asymmetric transbilayer distributions of amino lipids. *Biochemistry* 1994, *33*, 12573–12580.
29. Heyes, J.; Palmer, L.; Bremner, K.; MacLachlan, I. Cationic lipid saturation influences intracellular delivery of encapsulated nucleic acids. *J. Control. Release* 2005, *107*, 276–287.
30. Maier, M.A.; Jayaraman, M.; Matsuda, S.; Liu, J.; Barros, S.; Querbes, W.; Tam, Y.K.; Ansell, S.M.; Kumar, V.; Qin, J.; *et al.* Biodegradable lipids enabling rapidly eliminated lipid nanoparticles for systemic delivery of RNAi therapeutics. *Mol. Ther.* 2013, *21*, 1570–1578.
31. Allen, T.M. The use of glycolipids and hydrophilic polymers in avoiding rapid uptake of liposomes by the mononuclear phagocyte system. *Adv. Drug Deliv. Rev.* 1994, *13*, 285–309.
32. Ishida, T.; Harashima, H.; Kiwada, H. Liposome clearance. *Biosci. Rep.* 2002, *22*, 197–224.
33. Klibanov, A.L.; Maruyama, K.; Torchilin, V.P.; Huang, L. Amphipathic polyethyleneglycols effectively prolong the circulation time of liposomes. *FEBS Lett.* 1990, *268*, 235–237.
34. Webb, M.S.; Saxon, D.; Wong, F.M.; Lim, H.J.; Wang, Z.; Bally, M.B.; Choi, L.S.; Cullis, P.R.; Mayer, L.D. Comparison of different hydrophobic anchors conjugated to poly(ethylene glycol): Effects on the pharmacokinetics of liposomal vincristine. *Biochim. Biophys. Acta* 1998, *1372*, 272–282.

35. Woodle, M.C.; Lasic, D.D. Sterically stabilized liposomes. *Biochim. Biophys. Acta* 1992, *1113*, 171–199.
36. Wheeler, J.J.; Palmer, L.; Ossanlou, M.; MacLachlan, I.; Graham, R.W.; Zhang, Y.P.; Hope, M.J.; Scherrer, P.; Cullis, P.R. Stabilized plasmid-lipid particles: Construction and characterization. *Gene Ther.* 1999, *6*, 271–281.
37. Mok, K.W.; Lam, A.M.; Cullis, P.R. Stabilized plasmid-lipid particles: Factors influencing plasmid entrapment and transfection properties. *Biochim. Biophys. Acta* 1999, *1419*, 137–150.
38. Song, L.Y.; Ahkong, Q.F.; Rong, Q.; Wang, Z.; Ansell, S.; Hope, M.J.; Mui, B. Characterization of the inhibitory effect of PEG-lipid conjugates on the intracellular delivery of plasmid and antisense DNA mediated by cationic lipid liposomes. *Biochim. Biophys. Acta* 2002, *1558*, 1–13.
39. Monck, M.A.; Mori, A.; Lee, D.; Tam, P.; Wheeler, J.J.; Cullis, P.R.; Scherrer, P. Stabilized plasmid-lipid particles: Pharmacokinetics and plasmid delivery to distal tumors following intravenous injection. *J. Drug Target.* 2000, *7*, 439–452.
40. Tam, P.; Monck, M.; Lee, D.; Ludkovski, O.; Leng, E.C.; Clow, K.; Stark, H.; Scherrer, P.; Graham, R.W.; Cullis, P.R. Stabilized plasmid-lipid particles for systemic gene therapy. *Gene Ther.* 2000, *7*, 1867–1874.
41. Ambegia, E.; Ansell, S.; Cullis, P.; Heyes, J.; Palmer, L.; MacLachlan, I. Stabilized plasmid-lipid particles containing PEG-diacylglycerols exhibit extended circulation lifetimes and tumor selective gene expression. *Biochim. Biophys. Acta* 2005, *1669*, 155–163.
42. Judge, A.; McClintock, K.; Phelps, J.R.; Maclachlan, I. Hypersensitivity and loss of disease site targeting caused by antibody responses to pegylated liposomes. *Mol. Ther.* 2006, *13*, 328–337.
43. Semple, S.C.; Harasym, T.O.; Clow, K.A.; Ansell, S.M.; Klimuk, S.K.; Hope, M.J. Immunogenicity and rapid blood clearance of liposomes containing polyethylene glycol-lipid conjugates and nucleic acid. *J. Pharmacol. Exp. Ther.* 2005, *312*, 1020–1026.
44. Heyes, J.; Hall, K.; Tailor, V.; Lenz, R.; MacLachlan, I. Synthesis and characterization of novel poly(ethylene glycol)-lipid conjugates suitable for use in drug delivery. *J. Control. Release* 2006, *112*, 280–290.
45. Chonn, A.; Semple, S.C.; Cullis, P.R. Association of blood proteins with

large unilamellar liposomes *in vivo*. Relation to circulation lifetimes. *J. Biol. Chem.* 1992, *267*, 18759–18765.

46. Cullis, P.R.; Chonn, A.; Semple, S.C. Interactions of liposomes and lipid-based carrier systems with blood proteins: Relation to clearance behaviour *in vivo*. *Adv. Drug Deliv. Rev.* 1998, *32*, 3–17.
47. Mendez, A.J.; He, J.L.; Huang, H.S.; Wen, S.R.; Hsia, S.L. Interaction of rabbit lipoproteins and red blood cells with liposomes of egg yolk phospholipids. *Lipids* 1988, *23*, 961–967.
48. Rensen, P.C.; Schiffelers, R.M.; Versluis, A.J.; Bijsterbosch, M.K.; van Kuijk-Meuwissen, M.E.; van Berkel, T.J. Human recombinant apolipoprotein e-enriched liposomes can mimic low-density lipoproteins as carriers for the site-specific delivery of antitumor agents. *Mol. Pharmacol.* 1997, *52*, 445–455. [Google Scholar]
49. Bisgaier, C.L.; Siebenkas, M.V.; Williams, K.J. Effects of apolipoproteins A-IV and A-I on the uptake of phospholipid liposomes by hepatocytes. *J. Biol. Chem.* 1989, *264*, 862–866.
50. Yan, X.; Kuipers, F.; Havekes, L.M.; Havinga, R.; Dontje, B.; Poelstra, K.; Scherphof, G.L.; Kamps, J.A. The role of apolipoprotein e in the elimination of liposomes from blood by hepatocytes in the mouse. *Biochem. Biophys. Res. Commun.* 2005, *328*, 57–62.
51. Akinc, A.; Querbes, W.; De, S.; Qin, J.; Frank-Kamenetsky, M.; Jayaprakash, K.N.; Jayaraman, M.; Rajeev, K.G.; Cantley, W.L.; Dorkin, J.R.; *et al.* Targeted delivery of rnai therapeutics with endogenous and exogenous ligand-based mechanisms. *Mol. Ther.* 2010, *18*, 1357–1364.
52. Sapra, P.; Allen, T.M. Ligand-targeted liposomal anticancer drugs. *Progr. Lipid Res.* 2003, *42*, 439–462.
53. Cressman, S.; Dobson, I.; Lee, J.B.; Tam, Y.Y.; Cullis, P.R. Synthesis of a labeled RGD-lipid, its incorporation into liposomal nanoparticles, and their trafficking in cultured endothelial cells. *Bioconjug. Chem.* 2009, *20*, 1404–1411.
54. Allen, T.M.; Sapra, P.; Moase, E.; Moreira, J.; Iden, D. Adventures in targeting. *J. Liposome Res.* 2002, *12*, 5–12.
55. Di Paolo, D.; Ambrogio, C.; Pastorino, F.; Brignole, C.; Martinengo, C.; Carosio, R.; Loi, M.; Pagnan, G.; Emionite, L.; Cilli, M.; *et al.* Selective therapeutic targeting of the anaplastic lymphoma kinase with liposomal sirna induces apoptosis and inhibits angiogenesis in neuroblastoma. *Mol. Ther.* 2011, *19*, 2201–2212.

56. Li, S.D.; Huang, L. Targeted delivery of antisense oligodeoxynucleotide and small interference RNA into lung cancer cells. *Mol. Pharm.* 2006, *3*, 579–588.

57. Tam, Y.Y.C.; Chen, S.; Zaifman, J.; Tam, Y.K.; Lin, P.J.C.; Ansell, S.; Roberge, M.; Ciufolini, M.A.; Cullis, P.R. Small molecule ligands for enhanced intracellular delivery of lipid nanoparticle formulations of siRNA. *Nanomed. Nanotechnol. Biol. Med.* 2013, *9*, 665–674.

58. Li, S.D.; Chono, S.; Huang, L. Efficient oncogene silencing and metastasis inhibition via systemic delivery of siRNA. *Mol. Ther.* 2008, *16*, 942–946.

59. Chen, Y.; Sen, J.; Bathula, S.R.; Yang, Q.; Fittipaldi, R.; Huang, L. Novel cationic lipid that delivers siRNA and enhances therapeutic effect in lung cancer cells. *Mol. Pharm.* 2009, *6*, 696–705.

60. Cabral, H.; Matsumoto, Y.; Mizuno, K.; Chen, Q.; Murakami, M.; Kimura, M.; Terada, Y.; Kano, M.R.; Miyazono, K.; Uesaka, M.; *et al.* Accumulation of sub-100 nm polymeric micelles in poorly permeable tumours depends on size. *Nat. Nanotechnol.* 2011, *6*, 815–823.

61. Huo, S.; Ma, H.; Huang, K.; Liu, J.; Wei, T.; Jin, S.; Zhang, J.; He, S.; Liang, X.J. Superior penetration and retention behavior of 50 nm gold nanoparticles in tumors. *Cancer Res.* 2013, *73*, 319–330.

5

Gene Switches for Deliberate Regulation of Transgene Expression: Recent Advances in System Development and Uses

Nuria Vilaboa[1,2], Frank Boellmann[3] and Richard Voellmy[4,5]

[1]Unidad de Investigación, Hospital Universitario La Paz-IdiPAZ, 28046 Madrid, Spain
[2]CIBER de Bioingeniería, Biomateriales y Nanomedicina (CIBER-BBN), Madrid, Spain
[3]Pacific Biosciences, Menlo Park, CA 9402
[4]Dept. of Physiological Sciences, University of Florida, Gainesville, FL 32611
[5]HSF Pharmaceuticals S.A., 1814 La Tour-de-Peilz, Switzerland

ABSTRACT

It will be desirable in genetic therapies to exert temporal and/or spatial control over the expression of therapeutic genes. Temporal regulation may be achieved by placing the gene of interest under the control of one of several well-known gene switches. These gene switches comprise at least two elements, i.e., a ligand-dependent transactivator (or transinhibitor) and

Citation: Vilaboa N, Boellmann F, Voellmy R (2011) Gene Switches for Deliberate Regulation of Transgene Expression: Recent Advances in System Development and Uses. J Genet Syndr Gene Ther 2:107. doi:10.4172/2157-7412.1000107.

a transactivator responsive promoter that is functionally linked to the gene of interest. The systems respond to doxycycline, insect steroid hormones, steroid hormone antagonists, rapalogs, or various nonnatural ligands. Deliberate spatial regulation may be attained by the use of promoters that are activated by physical forces which can be directed to an area in need of therapy. Best known are heat shock promoters and radiationinduced promoters. More complex gene switches are capable of both spatially and temporally regulating genes of interest. The present article reviews improvements and further refinements of the latter gene switches as well as exemplifies uses of these gene switches in research and experimental therapy. It further reports on more recently developed gene switches, including complex systems involving multiple transactivators (or transinhibitors) and/or regulating gene product functionality at the transcriptional, posttranscriptional and/or activity level.

INTRODUCTION

Gene switches for deliberate regulation of transgenes typically comprise a transactivator or transinhibitor whose activity can be regulated and a transactivator-responsive or transinhibitor-susceptible promoter for controlling a gene of interest. The transactivators or transinhibitors typically are ligand-responsive, chimeric proteins comprising a DNA-binding domain, a ligand binding domain and a transcriptional activation domain or inhibition domain, respectively. Well known gene switches are based on tetracyclineresponsive transactivators and transinhibitors, mammalian or insect steroid receptor-derived transactivators and rapamycin-induced transactivators. Other gene switches make use of endogenous transcription factors that can be deliberately activated, e.g., by physical cues, and whose transient activation is tolerated by the cell. The best known systems of this kind make use of transcription factors that can be activated by heat or ionizing radiation.

It appears that gene switches were developed originally to regulate the timing, dosing and/or spatial definition of transgene expression in gene or cell therapy. Presumably, this development occurred in response to concerns that excessive concentrations of transgene therapeutic product could have adverse effects or that expression of certain therapeutics outside of the therapeutic time window or outside of a defined region in need of therapy may have unintended negative consequences. Many preclinical animal studies were performed with this aim in mind. However, gene switches were also employed in research, e.g., for studying gene function, facilitating expression of cytotoxic proteins in vitro and the like. We have last reviewed developments in the gene switch field in 2009 [1]. The present review is intended as an update that focuses on recent improvements of gene switches developed earlier as well as on discussing novel gene switches that were described subsequent to our review. It further reports on recent developments aimed at moving gene switch-regulated therapies into the clinic, discusses results of clinical trials that were undertaken and summarizes research uses of gene switches.

The main emphasis of this article is on transcription regulation. However, the same principles that were used to produce ligandregulated transactivators or transinhibitors also were employed increasingly to subject proteins to ligand regulation of their activity. Because a regulated biological effect or therapy may be provided either by regulating the amount or the activity of an effector protein, a number of examples for ligand regulation at the effector protein level are also described.

SMALL MOLECULE-DEPENDENT GENE SWITCHES

Gene Switches Comprising Tetracycline Repressor-Derived Transactivators and Transinhibitors

The E.coli Tn10 tetracycline resistance operon consists of a tetracycline repressor protein (TetR) and a specific DNA-binding site, the tetracycline operator (TetO, also referred to as TRE). TetR binds to TetO. Tetracycline or derivatives such as doxycycline bind to TetR, causing a reversible conformational change which results in dissociation of TetR from TetO. Several different tetracycline-responsive regulatory systems were developed. In the so-called TetOff systems target gene expression is on in the absence of tetracycline and is inactivated by the addition of the ligand (Figure 1A). In the so-called TetOn systems the target gene is off in the absence of ligand and is activated upon

its addition (Figure 1B). The original TetOff system consisted of a chimeric transactivator composed of TetR and an activation domain from herpes simplex virus protein VP16 (tTA), and a minimal CMV promoter supplemented with seven TRE sequences for driving a gene of interest. The typical TetOn systems are based on mutants of TetR whose TRE-binding ability is induced rather than inhibited by tetracycline. The systems include a transactivator assembled from mutant TetR and VP16 (rtTA) and a responsive promoter to which a gene of interest is linked. The original TetOn system was somewhat leaky, i.e., target gene expression also occurred at some appreciable rate in the absence of tetracycline. Over the years transactivators were improved to suppress this background activity. A much used second generation transactivator has been rtTA2s-M2. In a further strategy for reducing leakiness, expression of rtTA was no longer mediated by a constitutively active promoter but a transactivator-responsive promoter (Figure 1C). Hence, in the absence of tetracycline, transactivator levels are low. Addition of tetracycline activates available transactivators, which then enhance their own expression (a feature referred to below as "auto-activation") as well at that of the gene of interest. Alternative TetOn systems were produced that made use of the transrepressor domains (KRAB) of human zinc finger protein Kox1 or rodent zinc finger protein Kid-1. These transrepressor domains are known to inhibit all RNA polymerases II and III within a range of about 3 kb. The transinhibitors of these systems comprise TetR and a KRAB domain of Kox-1 or Kid-1 (referred to as tTS or tTS^{Kid}, respectively), and an active promoter supplemented with TRE sequences. In attempts to further reduce background activity, the two types of TetOn systems were combined. Hence, combined systems consist of an rtTA, a tTS and a transactivator-responsive promoter for controlling transcription of a gene of interest (Figure 1D). In the absence of tetracycline, tTS binds to the promoter of the gene of interest, further suppressing background expression. In the presence of tetracycline, tTS is inactivated and rtTA is activated. Consequently, tTS is replaced by rtTA on the promoter, resulting in induced transcription from the promoter. For additional information, the reader may also consult the 2009 review article by Stieger and colleagues [2].

System Development and Improvements

In a recent development, an auto-activated TetOn system was adapted for liver-specific gene therapy [3]. A bidirectional promoter unit was constructed comprising two albumin promoters in a headhead configuration separated by a sequence element containing 7 TetO sites. To one of the albumin promoters was functionally joined an rtTA-M2 gene [4] and to the other was linked a gene of interest. The unit was delivered by means of an AAV8 viral vector. Experiments with a luciferase reporter gene as the gene of interest showed >100 fold inducibility, which was maintained over multiple cycles of deactivation and reactivation. Expression was essentially confined to liver upon intravenous administration of the virus in a mouse model. When using an IL-12 gene as the regulated gene, the viral system was highly effective against hepatic tumors in a preventative as well as a therapeutic setting. Notably, mice "cured" of tumor developed an efficient T-cell memory response to tumor cells.

Leakiness was dramatically reduced in new "all-in-one" retroviral and transposon vectors that exhibited dynamic ranges (induced to uninduced expression of target gene) of up to 25000 fold [5]. "All-in-one" vectors refer to vectors that comprise genes for both a transactivator and a regulated gene. The preferred vectors contained a constitutively expressed rtTA-M2 gene and a transactivator-responsive reporter gene. The increases in dynamic range were primarily achieved through the development of improved transactivator-responsive promoters that allowed for nearly as strong doxycycline-induced expression but far lower background activity compared to the standard tetracycline repressor-responsive promoter [6].

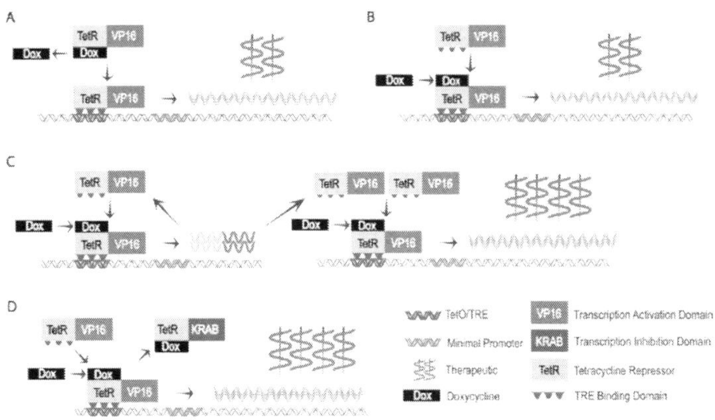

Figure 1: Tetracycline/doxycycline-regulated gene switches. A. TetOff system. B. TetOn system. C. Autoregulated TetOn system. D. Complex system comprising an rtTA and a tTS.

It may be desirable to irreversibly terminate transgene expression subsequent to a therapeutic regimen. This may be achieved by the use of a tet-O HAC–based vector that allows for regulation of transgene persistence in growing cells [7]. Tet-O HAC is a human artificial chromosome containing TetO sites inserted in its alphoid DNA array that replicates as a single-copy episome in human cells [8]. Expression of transinhibitor tTS resulted in a rapid loss of kinetochore function and loss of Tet-O HAC. Tet-O HAC was converted into a gene delivery vector by introduction of a loxP cassette for transgene insertion. The new vector was tested following introduction of an EGFP transgene cassette by Cre-loxP recombination (tet-O-EGFP HAC). Expression of the transgene could be detected in cells containing tet-O-EGFP HAC, which exhibited a high level of stability (50% retention after 2 months). An alternative approach to irreversibly terminating transgene activity was based on nonviral replicon vector pEPI that comprises an SV40 origin of replication and a chromosomal scaffold/matrix attached region (S/MAR) [9]. The vector further contains a CMV promoter driving a gene of interest, e.g., an eGFP reporter gene. Replication of this vector in mammalian cells apparently depends on transcription from the CMV promoter that extends into the S/MAR region. Insertion of an rtTA expression cassette (in a tail-to-tail arrangement with the gene of interest) as well as replacement of the CMV promoter with a transactivator-responsive promoter resulted in vector pEPI-Tet On [10]. When tested in CHO cells or in mice, eGFP expression from pEPI-Tet On was enhanced several fold in the presence of doxycycline. In the absence of doxycycline, the vector was gradually lost from growing cells.

A pair of high-capacity adenovectors for glioblastoma therapy were constructed, one containing a constitutively expressed gene for HSV thymidine kinase (HC-Ad-TK), and the other constitutively expressed rtTA and tTS genes as well as a gene encoding fms-like tyrosine kinase ligand 3 (Flt3L) controlled by a TetO-containing promoter (HC-AdTetOn-Flt3L) [11; 12 for methods]. The latter vector was capable of expressing Flt3L in a doxycycline-inducible fashion. Delivery of the vector pair into intracranial glioblastoma multiforme resulted in tumor shrinkage and enhanced survival in a large fraction of animals. None of the sham-treated animals survived.

An improved lentiviral vector for regulated expression of transgenes in neurons comprised several elements that had previously been optimized [13]. A first transcription unit consisted of a luciferase gene (example transgene) controlled by improved tetracycline repressorresponsive promoter TRE-tight1 [14]. The latter promoter was derived from second-generation promoter TRE_{tight} that contains a modified tetracycline response element consisting of seven repeats of a 36 bp sequence including the 19 bp TetO sequence (Clontech Laboratories). A third generation promoter with an even lower basal activity, P_{TRE3G}, has also been made available by Clontech. Following an observation of Osti and colleagues [15], chicken chromatin insulator cHS4 was used to separate this transcription unit from the second transcription unit, which consisted of a CMV promoter-driven modified rtTA (V16) gene [16,17]. The resulting vector showed excellent doxycycline regulation of luciferase expression in vitro in HeLa cells and neurons, and in vivo in spinal neurons. De-activation and re-activation of luciferase expression could be demonstrated. More recently, other lentiviral vectors, termed "polyswitch lentivectors", were described that also allow for tight doxycycline regulation of a gene of interest, contain acceptor sites for easy recombination cloning of a gene of interest and express living colors and a selectable marker [18].

Gene silencing through RNAi is an approach that makes it possible to carry out loss-of-function studies in mammalian organisms. A recent study reports the development of a new generation of lentivectors called pINDUCER for inducible RNAi in vivo [19]. Basic versions of these vectors comprise two expression cassettes, one expressed constitutively from a ubiquitin C promoter and containing genes for rtTA3 and a reporter joined by an IRES, and the other controlled by a TRE-containing promoter containing a fusion gene consisting of sequences coding for turboRFP and a sequence encoding a short hairpin RNA (shRNA) of choice embedded in microRNA-30 (miR30) sequences. Transactivator rtTA3 is an improved version of rtTA2s-M2 harboring further mutations F86Y and A209T. rtTA3 has improved doxycycline sensitivity and activity [20.21]. pINDUCER vectors were shown to be highly inducible in models of epithelial cancer and mammary development. A lentiviral vector of the series was also found capable of delivering BIRC4 shRNA in a doxycycline-induced fashion to BIRC4-dependent human breast tumor xenografts in nude mice, causing a substantial diminution of tumor growth. The pINDUCER system should be useful for large-scale in vivo genetic screening in models of human epithelial cancers. See also ref. 22 for another study that described similar technology. In a recent study retroviral vectors expressing shRNAs in a tetracycline-inducible fashion were used in a genetic screen that identified PTPN12 tyrosine phosphatase as a tumor suppressor in triple-negative breast cancer [23].

An all-in-one plasmid-based vector for expressing proteins of interest in a tetracycline-inducible fashion in pancreatic beta cells of rodents comprised a rat insulin promoter driving the expression of a gene of interest as well as a tTS gene linked to the gene of interest via IRES [24]. A TRE_{tight} sequence element was placed upstream from the insulin promoter. In the absence of tetracycline, residual transinhibitor was expected to bind to the TRE_{tight} element, thereby inhibiting insulin promoter activity. In the presence of tetracycline, transinhibitor should be released, resulting in enhanced expression of the gene of interest and of transinhibitor. The system was tested in Min6 and βTC mouse pancreatic beta cell lines and was found capable of conferring tight tetracycline regulation on the gene of interest.

Immune responses against Tet regulatory elements had been detected subsequent to intramuscular administration of vectors expressing Tet regulators. To investigate whether similar immune responses were induced in the brain, TetOff AAV vectors expressing potential therapeutic transgenes were administered to rats by intrastriatal injection [25]. Transgene expression could be demonstrated. Addition of doxycycline to the drinking water depressed expression by about 10 fold. Using a newly developed ELISA, antibodies against tTA could not be detected 5 or 9 weeks, respectively, after vector introduction in the brain. Intradermal vector injection served as a positive control in these experiments. Hence, TetOff AAV vectors did not elicit immune responses to Tet regulators in the brain and may therefore be safe vectors for the therapy of brain diseases such as Parkinson's disease. These conclusions are compatible with those of an earlier study, in which expression from an intrastriatally administered TetOn highcapacity adenovector was not affected by pre-existing immunity against rtTA [26]. It is noted that a similar absence of immune reactions against Tet regulators was observed in the eye [25,27].

A TetOn transgenic rat line was developed that expressed the rtTA-M2 transactivator from the ubiquitously active ROSA26 promoter [28]. Transactivator was detected in all major organs, including brain, heart, lung, liver, kidney and muscle. Doxycycline-inducible expression of an EGFP gene delivered by a lentivirus vector was observed in cells of testis, kidney and muscle tissues.

The transcriptomic effects of TetOn and mifepristone-inducible systems in the mouse liver in the presence and absence of ligand were examined [29]. Plasmids containing rtTA-M2 or GLp65 transactivator genes under the control of albumin promoters were introduced into the liver by hydrodynamic injection. The procedure resulted in transfection efficiencies of about 17%. The transactivator genes were expressed (one month after transfection) at a rate of about 1/6 of that of a liver reference gene (Mat1a), which is expressed at a medium-to-low level in hepatocyctes. RNA from livers was prepared one month after transfection. Analysis on 430A 2.0 microarrays (Affimetrix) revealed that expression of rtTA-M2 resulted in alterations in 69 gene probe sets and that of GLp65 in 1059 gene probe sets. (The Affimetrix chip contains 22690 probe sets corresponding to about 14000 genes.) Few additional changes were observed with RNA from animals that were also exposed to doxycycline or mifepristone, respectively. Most of the expression changes detected were relatively modest (equal to or less than three fold). Functional assignments indicated that alterations were mild and of little general significance. Therefore, it appeared that the two gene switches have little impact on the liver transcriptome profile and are safe for gene therapy applications. The analysis supports a slight preference for TetOn systems over mifepristone-inducible systems.

Hematopoietic stem cells (CD34+) can be efficiently transduced by high-capacity adenovectors with serotype 35 fiber knob domains (HD-Ad5/35). A vector designed for doxycycline-induced expression of transgenes in CD34+ cells, Ad5/35.Tet-1, contains a bicistronic arrangement of a GFP-coding sequence (example transgene) and a sequence encoding a tTS [30]. Expression is driven by a constitutively active (phosphoglycerokinase gene) promoter. Coding sequences and promoter are flanked by elements containing multiple TetO sites, polyA addition sequences and chromatin insulator cHS4. Another vector, Ad5/35.Tet-2, expresses an rtTA transactivator instead of a tTS. Furthermore, expression of the GFP and rtTA genes is driven by an rtTA-responsive promoter, i.e., a minimal CMV promoter supplemented by TetO sites. Both vectors were capable of expressing GFP in an inducible fashion in primary human CD34+ cells. Ad5/35. Tet-1 supported a lower induced level of transgene expression than Ad5/35.Tet-2. However, Ad5/35.Tet-2 was only active in a subset of CD34+ cells, whereas transgene expression occurred in all CD34+ cells transduced with Ad5/35.Tet-1.

A plasmid vector designed to allow phiC31 recombinase-mediated integration of a doxycycline-activated gene switch and associated regulated gene into the genome of a host cell contained a gene of interest controlled by a CMV promoter susceptible to inactivation by TetR due to the presence of two TRE sites located between the TATA box and the transcription initiation site [31]. Downstream from the gene of interest was a BGH polyA sequence. This gene cassette was followed by two copies of a cassette consisting of a CAG promoter (a combination of the CMV early enhancer element and a chicken β-actin promoter), a TetR gene functionally linked to the promoter and an SV40 polyA sequence. All three genes had the same transcriptional orientation. To minimize promoter interference (especially after integration in a host cell), cassettes 1 and 2 were each followed by a cHS4 chromatin insulator sequence. Finally, the plasmid also contained an attB recognition site for phiC31. Stable cell lines containing the system were isolated and examined: stable and robust transgene expression and high induction rates upon doxycycline addition were observed.

The usefulness of regulated expression systems is often hampered by the difficulty of identifying clonal cell lines in which the regulation of the transgene of interest is optimal. HeLa cell-derived lines were generated that carry an rtTA-M2 gene under the control of a human EF1alpha promoter and contain a locus retargetable by Flp-mediated cassette exchange [32]. Retargetable loci were selected for their ability to allow for maximal inducibility of an inserted luciferase reporter gene. The availability of these lines will facilitate a number of experiments that can be carried out in HeLa cells, e.g., gene knock-down experiments to examine gene function, etc.

Another study incorporated a tTA transactivator in a gene switch that is irreversibly triggered by a transiently activated receptor [33]. The switch should be useful as an alternative approach to measuring receptor activation that is likely to be more sensitive than assays that follow signaling events downstream from an activated receptor as well as should be largely independent from endogenous co-regulators. The assay, termed Tango assay, was developed for G protein-coupled receptors, receptor tyrosine kinases and steroid hormone receptors. As an example for a G protein-coupled receptor assay, a first construct was prepared that encoded a fusion protein consisting of a human arginine vasopressin receptor 2, a short sequence containing a cleavage site for viral proteinase TEV and a tTA transactivator. A second construct encoded an arrestin-TEV fusion protein. Upon co-introduction of these constructs into a cell containing a tTA-responsive reporter gene, ligand binding was expected to activate the fusion receptor and attract binding of the arrestin-TEV fusion protein. As a consequence, the tTA transactivator would be cleaved off the fusion protein and translocate to the nucleus to activate the reporter gene. In actual experiments, ligand exposure of cells containing the system resulted in a 200-fold increase in reporter activity.

Tetracycline Repressor-Based Systems Controlling Experimental Cancer Therapy

Human mammary tumor lines were stably transfected to express a gene encoding fusion protein CD8/caspase-8 under the control of a TetOn system [34]. Doxycycline-induced expression of the fusion protein caused apoptotic death of the cells. The cell lines were used to establish xenograft tumors in mice. Induction of apoptotic death created void spaces and channels in the tumors, which greatly facilitated distribution of oncolytic herpes virus throughout the tumor. This study outlines a new approach for enhancing the clinical effectiveness of oncolytic viruses.

To enhance the anti-cancer activity of a replicating adenovirus vector expressing IFN-alpha (KD3-IFN), a hepatocyte-derived tumor was co-infected with KD3-IFN and replication-defective adenovirus TetOn-TRAIL expressing tumor necrosis factor-related apoptosisinducing ligand (TRAIL) under the control of an rtTA [35]. Coinfection with TetOn-TRAIL was significantly more effective in slowing tumor growth in the presence of doxycycline than infection with KD3- IFN or TetOn-TRAIL alone. It is noted that ligand inducibility of the system was modest (tested by replacing TRAIL with a reporter gene): doxycycline increased expression by only about 5-10 fold. Another study tested the concept of using a brain sodium channel (ASIC2a) containing a mutation that renders it constitutively opened for cancer therapy in the brain [36]. A HSV/EBV hybrid amplicon vector was constructed that contained expressible genes for rtTA-M2 transactivator, tTS^{Kid} transinhibitor and the mutant channel, the gene for the mutant channel being controlled by an rtTA-responsive promoter. For delivery, the amplicon vector was packaged in vitro into HSV-1 virions. The system was successfully tested in vitro and in vivo in a subcutaneous mouse xenograft model of human glioma.

A TetOff system was used to time the expression of IL-4 in a study that tested the effect of IL-4 on myeloid-derived suppressor cell (MDSC) development at different stages of tumor growth [37]. MDSCs inhibit T cell activity and promote tumor growth. High-level expression of IL-4 early after subcutaneous administration of mouse fibrosarcoma cells prevented the generation of MDSCs and resulted in T cell-mediated tumor rejection. IL-4 was no longer effective, if first expressed after several days of tumor growth. For other cancer-related studies that made use of Tet regulators [38,39,40].

Tetracycline Repressor-Based Systems Controlling Experimental Therapy of Neurologic Disorders or Neurodegenerative Disease

Striatal expression of glial cell line-derived neurotrophic factor GDNF was proposed as a potential treatment for Parkinson's disease. A recombinant AAV2/5 vector was constructed that featured a gene for transactivator tTA2 (a VP16 miminal domain transactivator, Clontech) under the control of a CAG promoter and a human GDNF cDNA gene controlled by a transactivator-responsive promoter [41]. The vector was delivered to rats by nigrostriatal injection. GDNF expression in the substantia nigra and in serum was stringently dependent on doxycycline dose. Striatal GDNF induced weight alterations. These alterations were used to demonstrate the doxycycline dependence of GDNF expression over several cycles of repression and derepression.

Spinocerebellar ataxia type 3 is caused by expansion of a CAG repeat in the MJD1 gene that encodes the ataxia-3 protein. A transgenic mouse line was established that harbored a full-length human ataxin-3 cDNA gene containing 77 CAG repeats functionally linked to a Tet regulator-responsive promoter [42]. This line was crossed with a line containing a PrP promoter-driven tTA transactivator gene (TetOff) to generate a double transgenic line capable of expression of the diseasecausing protein under doxycycline control. The model was used to demonstrate that discontinuation of pathogenic ataxin-3 expression in an early disease state prevents the establishment of neuronal dysfunction characterized by reduced anxiety, hyperactivity, impaired Rotarod performance and lower body weight gain. These results suggest that early pharmacological intervention will be able to halt or at least slow disease progression.

Gene therapeutic safety is arguably enhanced by spatial restriction of therapeutic gene expression. The biodistribution of transgene expression in the CNS subsequent to stereotaxic injection of recombinant AAV2/1 vectors was compared [43]. The viral vectors comprised either a CMV promoter (the CMV vector) or a TetOn system (the TetOn vector) for regulating the activity of an EGFP gene. The latter vector included a bidirectional, TetO-containing promoter (derived from Clontech construct pBI-GL) driving an EGFP and an rtTA-M2 gene. To augment transactivator expression, the rtTA-M2-coding sequence was followed by a woodchuck hepatitis posttranscriptional regulator element sequence [44,45]. After intrastriatal injection (and doxycycline administration), EGFP was mainly expressed in neurons with both vectors. However, the two vectors differed in supporting EGFP expression in substantia nigra and the olfactory bulb, where expression was observed from the CMV vector but not the TetOn vector. The study [43] concluded that this represented a safety advantage of the TetOn vector in striatal gene therapy. Midbrain injection resulted in selective EGFP expression from the TetOn vector in tyrosine hydroxylasepositive neurons. Widespread midbrain expression of the marker gene was observed following administration of the CMV vector. Hence, the TetOn vector would appear to be better suited than the CMV vector for specific transgene expression in midbrain dopaminergic neurons.

Use of Tetracycline Repressor-Based Gene Switches In Other Experimental Therapies

Aminoglycoside antibiotics are used frequently to treat serious infections, e.g., septicemia, complicated intra-abdominal or urinary tract infections and nosocomial respiratory tract infections. Such treatments are associated with ototoxicity, in which cochlea and vestibule are attacked, and auditory and vestibular hair cells are destroyed. This results in irreversible hearing deficits that can manifest themselves months after antibiotic treatment. Sustained infusion of recombinant GDNF was known to protect the cochlear structure and function from noise- and drug-induced damage, although GDNF overdosing has contrary effects. An AAV1 vector was constructed that included a constitutively expressed gene for transactivator rtTA2s-S2 [4] and a GDNF gene controlled by a transactivator-responsive promoter [46]. GDNF was synthesized in a strictly doxycyclineinducible fashion in vector-infected cells. Introduction of the vector in rat cochlea and subsequent induced expression of GDNF successfully protected cochlear structure and function against damage induced by a course of treatment with kanamycin.

A lentivirus vector was prepared that contained a constitutively expressed rtTA (presumably rtTA2s-M2) gene and an expression cassette containing a gene for keratinocyte growth factor (KGF; also known as FGF7) and an IRES-linked EGFP reporter gene controlled by a transactivator-responsive promoter [47]. Mesenchymal stromal and hematopoietic stem cells were transduced with this lentivector to yield cells capable of doxycycline-induced EGFP expression. Transplantation of lentivector-transduced hematopoietic stem cells attenuated histological damage in a mouse model of bleomycininduced lung fibrosis.

An AAV2-based vector was developed comprising a basic fibroblast growth factor (bFGF) gene under TetOff control [48]. Vector-transduced mesenchymal stem cells could be shown to induce significant bone formation and angiogenesis in a rat calvarian defect model in the absence but not the presence of doxycycline. To develop a new modality of treatment of cartilage defects, a lentivirus was constructed comprising a gene for bone morphogenetic protein 2 (BMP-2) under the control of transactivator rtTA2s-M2 [49]. In the preferred vector, the BMP-2 gene was controlled by a PRE_{tight} promoter and the rtTA gene by a spleen focus-forming virus promoter. Stringent doxycycline inducibility of BMP-2 expression could be demonstrated in vector-transduced primary rabbit chondrocytes. Induced expression was sufficient to induce proteoglycan synthesis, a marker of chondrogenesis. In diabetes type 1, macrophages and T lymphocytes release cytokines that induce the generation of oxygen and nitrogen radicals in pancreatic islets which promote death of insulin-producing beta cells. Expression of mitochondrial peroxiredoxin III in rat insulinoma cells, under the control of a TetOn system, protected these cells against hydrogen peroxide, proinflammatory cytokines and streptozotocin stress [50].

Leber congenital amaurosis is a rare disease caused by a nonfunctional retinal pigment epithelium-specific gene (RPE65). Preliminary results from three small clinical trials suggested that expression of an RPE65 transgene can result in an improved visual function. Because it may be important to correctly titrate the amount of RPE65 provided to the retina, AAV2 vectors were developed [27] that expressed RPE65 under the control of either the rtTA-2M or the tTA2 transactivator [4]. Transactivators were controlled by CAG promoters. Induced expression of RPE65 from subretinally delivered AAV2 vectors in RPE65(-/-) Briard dogs resulted in a detectable recovery of retinal function, but no visual improvement. The retinal recovery was reversed upon cessation of RPE65 transgene expression. The study [27] argued that, while doxycycline-regulated RPE65 expression could be achieved, maximal levels of RPE65 produced by the fully active gene switches were below the level needed for the more substantial retinal recovery that is associated with improvement in visual function.

Two approaches were described that used doxycycline-regulated transgenes for opioids for the suppression of chronic pain. In the first approach, a three-plasmid system was assembled for achieving doxycycline-controlled expression of a proopiomelanocortin gene [51]. Proopiomelanocortin is a precursor protein that is processed in a tissue-specific fashion to yield different peptide hormones including beta-endorphin. The three plasmids were pTRE2 (Clontech) with an inserted proopiomelanocortin cDNA gene, pTet-On (Clontech) containing a CMV promoter-driven rtTA gene and pTet-tTS (Clontech) containing a CMV promoter-driven tTSKid gene. pTetOn had previously been described as pUHD17-neo 1 [52]. The three plasmid combination was introduced into rat dorsal root ganglia by electroporation. Stringent doxycycline regulation of beta-endorphin expression was observed in the transduced cells. To determine the anti-nociceptive effects of beta-endorphin therapy, the rat chronic constrictive injury (CCI) model was used. The three plasmids were introduced intrathecally, and spinal chord cells were transduced by electroporation. Robust reductions in thermal withdrawal latency and mechanical withdrawal threshold were observed in transduced animals subjected to a doxycycline regimen, but not in animals that did not receive doxycycline. Inducible expression of beta-endorphin could be measured in the cerebrospinal fluid. The second approach involved a cell therapy for reducing chronic pain [53]. A human preproenkephalin cDNA gene was subcloned into pRevTRE (Clontech). This plasmid and pRevTet-on (Clontech) were packaged using packaging cell line PT67. The resulting retroviruses were used to transduce an immortalized rat astrocyte cell line. The transduced cells were capable of expressing enkephalin in a doxycycline-induced fashion. The cells were introduced intrathecally into CCI rats, and changes in thermal and mechanical hyperalgesia were assessed. Substantial reduction of threshold values was observed. Reversible activation of enkephalin expression could also be demonstrated in vivo.

Other Uses of Tetracycline-Regulated Systems

To answer the question whether cardiovascular precursor cells are more effective than vascular precursor cells in cardiovascular cell therapy, i.e., whether transplantation of the appropriate cell type is critical for obtaining maximally effective stem cell therapy, a murine stem cell line was established that expressed Notch 4 under the control of a TetOn system [54]. Expression of Notch 4 in Flk1+ hematopoietic and vascular progenitors of this line re-specified the cells to a cardiovascular fate [55]. Transplantation of doxycycline-exposed cells in a mouse model of surgically induced myocardial infarction resulted in a substantially greater mean ejection fraction than that produced by cells that had not been doxycycline-activated [54]. These results imply that cardiovascular precursors are more effective than vascular progenitors in improving cardiovascular function after myocardial infarction.

Regeneration of mouse beta cells was thought to be largely dependent on induction of proliferation of terminally differentiated beta cells, although pancreatic progenitor cells were known to exist around the duct, which cells may contribute to regeneration. To answer the question whether non-beta cells can contribute to beta cell regeneration, a transgenic mouse model was established harboring an rtTA transactivator gene driven by a rat insulin II promoter and an rtTA-responsive p21 transgene [56]. Doxycycline-induced p21 expression inhibited proliferation of beta cells. Animals in which beta cell proliferation had been blocked showed better recovery from streptozotocin-induced diabetes than controls. Markers of early pancreatic development and pancreatic progenitors could be detected in islets of these mice. These results suggested that in the absence of beta cell proliferation, pancreatic progenitor cells contributed to recovery from induced diabetes.

A more recent study achieved nearcomplete beta cell ablation by means of diphtheria toxin [57]. A large fraction of regenerated beta cells were found to be derived from glucagon-producing alpha cells. The approach chosen for lineage tracing involved the use of a TetOn system to inducibly label glucagonproducing alpha cells.

Mouse and human fibroblasts can be re-programmed to a pluripotent state by co-expression of transcription factors Oct4, Sox2, Klf4 and c-Myc. Four lentiviral vectors were prepared, each of which contained a gene for one of these transcription factors functionally linked to a CMV minimal promoter supplemented with TRE sites [58,59]. Mouse embryo fibroblasts were engineered to constitutively express transactivator rtTA-M2. Cells co-transduced with the four lentivectors could be induced by doxycycline to undergo reprogramming. Selected pluripotent cells were used to produce chimeric mice. Cells derived from these mice could be reprogrammed by doxycyline addition with an efficacy that was 25-50 times better than direct reprogramming through lentiviral transduction and selection for pluropotency markers. This system should be well suited for studies of reprogramming as well as for genetic or chemical screens for factors involved in or capable of modulating reprogramming. In another study, the same technology was employed for demonstrating that embryonic and postnatal mouse fibroblasts could be rapidly and efficiently converted to neuronal cells by expression of transcription factors Ascl1, Brn2 and Myt11 [60]. These findings are expected to have an important impact on studies of neural development, disease modeling and human regenerative medicine. With regard to their potential future application in regenerative medicine, there has been concern that any one of the many viral insertions required to convert fibroblasts to inducible pluripotent stem cells may result in unpredictable genetic dysfunction. To reduce this possible risk, different methodologies for reprogramming cells to a pluripotent state were investigated. In one such study, human embryonic fibroblasts were co-transfected with two piggyback transposons, one comprising a doxycycline-inducible expression unit encoding transcription factors Oct4, Sox2, Klf4 and c-Myc, and the other a CAG promoter-driven rtTA gene [61]. Clonal lines that displayed embryonic stem cell morphology and were positive for alkaline phosphatase were obtained.

Gene regulatory networks may be analyzed by systematically disturbing them by means of overexpression of a transcription factor and then examining the genomic consequences. One such study focused on 50 selected transcription factors [62]. Transcription factor genes were introduced by recombinase-mediated cassette exchange into mouse embryo fibroblasts containing a tetracycline-repressed gene expression system at the Rosa26 locus [63]. This system that is referred to as ROSA-TET locus comprises a tTA gene driven by the Rosa26 promoter and, flanked by loxP and loxPV sites, a hygromycin resistance gene controlled by a TRE-containing mimimal CMV promoter. Cdx2 was identified as the factor that caused the most dramatic transcriptome perturbation. The cell lines generated will be a resource for future studies of biological networks in embryo fibroblasts and mice.

To facilitate studies of kidney diseases, a transgenic mouse line was generated that expressed transactivator rtTA2s-2M under a Pax8 promoter, which directed transactivator expression to renal tubular cells [64]. Inducible kidney disease models were generated by crossing Pax8-rtTA mice with TetO-MYC mice (model of polycystic kidney disease) or TetO-Tgf-beta1 mice (renal fibrosis). Triple transgenic mice, Pax8-rtTA/LC-1/Tsc1$^{flox/flox}$, when treated by doxycycline in utero, produced newborns with massive hyperproliferation in tubules and collecting duct epithelial cells, apparently as a consequence of the induced deletion of the tuberous sclerosis complex-1 (Tsc1) gene. These mice died a few weeks after birth with giant polycystic kidneys characteristic of autosomal recessive polycystic kidney disease. LC-1 mice contain a tetracycline-inducible Cre recombinase.

Sudden infant death syndrome (SIDS) is associated with robust morphological and biochemical deficits in serotonin neurons in the brain stem of SIDS brains, suggesting that a reduced serotonin function is a risk factor for the disease. Normal levels of serotonin neuron firing are maintained by a feedback mechanism involving serotonin autoreceptor Htr1a. A model for serotonin dysfunction was generated by crossing mice containing a Tet regulator-responsive endogenous Htr1a allele with mice expressing a TetOff transactivator under the control of the endogenous promoter of the serotonin transporter gene Slc6a4 [65]. In the absence but not in the presence of doxycycline, the double transgenic animals developed sporadic brachycardia and hypothermia, which frequently resulted in death. Interestingly, and in many ways reflective of the human condition, a time window could be defined (P25-80) within which the animals were particularly sensitive to Htr1a overexpression. This study suggests that a deficit in serotonin function is a sufficient cause for SIDS and should help to channel efforts to find a treatment for this condition.

Amplification of high-capacity adenovectors (HC-Ads) requires a helper virus (HV). Frequently, yields of HC-Ad are low, and preparations are contaminated with significant quantities of HV. Self-inactivating HV AdTetCre is an E1/E3-deleted Ad whose packaging signal is flanked by LoxP sites and that contains a constitutively expressed gene for transactivator rtTA-M2 as well as an rtTA-responsive gene for chimeric Cre-recombinase MerCreMer [66]. MerCreMer is a fusion of Cre-recombinase and two mutated estrogen receptor ligand-binding domains. The latter domains control the nuclear localization of the fusion protein. HV genome inactivation by removal of the packaging signal involves prior doxycycline-induced expression of chimeric Crerecombinase which translocates into the nucleus in the presence of ligand 4-hydroxytamoxifen. This system was used to produce at high yield HC-Ad that was only minimally contaminated by HV.

Tet systems were also used to produce packaging cell lines for deleted viruses. HEK 293 cells expressing a TetOff transactivator were engineered to express adenovirus L4 100K, 22K and 33K proteins under doxycycline control [67]. The cell lines were capable of packaging Ad5 vectors with novel late gene (L4) deficiencies. A producer line for efficient packaging of a non-primate lentivirus vector (based on equine infectious anemia virus, abbreviated EIAV) was derived from HEK 293 cells and included regulated Gag/Pol and VSV-G envelope protein genes [68]. Conditional expression of the latter transgenes was important because VSV-G is cytotoxic and Gag/Pol may be metabolically burdensome. Regulation was achieved by means of TetR expressed from a constitutively active gene and CMV promoters for driving the transgenes, into which promoters two TetO sequences had been inserted between TATA box and transcription start site (also known as T-Rex system; Invitrogen). To validate the line, it was stably transfected with an EIAV genome cassette encoding ProSavin, a lentivirus vector intended as a gene therapeutic for Parkinson's disease. Upon induction, ProSavin was produced with a titer that was comparable to that attained with the conventional transient packaging system. It is noted that Clontech Laboratories introduced their TetExpressTM Inducible Expression System in February 2011. At the core of this system is a membrane-permeable version of the previously marketed Tet-Off Advanced transcription activator protein. This transactivator is capable of entering cultured cells directly and activating a TetR-responsive gene of interest.

Judging from the number of recent publications, tetracycline repressor-derived gene switches are widely employed in research. To our knowledge, they are not being incorporated in vectors intended for human therapy. The study of Mättö and colleagues reminds us that tetracycline and doxycycline are antibiotics [69]. A 10-day course of antibiotic concentrations of doxycycline (150 mg/day/human subject) resulted in a dramatic increase in bifidobacteria resistant to the antibiotic. Bifidobacteria are normal constituents of the intestinal fauna. Hence, human use of Tet systems may not be unproblematic.

Gene Switches Comprising Ecdysone Receptor-Derived Transactivators

Insect ecdysone receptor (EcR) forms a dimeric transactivator with ultraspiracle protein (USP). USP is the insect homolog of vertebrate retinoid X receptor (RXR) [1]. The EcR/USP complex is stabilized by binding of ecdysteroids (insect steroids) to EcR, and the affinity of the complex for ecdysteroid response elements (EcRE) is enhanced. An early gene switch comprised chimeric receptor VpEcR and a responsive promoter. VpEcR is a fusion between Drosophila melanogaster EcR and an HSV VP16 transactivation domain. In a mammalian cell VpEcR forms a complex with RXR and, in the presence of an ecdysteroid, effectively binds to EcRE sequences in a promoter, resulting in activation of the promoter (Figure 2A). An improved gene switch comprised VgEcR, a mutated version of VpEcR capable of binding E/ GRE sequences, and a promoter supplemented with E/GRE sequences for mediating expression of a gene of interest. E/GRE is a hybrid between a glucocorticoid response element and a response element for a type II receptor (such as EcR). This gene switch performed well in certain cells (e.g., HEK 293 cells) but needed to be supplemented with RXR from exogenous genes in other cells. A system that eliminated this requirement for extra RXR was based on chimeric Drosophila/Bombyx ecdysone receptor DBEcR that combined activation and DNA-binding domains of VgEcR with hinge region and ligand-binding domain of Bombyx mori EcR. DBEcR was found to be a potent transactivator in the presence of insect steroids such as ponasterone A or of nonsteroidal ecdysone agonists such as GS-E, a synthetic diacylhydrazine (also known as RSL1). Ecdysone receptor-based gene switches were also formulated in the two-hybrid format. G:CfE (DEF) fusion receptor comprises the D, E and F domains of Choristoneura fumiferana EcR (CfEcR) and a yeast GAL4 DNA-binding domain. Fusion partner V:MmR (EF) includes the E and F domains of mouse RXR and a VP16 activation domain. The dynamic range of the chimeric receptor was found to be around 9000 (based on measurements of reporter gene expression in the presence vs. absence of ligand). Nonsteroidal ligands were more effective than ecdysteroids. Subsequent improvements further increased ligand sensitivity of the system. A system consisting of an RSL1-activated two-hybrid transactivator and a GAL4 binding site-containing target promoter for driving transgene expression is marketed by New England Biolabs under the trade name RheoSwitch (Figure 2B). [70,71,72] for detailed descriptions of selected ecdysone receptor-based gene switches.

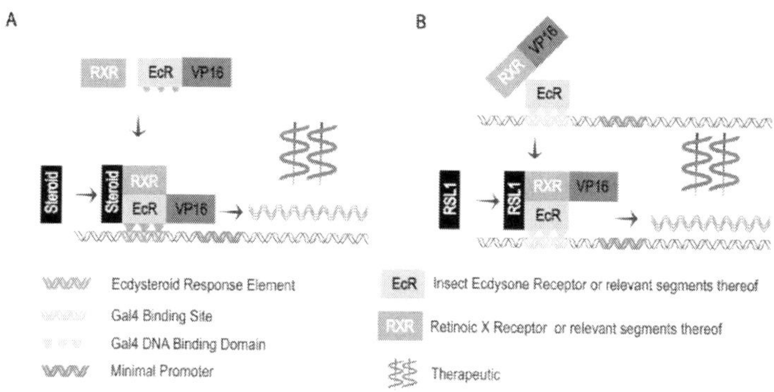

Figure 2: Ecdysone receptor-based gene switches. A."Classical" insect steroid-responsive gene switch. B. RheoSwitch. RSL: synthetic diacylhydrazine.

Research Concerning System Development & Improvements

The field of use of ligand-dependent transactivators and transinhibitors could be greatly expanded by the incorporation of zinc finger-based DNA-binding domains. Theroretically, any endogenous gene could be targeted by such transactivators and transinhibitors. A recent study reported the construction of a retroviral vector expressing a chimeric transactivator containing the CD54-31 zinc finger DNAbinding domain, a Drosophila melanogaster EcR ligand-binding domain, a human RXR ligand-binding domain separated from the EcR sequences by a 30-amino acid linker and the VP64 activation domain [73]. The CD54-31 domain is known to specifically interact with a sequence within the ICAM-1 promoter. Human cells transduced with this vector substantially upregulated ICAM-1 expression in response to ponasterone A.

The most potent ligands for ecdysone receptor-based transactivators are plant-derived ecdysteroids such as ponasterone A and synthetic diacylhydrazines such as GS-E (RSL1, New England Biolabs). Ecdysteroids contain multiple OH groups and have a less than ideal ADME profile. It was hypothesized that alkylation of one or more OH groups may be tolerated from an activity point of view and may reduce hydrophilicity, resulting in an improved ADME profile [74]. A number of alkylated ecdysteroids were synthesized and tested. A key finding of this study is that ponasterone A alkylated at the 22-position may be a better gene switch actuator than its parent as well as may have an improved ADME profile. Cell cycle dependence of induced expression of a target gene controlled by a RheoSwitch was investigated [75]. Induction of quiescence by either serum starvation or contact inhibition dramatically reduced induced gene expression in NIH3T3 cells. This phenomenon was observed regardless of whether non-steroidal inducer RSL1 or ponasterone A were utilized. It also could not be explained by a reduction of expression of the RheoSwitch transactivator components. In contrast, no marked difference in induced expression was observed between growing and quiescent NMuMG breast epithelial cells. These results indicate that activation of a RheoSwitch-controlled target gene is cell context-dependent. The findings would appear to be highly relevant for applications in which Rheoswitch-controlled gene expression is directed to normal tissues.

A novel gene switch entirely dispenses with the requirement for expression of insect EcR sequences and is based exclusively on human RXRalpha carrying Q275C, I310M and F313I mutations in the ligand-binding domain [76]. The mutant receptor is activated by synthetic compound LG335 but not natural ligand 9-cis-retinoic acid. A fusion transactivator consisting of a GAL4 DNA-binding domain and the mutant RXR was constructed. A retrovirus vector harboring an expressible gene for the latter transactivator and a GAL4-responsive gene of interest (GFP) was prepared. In experiments with NIH3T3 cells transduced with the viral vector, ligand addition to the cultures clearly enhanced GFP expression.

Ecdysone Receptor-Derived Systems Controlling Experimental Cancer Therapy

Recombinant IL-12 has been long known to have antitumor activity. Unfortunately, IL-12 treatments also have considerable adverse effects. Treatment-unrelated toxicity may be avoided by intratumoral genetic IL-12 therapy. Delivery to tumors of expressible IL-12 genes or dendritic cells expressing IL-12 has produced profound antitumor effects in animal models. Both to provide a tool for discovering the effective timing of IL-12 anti-cancer therapy and to create a vector with a safety advantage, a replication-defective adenovector was constructed that comprised a version of RheoSwitch termed "Rheoswitch Therapeutic System" (RTS) and a bicistronic IL12 target gene [77]. In RTS, coding sequences for VP16-RXR and Gal4- EcR, separated by the encephalitis virus internal ribosomal entry site, are transcribed from a human ubiquitin C gene promoter. Upstream from this transcription unit is placed a synthetic inducible promoter that drives expression of a gene of interest, in the instant case the bicistronic IL-12 gene unit, whereby the transcriptional orientation of the gene of interest is the same as that of the transactivator gene unit. C57BL/6 dendritic cells transduced with the adenovector were found to express IL-12 in a strictly ligand-dependent fashion. Transduced dendritic cells introduced into established B16 melanomas in C57BL/6 mice caused dramatic inhibition of tumor growth, provided that ligand was administered early after dendritic cell introduction and was maintained for at least about 5 days. Additional animal studies using the conditional IL-12-expressing adenovector were reported during the 2011 annual meeting of the American Society of Gene and Cell Therapy [78]. In the 2011 annual meeting of the American Society of Clinical Oncology, results of a first clinical trial were reported, in which trial safety and clinical activity of intratumoral administration of autologous dendritic cells transduced with the conditional IL-12- expressing adenovector in combination with oral ligand were tested [79]. The treatment was generally well tolerated. Of the 7 melanoma patients treated, partial or complete regression of injected and some uninjected lesions was observed in two patients. The same two patients had intratumoral changes in IL-12-associated gene expression as well as circulating CD8+ and/or CD4+ cells reactive against several melanoma-associated antigens.

Ecdysone Receptor-Derived Systems Controlling Experimental Therapy Of Neurologic Disorders Or Neurodegenerative Disease

Glutamate receptor subtype 5 is associated with disorders such as anxiety, depression or chronic pain. A cell line was described in which a newly cloned gene encoding mouse glutamate receptor type 5b had been placed under the control of the ponasterone A-responsive transactivator present in cell line HEK 293-EcR (Invitrogen) [80]. The cell line should be useful for in vitro characterization of receptorselective agonists and antagonists. Expansion of a polyglutamine repeat at the amino-terminal end of the huntingtin protein is the probable cause for Huntington's disease. A cell line that contained a RheoSwitch-controlled mutant huntingtin gene (comprising 72 Q polyQ length) was derived from HN10 cells [81]. (Vectors used were those available from Invitrogen Corp.) HN10 cells can be easily differentiated into post-mitotic neurons. Upon induction of mutant huntingtin expression, significant accumulation in intranuclear inclusions was observed as well as transcriptional changes resembling those occurring in animal models. Post-mitotic cells showed aconitase impairment, caspase 3 activation, neurite loss and reduced total protein characteristic of an ongoing neurodegenerative process. The cell line appears to represent a valid model of Huntington's disease and may be useful for drug screening applications.

Use of Ecdysone Receptor-Based Switches in Other Experimental Therapies

An AAV-RheoSwitch-GFP virus (obtained from Intrexon Corp., Blacksburg, VA) was used to transduce rabbit intervertebral disc cells [82]. Ligand-induced, reversible GFP expression was observed. Induced expression could be maintained for a period of more than a week. The virus (either 108 or 1.5 x 10^{11} particles per disc) was subsequently introduced into the L2-L3, L3-L4 and L4-L5 intervertebral discs of New Zealand white rabbits. Ligand was administered for 5 days by subcutaneous injection. Animals were sacrificed either 2 h after the final ligand dose or 5 days later. Immunofluorescence analysis revealed GFP expression in the disks of all animals that had received ligand. No GFP expression was detected in adjacent untreated discs, spinal cord, dura, bone, liver or brain. The study concluded that the RheoSwitch system should be a useful tool for controlling genetic therapies of intervertebral disc degeneration.

Another study made use of the RheoSwitch system for controlling expression of short hairpin RNAs (shRNAs) in stable cell lines [83]. Initially, a genomic microRNA (miR-122) fragment was expressed under RheoSwitch control. Correctly processed miR-122 was obtained upon induction with ligand RSL1. Downregulation of reporter genes as well as of endogenous miR-122 target genes could be demonstrated. The system was adapted to express any desired small interfering RNA (siRNA) under RheoSwitch control.

Mammalian Steroid Receptor-Based Gene Switches

Mifepristone-activated gene switches were developed subsequent to the identification of a mutant human progesterone receptor lacking 42 carboxy-terminal amino acids that failed to bind progesterone but instead interacted with progesterone antagonists such as mifepristone and enhanced transcription of reporter genes [84]. Mifepristone is an orally available approved drug substance also known as RU-486. An early chimeric transactivator contained an N-terminal transcriptional activation domain obtained from HSV VP16, the ligand-binding domain of the mutant receptor and a C-terminal GAL4 DNA-binding domain [85]. Improvements were made, which included the use of a transcriptional activation domain from human protein NF-κB in lieu of the VP16 sequence. This resulted in a partially humanized transactivator (GLp65) [86], in which the GAL4 DNA-binding domain remained the only nonhuman element. An autoinducible gene switch in which the transactivator gene was controlled by a basal promoter supplemented with GAL4 response elements was developed to further reduce undesired transactivation in the absence of mifepristone [87]. Shortening of the GAL4 DNA-binding domain resulted in transactivator GS4 which has a reduced propensity for spontaneous homodimerization and, consequently, a dimished ability to transactivate in the absence of ligand [88]. The commonly used concentrations of mifepristone for inducing gene expression are in a subclinical range (i.e., 0.1-10 nM). Still, there remain concerns that even these low doses may affect the ovarian cycle. Implanted mifepristone timed release pellets were utilized to avoid repetitive administration of the ligand [89]. A study that compared cytotoxicity of ligands of regulatable systems in a panel of normal and cancer cell lines found no toxicity of mifepristone and muristerone A but reported some detrimental effects for doxycycline [90]. Mifepristone-regulated systems have been tested extensively. A transgenic mouse was created that harbored a gene for a mifepristone-dependent transactivator controlled by an enhancer/promoter region from the transthyretin gene to restrict expression to the liver and a human growth hormone (hGH) reporter gene linked to a GAL4-responsive promoter [91]. Serum levels of hGH were very low in the absence of mifepristone, and administration of a single dose of ligand resulted in a transient increase. Transgene expression could be reinduced by administration of a second dose of the ligand. A similar approach was taken to express inhibin A in a mifepristone-dependent fashion in mouse liver [89]. The human keratin promoter [92], the human surfactant protein-C (SPC) promoter [93], the cardiac-specific α-myosin heavy chain (αMHC) promoter [94-96] and the mouse mammary tumor virus (MMTV) promoter [97] are among the promoters used in bigenic mice for tissuerestricting expression of mifepristone-dependent transactivator and, consequently, target transgenes. A HSV vector was constructed from which a beta-galactosidase reporter was expressed in a mifepristonedependent fashion in the brain [98]. Adenoviral vectors were also used to achieve mifepristone-regulated expression in vivo of hGH [86], human and mouse IL-12 [99] and vasodilatadory atrial natriuretic peptide [100].

Progesterone Receptor-Derived Gene Switches: System Development

An all-in-one mifepristone-regulated system (pBRES) was developed that contains a transactivator as well as a transgene cassette [101]. Plasmids having the two cassettes in the four possible relative orientations were constructed. The transactivator, a fusion protein consisting of the yeast GAL4 DNA-binding domain, the transcriptional activation domain of the human NF-κB p65 subunit and a truncated ligand binding domain of the human progesterone receptor was expressed under the control of a chicken skeletal muscle alpha-actin promoter and the transgene was controlled by a minimal promoter containing six copies of 17-bp GAL4 binding sites. The in vitro performance of the four pBRES plasmids was tested using different inserted genes, i.e., genes for human interferon-beta, human erythropoietin, human and murine secreted alkaline phosphatase, and firefly luciferase. The system could be improved by replacement of the chicken alpha-actin promoter with a human ubiquitin B promoter which resulted in lower background levels in the absence of ligand. The pBRES plasmids were also tested in vivo following introduction into murine hind limb muscles by electroporation and intraperitoneal administration of mifepristone. From both in vitro and in vivo experiments it appeared that the head-to-head orientation of gene cassettes resulted in elevated background activity and, consequently, lower induction. The human interferon-beta gene in a pBRES system having the two cassettes arranged in a tail-to-tail orientation was transferred to an AAV shuttle vector. Recombinant AAV1 particles were injected into the hind limb of mice. During a one-year observation period the therapeutic gene remained capable of being switched on by administration of mifepristone. In a subsequent study, a murine interferon-beta gene (mIFNb) was introduced into the four different pBRES plasmids, which were electroinjected into the hind limb muscles of mice [102]. Transgene expression was followed by measuring the induction of IP-10, a chemokine that is specifically upregulated by type I interferons. All four pBRES mIFNb vectors exhibited approximately the same level of induction, and one preferred plasmid having a tailto-tail arrangement of gene cassettes was used for subsequent studies. Inducible expression of IP-10 could be demonstrated over a period of more than three months. Chemokine expression was proportional to plasmid and mifepristone dose. The same plasmid was also tested in a murine model of acute experimental allergic encephalomyelitis that mimics the symptoms of multiple sclerosis. Mice were immunized by subcutaneous injection of proteolipid protein (in complete Freund's adjuvant supplemented with M. tuberculosis H37Ra) followed by intraperitoneal injection of pertussis toxin. Plasmid administration and concomitant implantation of mifepristone pellets resulted in a significant delay in the onset of disease and a decrease in the overall severity of disease scores compared to animals that had received plasmid but not ligand.

A library of mifepristone analogs was screened to identify a ligand that had reduced anti-progesterone receptor activity [103]. Compound BLX-913 was not able to induce the pBRES system and had a 30 fold lower antiprogestin activity than mifepristone. A homology model of the pBRES transactivator ligand-binding pocket was developed to search for steric conflicts between the inactive compound and amino acids lining the binding pocket. Site-directed mutagenesis was used to create a transactivator that can be triggered by BLX-913 and that is similarly effective in target gene transactivation as the original mifepristone-activated factor.

To reduce the likelihood that mifepristone-regulated transactivator binds to eukaryotic promoters, an alternative transactivator was developed that contained bacterial repressor LexA sequences instead of a yeast GAL4 DNA-binding domain [104]. The operator-promoter sequence consisted of a synthetic LexA operator containing four ColE1 operator sequences that was joined to a minimal 35S promoter from Cauliflower Mosaic Virus. The system was successfully tested in transgenic zebrafish, in which EGFP reporter expression was induced upon administration of mifepristone in the water and downregulated upon ligand withdrawal. The system also performed well in transgenic zebrafish lines carrying an inducible cell-toxic K-ras^{V12} oncogene.

To regulate the activity of the endogenous gene for vascular endothelial growth factor A (VEGF-A), a fusion gene was assembled that encoded a chimeric transactivator consisting of a synthetic zinc finger protein DNA-binding domain capable of targeting the VEGF-A gene promoter, a truncated ligand-binding domain from the human progesterone receptor and an activation domain from the p65 subunit of human NFkB [105]. Transient and stable transfection experiments employing different cell lines demonstrated induced, dose-dependent secretion levels of VEGF-A in the presence of mifepristone and low background expression in its absence. Induction of VEGF-A expression was reversible; VEGF-A level returned to basal within 48 hours after removal of inducer.

Progesterone Receptor-Derived Gene Switches: Experimental Cancer Therapy

An adenovirus capable of expressing TRAIL under the control of a mifepristone-inducible system was constructed [106]. Human lung carcinoma cells transduced with the adenovirus did not express detectable levels of TRAIL in the absence of mifepristone, but the transgene could be induced by mifepristone in a dose-dependent manner. The antitumor efficacy of the virus was tested in a mouse xenograft model of human lung tumor. Virus was injected in the tail vein, and mifepristone was administered intraperitoneally. This treatment resulted in significant suppression of tumor development when compared to animals that only received virus but not ligand.

Soluble VEGF receptor 1 (sflt1) was expected to neutralize vascular endothelial growth factor (VEGF) function and inhibit tumor growth. Two high capacity adenovectors were prepared, of which one contained a TTRB promoter (liver-specific)-controlled gene for a mifepristone-regulated transactivator and a transactivator-responsive sflt1 gene and the other a constitutively active sflt1 gene driven by an EF1alpha promoter [107]. Functional assays demonstrated that soluble receptor expressed from both vectors inhibited VEGF function. The two vectors were injected in tumor-bearing C57BL/6J mice via the tail vein; transient suppression of aggressive tumor growth was observed as a result of either constitutive or inducible expression of soluble receptor. Unexpectedly, mice receiving the virus that constitutively expressed sflt1 developed kidney damage, ascites and died thereafter. Similar outcomes were observed in mice that were injected with the virus carrying the inducible system and received mifepristone pellets. However, when inducer was administered intraperitoneally on a daily basis, which resulted in transiently elevated levels of soluble receptor, animals were free of ascites and survived. These results indicate that intermittent inhibition of VEGF function can be an effective antitumor strategy.

The immunomodulatory cytokine interleukin-12 (IL-12) is a potent inducer of interferon (IFN)-gamma and exerts antitumor effects against liver cancer. To be able to modulate the intensity and duration of cytokine expression, the two murine IL-12 chains were placed under the control of either a TetOn gene switch (using transactivator rtTA2s-M2) or a mifepristone-inducible system (using transactivator GLp65) [108]. To restrict production of IL-12 to the liver, transactivator expression was directed by liver-specific promoters. Plasmid containing the doxycycline-inducible IL-12 expression system was administered to immunocompetent mice by hydrodynamic injection. The mifepristoneinducible system was delivered by intravenous or intrahepatic injection of a high-capacity adenovector. Upon administration of doxycycline or mifepristone, respectively, IL-12 levels in the serum of the animals initially rose but then experienced a time-dependent substantial reduction. Inducibility could be restored after a resting period. Inhibition of murine IL-12 production was associated with diminished transactivator expression, an effect that was caused by interferongamma-mediated downmodulation of liver-specific promoters. To neutralize this problem, the mifepristone administration regimen was optimized [109]: mifepristone was administered intraperitoneally with stepwise increases that maintained sustained levels of IL-12 for more than 10 days in mice that had been administered high capacity adenovector intrahepatically. The induction protocol could be repeated after a resting period. The improved mifepristone dosing regimen was tested in experiments that assessed the efficacy of IL-12 gene therapy either as a monotherapy or in combination with oxaliplatin. Liver tumors were produced by direct administration of MC38Luc1 cells to the livers of syngeneic mice. Vector was introduced in the proximity of tumor nodules. Compared to animals treated only with oxaliplatin or only with IL-12 gene therapy, the combination of both strategies achieved the best therapeutic effects and long term survival. Combination of IL-12 gene therapy and oxaliplatin chemotherapy also was most effective in protecting animals against tumor rechallenge. The same adenovector expressing murine IL-12 was also tested in a model of HBV infection in woodchucks chronically infected with woodchuck hepatitis virus (WHV) [110]. Repression of liver-specific promoters appeared to be less dramatic in woodchuck than in mice. Direct intrahepatic injection of the vector in animals with chronic WHV infection and stable viremia followed by intraperitoneal administration of mifepristone resulted in high-level expression of IL-12 and IFNgamma around the injection site. Woodchucks with low basal viremia levels (below 10^{10} viral genomes/ml) exhibited an intense and sustained decrease of serum WHV DNA upon IL-12 induction. This antiviral effect was associated with induction of T-cell immunity against viral antigens and reduction of hepatic expression of transcription factor Foxp3, a key marker for regulatory T cells.

Progesterone Receptor-Derived Gene Switches In Other Disease Models

An intestinal cell line capable of insulin processing and glucose sensing was derived from mouse GTC-1 cells. The line contained a human preproinsulin gene that was controlled by a mifepristonedependent transactivator, which was expressed from a glucosedependent insulinotropic polypeptide (gip) promoter [111]. A second cell line was generated, in which the transactivator gene was controlled by a minimal thymidine kinase promoter supplemented with GAL4 binding sites to constitute an autoregulatory loop. While in the two cell lines human insulin content increased with mifepristone dose and was negligible in the absence of the ligand, total insulin production was higher when feedforward control of transactivator expression was utilized. The better system was then tested in a mouse model of streptozotocin-induced diabetes. Engineered cells were encapsulated and transplanted into the peritoneal cavity of mice. Upon administration of mifepristone pellets blood glucose in streptozotocintreated animals decreased to levels indistinguishable from those in streptozotocin-naive controls. Glucose levels in mifepristone-treated, transplanted diabetic mice remained lower than those in untreated animals during oral glucose tolerance tests.

Another study that related to weight management utilized a mifepristone-regulated transactivator expressed under the control of the gip promoter for dosing leptin [112]. A GTC-1 cell line was generated in which leptin production was stringently dependent on mifepristone dose. Transplantation of the conditionally leptinproducing gut cells under the kidney capsule of ob/ob leptin-deficient mice followed by implantation of mifepristone pellets led to reduced food intake and steady loss of body weight when compared to animals that were transplanted with engineered cells lacking the transgene component. Reduction in food intake and loss of body weight correlated with the leptin level in the circulation as revealed by experiments that compared two different mifepristone doses. Plasma insulin levels also decreased in a mifepristone dose-dependent fashion and remained significantly reduced in mice that had received leptin-producing cells. Interestingly, blood glucose levels remained lower in the group of mice transplanted with leptin-producing cells than in the control group, with the leptin-treated group remaining normoglycemic and the controls hyperglycemic. This effect lasted for 38 days after transplantation, i.e., much longer than the leptin-induced reduction of food intake and loss of body weight. However, the mifepristone-induced leptin cell therapy did not appear to induce weight loss in C57Bl/6J mice maintained on a high-fat diet.

Other Uses of Progesterone Receptor-Derived Gene Switches

A mifepristone-inducible system was also used for preventing expression of cytotoxic and anti-HIV genes during production of lentivirus particles [113].

A transactivator-responsive promoter was subcloned into a self-inactivating lentiviral vector plasmid in either the forward or the reverse orientation with respect to the direction of viral genomic RNA. The activities of several transgenes were examined in lentivirus-infected cells that expressed a mifepristone-dependent transactivator. Infectious lentiviral particles could be produced with a human CD14-coding sequence inserted in lentiviral vector plasmid in either direction. However, the forward orientation resulted in high levels of background expression in the absence of the inducer, which was likely due to transcriptional interference by a heterologous promoter required for Tat-independent transcription of the viral RNA genome. When a cytotoxic gene was assayed, i.e., the vesicular stomatitis virus M gene, the reverse orientation but not the forward orientation resulted in high titers of virus. The reverse vector was able to overcome self inhibition caused by an anti-HIV transgene, a dominant negative mutant of human VPS4B that blocks HIV-1 release and infectivity. The study demonstrated that a lentiviral vector that transduces a harmful gene can be produced, provided that a mifepristone-inducible expression unit comprising the harmful gene is introduced in the virus in the reverse orientation.

Estrogen Receptor-Derived Systems

Estradiol-regulated expression systems are based on a fusion protein that includes a human estrogen receptor (ER) ligand-binding domain and is transcriptionally active in the cell nucleus. The ER domain sequesters the fusion protein in the cytoplasm until ligand binds to the ER domain and translocation to the cell nucleus can occur. A completely humanized transactivator named HEA was created that incorporated a DNA-binding domain from hepatocyte nuclear factor-1, the G521R mutant of the ligand-binding domain of human ER-alpha that binds anti-estrogen 4-hydroxytamixofen but not estradiol, and the activation domain from the p65 subunit of human NFκB [114]. The system exhibited robust and sustained inducibility of an erythropoietin transgene in muscle of mice that received tamoxifen, which is converted to 4-hydroxytamoxifen in vivo. Predictably, administration of this inducer in the context of long-term human gene therapy will raise safety concerns, because the doses needed to activate HEA (in mice) are comparable to those used in clinical practice. Prolonged treatment with tamoxifen is associated with an increased risk for endometrial cancer due to its partial agonist activity on endogenous ER-alpha. These concerns were met by the identification of an ER ligand-binding domain double mutant (L384M, M421G) with a decreased affinity for estradiol and enhanced binding to compounds that do not interact with estrogen receptors [115]. A triple mutant including the G521R substitution was insensitive to physiological concentrations of estradiol and had nanomolar affinity for novel ligand CMP8. The latter mutated ER ligand-binding domain was grafted onto chimeric transactivator HEA. The improved transactivator induced a human secreted alkaline phosphatase reporter gene by about 2000 fold in the presence of CMP8, but remained inactive in the presence of estradiol.

Regulation of Protein Activity Based On A Steroid Receptorligand Interaction

Regulation by 4-hydroxitamoxifen has been exploited in inducible gain-of-function and loss-of-function studies using transgenic animals [116]. In these systems, one transgenic mouse strain carries a gene for a 4-hydroxytamoxifen-regulated Cre or Flp recombinase, i.e., an estrogen receptor ligand-binding domain-recombinase fusion protein. The second strain carries a conditional allele of a gene of interest. When a mouse inherits both transgenes and is exposed to 4-hydroxytamoxifen, the recombinase is translocated to the nucleus where it inactivates the gene of interest by removing an essential exon flanked by LoxP or FRT sites, respectively, from the gene or activates gene expression by removing a stop cassette. Using the same technology, temporal regulation of reporter genes was obtained upon electroporation of plasmids carrying a reporter gene into the embryonic mouse brain or postnatal retina [117]. Spatio-temporal control of transgene expression could be achieved by means of ligand-regulated recombinase genes that were driven by cell- or tissue-specific promoters; transgene expression could be so confined, e.g., to rod photoreceptors, bipolar cells, amacrine cells and glia [43], melanocytes [118], epithelial cells [119], osteoblasts [120] and podocytes [121]. A mifepristone-inducible Cre recombinase was obtained by fusion with the ligand-binding domain of a mutant human progesterone receptor [122]. More recently, high-efficiency fusion variants of Flp and PhiC31 recombinases were engineered. These fusion recombinases mediate site-directed gene insertion in thepresence of mifepristone but exhibit low levels of activity in the absence of ligand [123].

A recent study attempted to direct inducible Cre recombinase activity to keratin-positive cells of the airways [124]. Bigenic reporter mice were created by crossing a strain that expressed a truncated, mifepristone-inducible progesterone receptor-Cre recombinase fusion protein under the control of a keratin 5 or 14 promoter with a Rosa26 reporter strain in which a loxP-neomycin-loxP sequence separated constitutively active Rosa26 promoter and beta-galactosidase reporter gene (R26-LacZ). Other bigenic reporter mice generated expressed a Cre fusion protein that contained a shortened progesterone receptor sequence to reduce background recombinase activity. A solvent system of 20% acetone/80% sesame oil with isoflurane anesthesia was established for delivering mifepristone to the trachea. Ligand administration induced significant tracheal Cre activity as shown by beta-galactosidase staining. The fusion protein with the shortened progesterone receptor segment exhibited the most tightly regulated and robust recombinase activity. In another study a gene for a fusion protein was assembled that consisted of a green fluorescent protein, Cre recombinase and a progesterone receptor ligand-binding domain [125]. Because of the low background activity and natural sensitivity to mifepristone of the lactoferrin promoter, the fusion protein gene was placed under the control of a truncated lactoferrin promoter. A transgenic mouse line was created, and crossed to Cre reporter strain R26-LacZ. Only the salivary glands of bigenic mice that had received implants of mifepristone pellets displayed X-gal staining, suggesting that the fusion protein was expressed and activated specifically in salivary glands.

To minimize basal transgene expression, another study combined ligand regulation at the transcriptional and protein activity levels [126]. Bigenic mouse lines were generated by breeding a transgenic mouse line that expressed chimeric factor Glp65 under the control of an epidermal-specific keratin 14 promoter (which was used to direct gene expression to the keratinocytes of the basal epidermis and hair follicles) with a transgenic mouse line that carried a GLp65-responsive gene for inducible caspase-3 or caspase-9 precursors. The Glp65 fusion protein was exclusively expressed in skin keratinocytes, which in the presence of mifepristone initiated caspase precursor expression. Administration of lipid-permeable rapalog AP20187 (see below) caused dimerization of inactive caspase precursor molecules, which dimerization resulted in their self activation. The double induction system was tested in neonates by delivering mifepristone in utero, along with progesterone to counter the abortion side effect, followed by topical administration of AP20187. The skin of the so treated neonates showed peeling, appeared dehydrated and exhibited strong induced expression of caspase 3 and 9 when compared to the skin of control littermates.

Dimerizer-Activated Gene Switches

Dimerizer-activated gene switches make use of "heterodimeric" transcription factors in which DNA-binding and activation domains are on separate polypeptides that have no intrinsic affinity for one another but can be held together by a small dimerizer molecule. The original gene switch made use of small immunosuppressive molecule rapamycin which interacts with immunophilin FKBP12 and as rapamycinFKBP12 complex with mTOR/FRAP. FRAP is a phosphoinositide3-kinase homolog that is involved in the control of cell growth and division. A fully "humanized" dimerizer-activated transactivator was engineered (Figure 3A): a first fusion protein comprised human zinc finger and homeodomain-binding domain ZFHD1 and three copies of FKBP12, and a second fusion protein included a 100-residue domain of FRAP (FRB) and a transcription activation domain from the p65 subunit of human NFkB. Subsequently, an activation domain from human heat shock factor HSF1 was added. A minimal IL-2 promoter supplemented with 12 ZFHD1 binding sites served to control expression of a chosen target gene. Stringent regulation by rapamycin of target gene expression was demonstrated in various in vitro and in vivo models. The system was made more attractive by the development of rapalogs that are non immunosuppressive analogs of rapamycin which retained the ability to function as dimerizers. The technology is presently commercialized under the trade name ARGENT by Clontech Laboratories. A second dimerizer-activated gene switch uses small molecule FK506 as dimerizer [1,127,128,129].

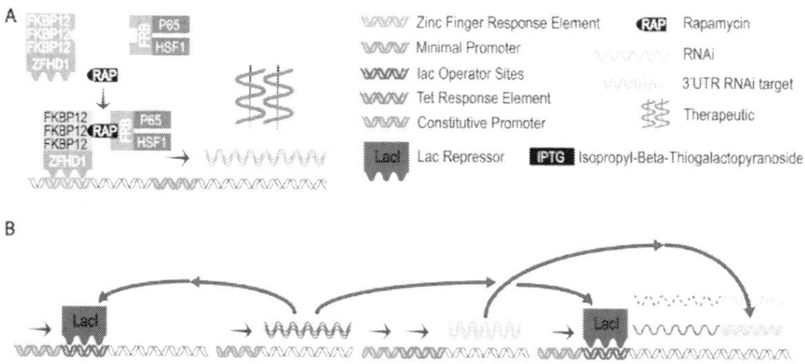

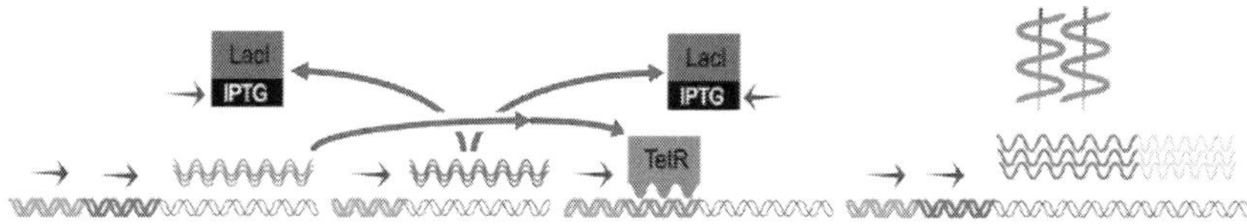

Figure 3: Dimerizer-induced gene switch and complex tunable gene switch. A. Rapamycin/rapalog-inducible gene switch. B. Lac-Tet-RNAi gene switch. Gene expression patterns in the absence (upper portion) or presence of inducer IPTG (lower portion) are outlined.

To further improve control of expression of secreted therapeutic proteins, a system was devised in which not only transcription but also exocytosis was rapalog-regulated [130]. The system consisted of two lentiviral vectors. Both vectors contained the central Flap sequence to improve transduction of non dividing cells [131]. The first vector, the transactivator expression vector, comprised a CMV promoter for directing expression of a bicistronic cassette containing coding sequences for the two subunits of dimerizer-activated transactivator separated by an internal ribosome entry sequence derived from the encephalomyocarditis virus. The second vector, the trophic factor secretion vector, included a transactivator-responsive promoter that mediated expression of a SS-4xFKBP-36M-FCS-target polypeptide fusion protein. The fusion protein-coding sequence was followed by a woodchuck hepatitis virus responsive element to enhance production of the fusion protein. SS refers to a signaling sequence, 4xFKBP36M to four tandem repeats of FKBP12 carrying the F36M mutation (transforming the FKBP into a conditional aggregation domain), and FCS to a furin cleavage site [132]. As target polypeptide either GFP or neurotrophic factor GDNF was employed. In vitro, cotransduction of the two vectors resulted in robust expression of target proteins in the presence of rapalog AP21967, which proteins were effectively secreted. The proteins were essentially undetectable in the absence of rapalog. Induction of target protein expression was shown to be reversible. Analogous results were observed when the vectors were co-introduced into the striata of mice.

It is noted that in recent studies rapalogs appear to have been used primarily to regulate protein activity rather than gene expression. In one such study, an adenovector was prepared that expressed iCaspase under the control of an endothelial cell-specific vascular endothelial growth factor receptor-2 (VEGFR2) promoter [133]. iCaspase is a fusion protein containing two copies of FKBP12 variant V36 and procaspase-9 [134].

The adenovector caused apoptosis in the presence of rapalog AP20187 in human dermal microvascular endothelial cells but not in several human tumor cell lines. In vivo experiments showed that intra-tumoral delivery of the vector followed by i.p. administration of rapalog resulted in ablation of tumor microvessels and inhibition of tumor growth. Another study tackled the question whether activation of protein kinase Akt2 is sufficient to cause translocation of glucose transporter 4 to the cell surface, thereby enhancing insulin-regulated glucose uptake into muscle and fat cells [135]. 3T3-L1 adipocytes were engineered to express a fusion protein containing full-length Akt2 and rapamycin-binding domain FRB. To recruit the Akt2 fusion protein to the cell membrane, the cells also expressed a fusion protein containing two copies of FKBP12 to which was added N-terminally a Src myristoylation signal. Rapalog addition resulted in activation of the fusion Akt2 as evidenced by phosphorylation of the factor at characteristic sites as well as phosphorylation of downstream substrates. Transient activation of the fusion Akt2 also increased glucose transport and glucose transporter 4 translocation to the plasma membrane, validating the hypothesis. An alternative insulin therapy involving insulin-secreting keratinocytes was evaluated in yet another study [136]. Keratinocytes were transduced with a lentivirus vector that expressed a fusion protein containing several copies of a selfdimerizing variant of FKBP12 and proinsulin preceded by a furin cleavage signal. The transduced keratinocytes could be induced by rapamycin to secrete insulin. Insulin secretion was observed within 30 min of rapamycin addition and ceased 2-3 hours after rapamycin removal. Transplanted lentivector-transduced keratinocytes secreted active insulin for weeks and were capable of repeated rapamycin stimulation of insulin secretion. Induced insulin secretion reversed hyperglycemia in athymic mice made "diabetic" by exposure to streptozotocin. These results showed that insulin produced in a regulated fashion by skin cells can control hyperglycemia. Finally, dimerization technology was employed to study the function of dynein light chains LC8 and TcTex1 [137]. The latter LCs bind as dimers to dimers of dynein intermediate chains (IC). The authors prepared LC traps which are fusion proteins between a selfdimerizing variant of FKBP12 and an LC-interacting sequence from IC. Evidence was obtained that these LC trap proteins dimerize and bind to LCs in the presence of rapalog. The system was employed to demonstrate that LCs are involved in lysosomal, endosomal and Golgi organization, but apparently are not critical to mitotic progression despite the central role of dynein in cell division.

Regulation or Coregulation by siRNAs and miRNAs

A problem common to most mammalian gene regulation systems is residual target gene expression when the systems are turned off. This is due to basal promoter activity as well as to random interactions of transactivators with their binding sites. Recently developed strategies make use of RNAi to more stringently control transgene expression. A tunable genetic switch termed LTRi (Lac-Tet-RNAi) couples repressor proteins and RNAi target modules to deactivate a transgene [138] (Figure 3B). In this system, bacterial repressor LacI is expressed constitutively from a cytomegalovirus promoter. In the absence of ligand isopropyl-beta-thiogalactopyranoside (IPTG), LacI binds to lac operator sites in the transgene, repressing transgene transcription. LacI also binds to lac operator sites located in a tetR gene, repressing its expression. In the absence of TetR, a short hairpin RNA (shRNA) is transcribed from a tetO site-containing active promoter. This shRNA binds to a complementary synthetic sequence element (originating from the E.coli beta-galactosidase gene) present in the 3′UTR of transgene mRNA, targeting for degradation any mRNA molecule made by leaky transcription. In the presence of IPTG, LacI is inactivated, and target and TetR genes are transcribed at their respective maximal rates. TetR so made binds to the TetO sites upstream of the RNAi gene, repressing transcription of the shRNA. The robustness of the LTRi system was tested in stably transfected cells, using an EGFP reporter gene as the transgene. The circuit showed greater than 99% repression in the off state and could be repeatedly switched between induced and uniduced states. The level of transgene expression was dependent on IPTG dose. To demonstrate how tightly transgenes can be regulated by the LTRi system, genes such as the highly toxic alpha chain of diphtheria toxin, the Cre recombinase and the proapoptotic gene bax were placed under its control.

The induction profile of several gene control switches was improved by a strategy that relied on the fact that the effectiveness of silencing increases with siRNA-to-target ratio [139]. siRNA was expressed from an intronic sequence within a gene for a ligand-activated transactivator, and siRNA sequence specific tags were placed in the transgene to enable silencing by bound siRNA. This approach succeeded in reducing the leakiness of an erythromycin and a tetracycline-responsive gene switch. Transfection experiments demonstrated the ability of the siRNA-based silencing system to control expression of two highly toxic gene products, the apoptotic RIP death domain and M. esculenta derived linamarase, which converts the innocuous linamarin into gaseous cyanide. Autoregulated silencing vectors were constructed by introducing an RNAi-coding sequence into a transactivator gene and by placing this gene under the control of a promoter that is responsive to the transactivator. Intronically encoded siRNA was also used as an element of a three component mammalian oscillator circuit that is capable of mimicking circadian regulation [140]. The first component consisted of a tTA gene controlled by a tetracycline-responsive promoter. An untranslated Photinus pyralis firefly luciferase-derived sequence tag was added at the 5' end of the tTA-coding sequence. The presence of this sequence tag enabled interference-based silencing of the tTA transcript by a luciferase-specific siRNA. The second component comprised a gene for a macrolide-responsive transactivator expressed from a tTA-responsive promoter. The latter transactivator gene included an intronic sequence that encoded luciferase-specific siRNA. The third component of the system consisted of a gene for a destabilized variant of yellow fluorescent protein d2EYFP controlled by a macrolide-responsive promoter. The activity of the system can be modulated by tetracycline and erythromycin. Leaky transcription from the tetracycline-responsive promoter leads to positive feedback, resulting in high levels of tTA as well as in increased expression of siRNA and macrolide-responsive transactivator. The macrolide-responsive transactivator will activate the d2EYFP gene. The siRNA mediates the breakdown of tTA, turning off the network. As a consequence, levels of d2EYFP and siRNA decrease, allowing the circuit to restart. Cotransfection assays indicated variations in cell fluorescence intensity that displayed an oscillatory behavior with a period of 26 hours.

New systems were recently described that exploit microRNAs (miRNAs) for transgene regulation [141]. miRNAs are small single-stranded, non-coding RNAs transcribed from DNA that is not translated, which RNAs function as regulators based on complementarity to mRNA transcrips of coding genes. Binding of a miRNA to a target mRNA results in translational repression and "gene silencing". By inserting complementary miRNA target sites into the 3′ untranslated region of transgene cDNAs, transgene expression can be subjected to endogenous miRNA regulation. Intrinsic differences in miRNA expression patterns can be exploited to finely control transgene expression [142]. However, a threshold level of miRNA must be reached for target mRNA suppression to occur.

Certain miRNAs are expressed broadly in the human body, whereas others display differential expression in developmental, tissue, or cell type-specific manners. A baculoviral vector for glioma suicide therapy was constructed harboring a herpes simplex virus thymidine kinase (HSVtk) gene controlled by a glial fibrillary acidic promoter for restricting transgene expression to the glial cell lineage [143]. The vector contained target sequences for three miRNAs that are enriched in astrocytes but downregulated in glioma cells. This strategy resulted in a significant improvement of in vivo selectivity over a control vector lacking the target sequences as observed in experiments involving injection of vector into the striatum of glioblastoma-bearing nude mice followed by daily intraperitoneal injection of ganciclovir for 5 days. In another study, the specificity of hepatic miRNA miR-122 was used to suppress transgene expression in the liver after intravascular administration of an AAV vector [144]. An analogous approach was used to restrict transgene expression to the central nervous system, detargeting liver, heart, and skeletal muscle, i.e., the tissues that are most efficiently transduced by systemically administered AAV vectors [145].

Overexpression of hypothalamic brain-derived neurotrophic factor (BDNF) results in weight loss. In a study aimed at developing a system for sustainable weight management, an AAV vector was constructed that contained a gene for BDNF controlled by a constitutive promoter [146]. The construct also expressed a specific miRNA directed against BDNF transcripts under the direction of the BDNF-responsive Agrp promoter. BDNF expression led to weight loss, increased Agrp activity and BDNF miRNA expression. The therapeutic efficacy of the dual cassette vector was tested in db/db mice, a genetic model of diabetes and obesity. Administration of the vector resulted in an initial weight reduction which was subsequently maintained, indicating efficient autoregulation of the BDNF transgene. A vector encoding BDNF and a scrambled miRNA caused progressive weight loss. The abovediscussed studies demonstrate that incorporation of a miRNA-based component can significantly contribute to transcriptional targeting or control of transgene expression.

TRANSGENE ACTIVATION BY FOCUSED HEAT OR IONIZING RADIATION

Heat Shock Promoters and Heat Shock Promoter-Based Systems

A common response of all living organisms to abnormally high temperatures or other environmental or metabolic stresses is the induction of a group of highly conserved proteins, termed heat shock proteins (HSPs). Stimuli that induce HSP expression include radiation, heavy metals, ischemia, nitric oxide radicals and the like, whereby elevated temperature arguably is the most robust inducer of hsp gene expression. All the stimuli share the ability to activate heat shock transcription factors (HSFs). Different hsp promoters differ in their inducibility. Based on reporter assays, the promoter of the human hsp70B gene (HSP A7) is one of the most highly heat-inducible promoters [1,147,148]. It is characterized by a very low basal activity that increases several thousand fold in a cell subjected to a sublethal heat treatment (heat shock) [149]. It is noted that the promoter of the highly related hsp70B' gene (HSP A6) has closely similar properties to the human hsp70B promoter. The architecture of hsp promoters shows considerable variation, but all heat-inducible promoters contain one or more sequence elements termed heat shock elements (HSEs) that are binding sites for heat shock transcription factors (Figure 4A). HSEs are the elements that confer heat regulation on the promoters and are comprised of three or more modules of the pentamer sequence NGAAN arranged in alternating orientation. Mammalian cells express several types of HSF, but heat induction of hsp genes cannot be observed in the absence of HSF1 [150]. In unstressed cells, HSF1 is largely present in an inactive form. Upon stress exposure, HSF1 is induced to homotrimerize and acquires the ability to bind to HSEs. Concentration of HSF1 in the nucleus and acquisition of transcriptional competence are further events that occur nearly simultaneously with homotrimerization. Activation of HSF1 is transient. HSF1 and, consequently, hsp promoters are inactivated within a few hours after an activating heat treatment.

Simple methods for administering heat to a cell or an experimental animal involve contacting with a heated surface or liquid. More sophisticated methods rely on microwave or infrared radiation, or ultrasound. As heat can be administered in a focused fashion, hsp promoters can be potentially used for spatially controlling transgene expression. In fact, several studies demonstrated that spatial accuracy of tissue heating can be achieved by focused ultrasound radiation guided by online MR thermometry [151]. Focused ultrasound induced hyperthermia was found to be capable of precisely inducing expression of a target gene under the control of an hsp70B promoter in the liver, a deep seated organ [152].

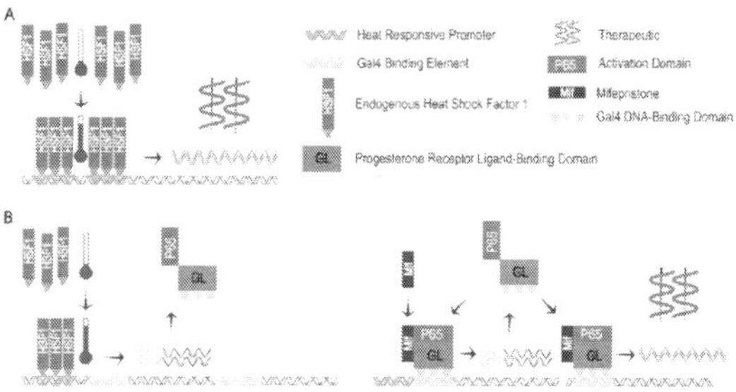

Figure 4: Heat-triggered gene switches. A. Heat shock promoter-based system. B. SafeSwitch system. Heat-activated expression of inactive transactivator is depicted on the left. On the right is shown that autoactivated synthesis of transactivator as well as induced expression of transgene product occurs subsequent to ligand addition.

HSF1 is expressed ubiquitously, and the mechanism for heat induction of hsp promoters exists in virtually every cell type. Thus, it is possible to use these promoters to uniformly control therapeutic gene activity throughout an organ comprised of different tissues or a tumor. In addition, peak induced activity of a hsp promoter is a function of both temperature and length of temperature exposure, allowing modulation of a therapeutic outcome by changing parameters of the activating treatment. As a generality, hsp promoters appear to be more efficient than tissue-restricted promoters.

Several groups developed in vivo models of localized activation of human hsp70B promoter-controlled transgenes [153,154,155,156]. Most of these studies related to cancer therapies in which the hsp70B promoter was employed for directing expression of cytokine, suicide or membrane fusion-inducing genes.

In a recent study, the human hsp70B’ promoter, an artificial promoter comprising five HSEs arranged in tandem and a combination promoter, in which the five HSEs were situated upstream from the hsp70B’ promoter, were compared [157]. A green fluorescent reporter gene was linked to each of the promoters, and activities were tested in transfected human pancreatic tumor cell lines. The combination promoter produced the highest reporter activity and was subsequently employed for controlling the expression of a mildly toxic variant of diphtheria toxin A in stably transfected cells. Heat shock completely arrested cell growth. The construct was also tested in human PSN1 tumor cells that were inoculated subcutaneously in nude mice. Tumor heating was by directly contacting the skin with a heated surface;

animals underwent two 15-min heat treatments at 42°C administered 72 hours apart. The heat treatments induced significant reductions in tumor volumes compared to unheated tumors, indicating that the engineered promoter is useful for controlling expression of toxic/ therapeutic products in vivo.

Heat inducibility of an hsp70B' promoter-driven luciferase reporter gene present either in a replication-deficient adenovirus or in a melanoma-targeted oncolytic adenovirus (Ad2xTyr) was analyzed in another study [158]. Heat treatment induced reporter expression from the replication-deficient adenoviral vector by about 8000-fold. Heat inducibility of luciferase expression was also strong in Ad2xTyr-infected permissive cells, but regulation of hsp70B' promoter activity was all but lost in the course of viral replication. Insertion of upstream insulator elements did not protect the promoter against this dysregulation. These results caution against an unquestioned use of inducible promoters in replicating systems, even if these promoters had worked well in replication-deficient vectors.

During recent years, efforts were made to develop systems capable of codelivering thermal energy and heat-inducible vectors. One approach involved the use of Mn-Zn ferrite magnetic nanoparticles that generate thermal energy when exposed to an alternating magnetic field (AMF) [159]. The nanoparticles were surface-modified with poly(ethylenimine) and loaded with plasmid DNA containing a hsp70B promoter-controlled bacterial beta-galactosidase gene. Cultured cells were subject to magnetofection with the resulting complexes and were subsequently exposed to AMF (230 kHz) for 60 min. Medium temperature increased as a function of nanoparticle dose. Induced beta-galactosidase activity could be observed in AMFtreated cells. The complexes were also injected into human tumors in nude mice; AMF treatment induced reporter expression, although this induction was less dramatic than that previously observed in cultured cells. While the feasibility of the approach was established by the study, concerns remain regarding the potential systemic toxicity of inorganic nanoparticles.

A self-inactivating retroviral vector was constructed that comprised IRES-linked human sodium iodide symporter (hNIS) and GFP cDNA genes controlled by a hsp70A promoter [160]. Heat treatment (42°C for 60 min) of stably vector-transduced human RG2 glioma cells produced a 39 fold increase in GFP-positive cells. Uptake experiments with $Na^{131}I$ and $^{99m}TcO4$-indicated accumulation of radiotracers after heat induction. Subcutaneous xenografts of stably vectortransduced RG2 cells in nude mice were subjected to heat treatment in a circulating water bath at 42°C for 60 min, and fluorescence, singlelabel scintigraphic imaging and ^{124}I-iodide micro-positron emission tomography (microPET) were performed 24 hours later. Low levels of GFP fluorescence and ^{99m}TcO4- accumulation were recorded prior to heat. A marked increase in both fluorescence and 99mTcO4- uptake was observed after heat exposure. microPET imaging undertaken subsequent to intravenous injection of ^{124}I-iodide revealed induction of hNIS expression.

Thymidine kinase activity produced by a replication-deficient type 5 adenovector containing an hsp70B promoter-linked gene for a HSV1-TK-EFGP fusion protein was evaluated by a tritiatedpenciclovir uptake assay in human head and neck squamous cell carcinoma SCC9 cells [161]. Increased cell-associated radioactivity was observed one day after heating at 41°C for 1 hour. Analogous results were obtained when EGFP expression was analyzed. Thymidine kinase activity was also examined in vivo in mice bearing subcutaneous SCC-9 xenografts that were directly injected with the adenovector. A temperature probe was placed in the tumor, and hyperthermia was delivered by means of an ultrasound transducer. A biodistribution study showed specific tumor uptake of tritiated penciclovir after heating at 41°C for 1 hour. In vivo expression of thymidine kinase was monitored noninvasively by microPET imaging after injection of 9-(4-[^{18}F]-fluoro-3- hydroxymethylbutyl) guanine substrate. Significant differences were observed between heated and non-heated tumors. These experiments demonstrated that PET imaging can be used to monitor heat induction of thymidine kinase gene therapy.

As also alluded to before, heat has become an attractive means for controlling gene expression due to the availability of technology for externally inducing local hyperthermia by high-intensity focused ultrasound radiation (HIFU) and measuring temperature distribution in the targeted tissues by magnetic resonance temperature imaging (MRI). In a recent study that was aimed at comparing local temperature distribution and heat-induced transgene expression, thigh muscles of transgenic NFL-1 mice that harbor a firefly luciferase gene driven by a mouse hsp70 promoter (HSPa1b) were heat-treated by MRI-guided HIFU [162]. Control of reporter gene expression was achieved by adjustment of the HIFU power based on MR thermometry. Luciferase activity was followed by bioluminescence imaging. These experiments revealed a close correspondence between spatial distribution of temperature and reporter gene expression. The noninvasiveness of the heating approach was evaluated by histological analysis of the heated tissue. No damage was observed for a 43°C/2 min hyperthermia protocol, whereas increasing the duration of the HIFU treatment to 8 min resulted in significant tissue alterations.

Owing to autoregulation, hsp promoters do not remain active for periods exceeding a few hours, and gene products typically reach peak concentration about one day after the activating heat treatment. Sustained expression of a hsp promoter-controlled therapeutic gene in a patient would require daily or twice daily heat treatment, resulting in a regimen that is impractical. Inadvertent activation of hsp promotercontrolled therapeutic transgenes in non-target tissues could be an important problem, e.g., in the context of cytotoxic or angiogenic therapies. Such activation could arise from an elevation in body temperature caused by disease, strenuous exercise, intoxication by heavy metals and other toxicants, certain pharmacological interventions, or ischemic events [163-171]. As exclusive delivery of a transgene to a desired tissue or body region it is not possible with currently available technology [172], this could result in undesired transgene expression and adverse effects in major organs. To activate a therapeutic gene only at the time and only for the duration required for effective therapy and to restrict expression of the gene to the tissue/organ in need of therapy, we designed regulatory circuits that combined an hsp70B promoter and a small molecule-activated gene switch [173]. These gene switches consist of (a) a ligand-activated transactivator expressed under the dual control of a promoter or promoter cassette that is responsive to heat and the transactivator and (b) a promoter responsive to the transactivator for controlling a gene of interest. In a cell containing such a gene switch and a target gene neither transactivator nor target gene are expected to be expressed in the absence of ligand and/or without an activating heat treatment. When the cell is exposed to a transient elevation of ambient temperature, endogenous HSF1 will be activated and will transactivate the ligand-activated transactivator gene. Subsequent to the heat exposure, HSF1 activity will decline and return to basal levels. In the absence of ligand, the transactivator will remain inactive, and the target gene will remain silent. In the presence of ligand, transactivator synthesized in response to the transient heat treatment will be activated and will begin transactivating any gene that is controlled by a transactivator-responsive promoter, i.e. the transgene will be expressed, and the transactivator will autoactivate its own gene. This feedforward mechanism is expected to sustain transactivator levels after HSF1 has been inactivated. Rate of synthesis and half life of the transactivator as well as the doubling time of the cell will determine the concentration of transactivator present in the cell, which concentration will eventually stabilize. In the presence of ligand, expression of the transgene will be maintained for an extended period. The system can be turned off by withdrawal of ligand. The main advantage of these gene switches is that they cannot be activated in the absence of the specific ligand, providing a solution to the problem that hsp promoters are subject to inadvertent activation in nontarget locations under certain physiological or pathological conditions.

Gene switches of this kind employing different ligand-activated transactivators were built and tested [173]. A particular two component system, termed SafeSwitch, employed mifepristone-activated chimeric transactivator GLp65 (Figure 4B). The first component consisted of a GLp65 gene under the control of a promoter cassette that combined a human hsp70B promoter and a GAL4-responsive promoter. This promoter cassette conferred stringent heat and autoactivated expression on the linked transactivator gene. The second component was a GAL4-responsive promoter that was linked to a luciferase reporter gene. Transient and stable transfection experiments established that SafeSwitch remains stably in the inactive state prior to heat activation or in the absence of mifepristone, is highly active for an extended period subsequent to a single transient heat treatment in the presence of mifepristone, allows modulation of target gene expression as a function of the intensity of the activating heat treatment and is inactivated by withdrawal of mifepristone. Plasmids encoding the two components of SafeSwitch were introduced into gastrocnemius muscles of adult C57BL/6 mice by electroinjection. Heat activation combined with administration of mifepristone resulted in about a 1000-fold induction of reporter activity one day after heat treatment. In animals that received a daily dose of mifepristone, reporter activity induced by a single heat treatment could be maintained at a high level for at least 6 days. Discontinuation of mifepristone administration deactivated the gene switch. More recently, we built another heattriggered, small molecule-activated gene switch that employs a dimerizer-controlled chimeric transactivator comprising modules derived from human proteins FKBP12 and FRAP for inducibly joining DNA-binding and activation domains [174]. The system was tested in transfection experiments using luciferase as the target gene. The gene switch showed the desired properties of only being activated by heat in the presence of rapamycin or rapalog AP21967, of maintaining high level reporter expression for several days after heat activation, and of being silenced by removal of ligand. Cytotoxicity from HSV thymidine kinase (in the presence of ganciclovir) or fusogenic GALV expression could be adequately controlled by the heat and rapamycin-dependent gene switch. Heat-activated, rapamycin-dependent gene switch and SafeSwitch supported comparable levels of target gene activity. The two heat activated gene switches may be used in combination for precisely controlling transgenes that need to be expressed in a sequential fashion.

Radiation-Inducible Promoters

A group of genes, the so-called "immediate early genes", react rapidly to ionizing radiation (IR). These genes encode transcription factors such as c-FOS, c-JUN, AP-1, NFkB and EGR-1. IR results in upregulation of the egr-1 promoter within 15 min; the promoter remains active thereafter for about three hours. The egr-1 promoter contains sequence elements referred to as CarG sequences that are responsible for the enhanced promoter activity subsequent to IR [175]. As induction by IR can be abrogated by antioxidants, it is believed that the promoter senses reactive oxygen species. Consistent with this hypothesis, the egr-1 promoter is also induced by neutrons, radioisotopes, photons and various chemotherapy agents including cisplatin, 5-fluorouracil, gemcitabine, paclitaxel, doxorubicin, cyclophosphamide and temozolomide. Subsequent to the proposal that a radiation-inducible promoter such as the egr-1 promoter could be utilized for targeted cytotoxic tumor therapy, a number of preclinical studies were carried out, many of which employed the TNF-alpha gene as the cytotoxic gene. These studies suggested that intratumoral delivery of a radiation-inducible TNF-alpha gene (by means of a replication-defective adenovector) substantially enhances anti-tumor activity of IR. This therapeutic approach was subsequently taken to the clinic. The vector used in the clinical experiments, which was named TNFerade, was a type 5 adenovector with E1, E3 and E4 deletions containing an egr-1 promoter-controlled TNF-alpha gene. Several phase I trials of combination IR and TNFerade therapy were carried out on subjects with various types of tumors. Generally, the results showed inproved tumor responses compared to historical controls. Side effects were fever, chills and pain at the injection site. A successful phase II trial of IR, TNFerade and 5-fluorouracil was subsequently carried out in patients with advanced non-metastatic pancreatic cancers. A maximal tolerated dose of $4x10^{11}$ pfu TNFerade was defined. Based on the encouraging results of this trial, a phase III randomized trial was intitiated. Unfortunately, the trial had to be stopped in March 2010 by GenVec, the company that had underwritten clinical development, subsequent to an interim analysis indicating that the required statistical significance for registration could not be attained. As TNFerade had previously also produced positive results in early trials of esophageal cancer, head and neck cancer, rectal cancer and soft tissue sarcomas, pivotal studies in these cancer types may be warranted.

OTHER INDUCIBLE SYSTEMS

Phloretin, a non-toxic flavonoid found in apples, binds to the TtgR-operator complex releasing cognate repressor TtgR from the specific operator O_{TtgR}. This results in induction of TtgABC production and effective pump-mediated efflux of the flavonoid from Pseudomonas putida. Based on the phloretin-sensitive TtgR-O_{TtgR} interaction in P. putida, a synthetic mammalian phloretin-adjustable control system named PEACE was assembled [176]. A transactivator was created by fusing TtgR to the VP16 transactivation domain. In the absence of phloretin, the synthetic transactivator was able to bind to and activate a promoter containing O_{TtgR} sequences linked to a minimal cytomegalovirus promoter. Addition of phloretin releases the transactivator from the operator sequences. The system was tested extensively in different mammalian cell lines and human primary cells and was found to be capable of stringently controlling a secreted alkaline phosphatase (SEAP) reporter gene. Phloretin was tested as a potential transdermal transgene inducer. A CHO cell line harbouring a SEAP reporter gene under the control of the PEACE system was microencapsulated in alginate-poly(L-lysine)-alginate beads and implanted subcutaneously in mice. Phloretin-containing cream was applied to the shaved skin area near the implantation site. After 72 hours, reporter gene activity was high in animals that had not received phloretin and was reduced, to a degree depending on phloretin dose, in animals that had been treated with the flavonoid.

The ability of Chlamydia pneumoniae to sense physiological signals and synchronize its metabolism with the host cell was exploited for the design of an L-arginine-regulated gene switch [177]. ArgR, an arginine-binding apo-repressor that functions as a biosensor, specifically interacts with arginine-responsive transactivator (ARG box) operators and represses the glnPQ operon that encodes a putative arginine transport system. The N-terminal domain of ArgR contains the DNA-binding domain, whereas the C-terminal half specifically binds L-arginine. A two-component L-arginine-regulated gene switch (ART) was assembled. The first component, encoded by a constitutively active gene, was a chimeric transactivator comprising ArgR and a transactivation domain from human NFkB subunit p65 or viral protein VP16. The second component was constructed by cloning an ArgRspecific operator module comprising two ARG boxes 5′ of a minimal version of the human cytomegalovirus promoter. Cotransfection experiments showed that at a low L-arginine concentration, the transactivator was in a low-affinity DNA-binding state and did not interact with its specific operator sequence. At an elevated L-arginine concentration, the transactivator assumed a high-affinity conformation and activated transcription. The gene switch was tested in mice that had been transplanted i.p. with microencapsulated engineered cells harboring ART as well as a SEAP reporter gene. The system performed well as SEAP levels in the serum correlated with administered arginine dose.

Other bacterial transcriptional repressors were also employed for engineering synthetic circuits. A uric acid–responsive expression network (UREX) was built around bacterial repressor HucR from Deinococcus radiodurans that in the absence but not in the presence of uric acid binds to operator sequence hucO [178]. A transinhibitor construct encoded a fusion protein consisting of HucR and a KRAB transsilencing domain. A reporter construct included a SEAP reporter gene that was controlled by a simian virus 40 promoter containing multiple hucO modules. Cultures co-transfected with transinhibitor and reporter constructs exhibited very low basal expression of reporter in the absence of uric acid; uric acid treatment derepressed reporter expression in a dose-dependent manner. The sensitivity of the UREX circuit to uric acid could be increased by addition of an expression construct for human urate-anion transporter URAT1. A mammalian cell line stably transfected with the latter three constructs was used to demonstrate that the UREX gene switch was capable of responding to pathological changes in urate levels in urate oxidase-deficient mice which develop hyperuricemia with human-like symptoms. (Urate oxidase converts urate into the more soluble and renally secretable allantoin.) Urate oxidase-deficient mice that had received transgenic cells intraperitoneally exhibited high levels of SEAP. SEAP activities were dramatically lower in animals that had also been treated with urate-reducing agent allopurinol. To autoregulate uric acid levels, a codon-optimized uricase/urate oxidase (Uox) gene from Aspergillus flavus was placed under the control of the three component UREX system. In vitro experiments indicated that feedback control of uric acid levels could be achieved and suggested that the approach could be utilized for reducing pathologic urate levels to physiological levels in an autocontrolled fashion.

Melanopsin, a member of the opsin family, triggers an intracellular calcium increase in response to blue light. This increased calcium level leads to calcium-dependent activation of calcineurin and calcineurinmediated mobilization of transcription factor nuclear factor of activated T cells

(NFAT). A synthetic mammalian light-controlled transcription system was recently constructed that consisted of a constitutive melanopsin expression vector and a NFAT-responsive SEAP reporter construct [179]. Cotransfection experiments showed that bluelight induced reporter expression. Reporter levels were dependent on the intensity and duration of light pulses administered. Reporter expression could also be light-controlled in a standard bioreactor setting. In mice containing intraperitoneal implants of engineered cells capable of light-triggered SEAP expression, blue light was delivered to the implanted cells by means of optical fibers. Measurements of SEAP levels in sera from irradiated and non-irradiated mice indicated that light-triggered expression was achieved in vivo. The system was also tested in a therapeutic setting in which expression of a glucagon-like peptide-1 variant, shGLP-1, was controlled by a NFAT-responsive promoter. Light-triggered expression of shGLP-1 was able to induce secretion of insulin in vitro. Transgenic cells harboring a shGLP-1 gene controlled by the light-inducible system were microencapsulated and implanted subcutaneously into diabetic db/db mice. After the animals were illuminated, insulin levels increased significantly. Furthermore, blood glucose levels remained well controlled after intraperitoneal administration of glucose.

CONCLUDING REMARKS

Today, systems responding to chemical or physical stimuli are available for effectively timing and spatially restricting expression of a transgene. During recent years, some of the "classic" gene switches, including those based on tetracycline repressor-, ecdysone receptoror mammalian steroid receptor-derived transactivators, have been refined to reduce leakiness in the absence of ligand. Some of these gene switches were introduced in high capacity vectors to avoid the need for using multiple separate components, were incorporated in self-inactivating vectors capable of irreversibly terminating transgene expression subsequent to a therapeutic intervention or were engineered to target specific tissues or cells. Use of zinc finger-based DNA-binding domains in regulatable transactivators or transinhibitors made it possible to target any endogenous gene. Recently developed strategies incorporated RNAi or miRNA regulation into transcriptional targeting vectors for improving stringency of transgene regulation. Ligands were modified to elevate sensitivity of regulation or achieve better ADME profiles, and transactivators that do not interact with natural ligands were constructed and successfully tested. Furthermore, novel gene switches were constructed that are controlled by inducers such as phloretin, arginine, uric acid or blue light. Inducible systems were incorporated in experimental cancer and neurodegenerative disease therapies to confine transgene expression to the therapeutic time window or avoid expression in locations in which transgene expression is not desired. The results of a first clinical trial using a replication-defective adenovector harboring a RheoSwitch-controlled cytokine gene were reported recently, in which trial regression of melanoma lesions was seen. Hence, regulatable systems have reached the clinic and may be on their way of finding their place in human genetic medicine. Other disorders that have been experimentally controlled by inducible systems include diabetes, multiple sclerosis, intervertebral disc degeneration, lung fibrosis and hyperuricemia.

In summary, a tremendous research effort has been dedicated to the development of regulatable switches. As a result, a multitude of promoters, transactivators, transinhibitors and other elements have been validated and are available not only for the design of gene therapy strategies but also for exploration of the functional roles of genes involved in the onset as well as progression of many diseases/disorders.

REFERENCES

1. Vilaboa N, Voellmy R (2009) Deliberate regulation of therapeutic transgenes: Gene and Cell Therapy: Therapeutic Mechanisms and Strategies. (3rdedn) CRC Press, Boca Raton, FL, pp. 619-636.
2. Stieger K, Belbellaa B, Le Guiner C, Moullier P, Rolling F (2009) In vivo regulation using tetracycline-regulatable systems. Adv Drug Deliv Rev 61: 527- 541.
3. Vanrell L, Di Scala M, Blanco L, Otano I, Gil-Farina I, et al. (2011) Development of a liver-specific Tet-On inducible system for AAV vectors and its application in the treatment of liver cancer. Mol Ther 19: 1245-1253.
4. Urlinger S, Baron U, Thellmann M, Hasan MT, Bujard H, et al. (2000) Exploring the sequence space for tetracycline-dependent transcriptional activators: novel mutations yield expanded range and sensitivity. Proc Natl Acad Sci USA 97: 7963-7968.
5. Heinz N, Schambach A, Galla M, Maetzig T, Baum C, et al. (2011) Retroviral and transposon-based tet-regulated all-in-one vectors with reduced background expression and improved dynamic range. Hum Gene Ther 22: 166-176.
6. Gossen M, Bujard H (1992) Tight control of gene expression in mammalian cells by tetracycline-responsive promoters. Proc Natl Acad Sci USA 89: 5547- 5551.
7. Iida Y, Kim JH, Kazuki Y, Hoshiya H, Takiguchi M, et al. (2010) Human artificial chromosome with a conditional centromere for gene delivery and gene expression. DNA Res 17: 293-301.
8. Nakano M, Cardinale S, Noskov VN, Gassmann R, Vagnarelli P, et al. (2008) Inactivation of a human kinetochore by specific targeting of chromatin modifiers. Dev Cell 14: 507-522.
9. Piechaczek C, Fetzer C, Baiker A, Bode J, Lipps HJ (1999) A vector based on the SV40 origin of replication and chromosomal S/MARs replicates episomally in CHO cells. Nucleic Acids Res 27: 426-428.
10. Rupprecht S, Hagedorn C, Seruggia D, Magnusson T, Wagner E, et al. (2010) Controlled removal of a nonviral episomal vector from transfected cells. Gene 466: 36-42.
11. Muhammad AK, Puntel M, Candolfi M, Salem A, Yagiz K, et al. (2010) Study of the efficacy, biodistribution, and safety profile of therapeutic gutless adenovirus vectors as a prelude to a phase I clinical trial for

glioblastoma. Clin Pharmacol Ther 88: 204-213.

12. Curtin JF, Candolfi M, Puntel M, Xiong W, Muhammad AK, et al. (2008) Regulated expression of adenoviral vectors-based gene therapies: therapeutic expression of toxins and immune-modulators. Methods Mol Biol 434: 239-266.
13. Tian X, Wang G, Xu Y, Wang P, Chen S, et al. (2009) An improved tet-on system for gene expression in neurons delivered by a single lentiviral vector. Hum Gene Ther 20: 113-123.
14. Sipo L, Hutado Pico A, Wang X, Eberle J, Petersen I, et al. (2006) An improved Tet-On regulatable FasL-adenovirus system for lung cancer therapy. J Mol Med 84: 215-225.
15. Osti D, Marras E, Ceriani L, Grassini G, Rubino T, et al. (2006) Comparative analysis of molecular strategies attenuating positional effects in lentiviral vectors carrying multiple genes. J Virol Methods 136: 93-101.
16. Das AT, Zhou X, Vink M, Klaver B, Verhoef K, et al. (2004) Viral evolution as a tool to improve the tetracycline-regulated gene expression system. J Biol Chem 279: 18776-18782.
17. Zhou X, Vink M, Klaver B, Berkout B, Das AT (2006) Optimization of the Tet-On system for regulated gene expression through viral evolution. Gene Ther 13: 1382-1390.
18. Laterriere MG, Cherpin O, Kim Y-S, Jensen J, Salmon P (2011) Polyswitch lentivectors: "all-in-one" lentiviral vectors for drug-inducible gene expression, live selection and recombination cloning. Hum Gene Ther Jul 15 Epub ahead of print] Hum Gene Ther 22:1255-1267.
19. Meerbrey KL, Hu G, Kessler JD, Roarty K, Li MZ, et al. (2011) The pINDUCER lentiviral toolkit for inducible RNA interference in vitro and in vivo. Proc Natl Acad Sci USA 108: 3665-3670.
20. Shin K-J, Wall EA, Zavzavadjian JR, Santat LA, Liu J, et al. (2006) A single lentiviral vector platform for microRNA-based conditional RNA interference and coordinated transgene expression. Proc Natl Acad Sci USA 103: 13759-13764.
21. Das AT, Zhou X, Vink M, Klaver B, Verhoef K, et al. (2004) Viral evolution as a tool to improve the tetracycline-regulated gene expression system. J Biol Chem 279: 18776-18782.
22. Zuber J, McJunkin K, Fellmann C, Dow LE, Taylor MJ, et al. (2011)

Toolkit for evaluating genes required for proliferation and survival using tetracyclineregulated RNAi. Nat Biotechnol 29: 79-85.

23. Sun T, Aceto N, Meerbrey KL, Kessler JD, Zhou C, et al. (2011) Activation of multiple proto-oncogenic tyrosine kinases in breast cancer via loss of the PTPN12 phosphatase. Cell 144: 703-718.
24. Bulat N, Widmann C (2008) Generation of a tightly regulated all-cis beta cellspecific tetracycline-inducible vector. Biotechniques 45: 411-420.
25. Han Y, Chang QA, Virag T, West NC, George D, et al. (2010) Lack of humoral immune response to the tetracycline (Tet) activator in rats injected intracranially with Tet-off rAAV vectors. Gene Ther 17: 616-625.
26. Xiong W, Candolfi M, Kroeger KM, Puntel M, Mondkar S, et al. (2008) Immunization against the transgene but not the TetON switch reduces expresson from gutless adenoviral vectors in the brain. Mol Ther 16: 343-351.
27. Lhériteau E, Libeau L, Mendes-Madeira A, Deschamps JY, Weber M, et al. (2010) Regulation of retinal function but nonrescue of vision in RPE65-deficient dogs treated with doxycycline-regulatable AAV vectors. Mol Ther 18: 1085- 1093.
28. Sheng Y, Lin CC, Yue J, Sukhwani M, Shuttleworth JJ, et al. (2010) Generation and characterization of a Tet-On (rtTA-M2) transgenic rat. BMC Dev Biol 10: 17.
29. Reboredo M, Kramer MG, Smerdou C, Prieto J, De Las Rivas J (2008) Transcriptomic effects of Tet-on and mifepristone-inducible systems in mouse liver. Hum Gene Ther 19: 1233-1247.
30. Wang H, Cao H, Wohlfahrt M, Kiem HP, Lieber A (2008) Tightly regulated gene expression in human hematopoietic stem cells after transduction with helperdependent Ad5/35 vectors. Exp Hematol 36: 823-831.
31. Inoue K, Sone T, Oneyama C, Nishiumi F, Kishine H, et al. (2009) A versatile nonviral vector system for tetracycline-dependent one-step conditional induction of transgene expression. Gene Ther 16: 1383-1394.
32. Weidenfeld I, Gossen M, Löw R, Kentner D, Berger S, et al. (2009) Inducible expression of coiding and inhibitory RNAs from retargetable genomic loci. Nucleic Acids Res 37: e50.

33. Barnea G, Strapps W, Herrada G, Berman Y, Ong J, et al. (2008) The genetic design of signaling cascades to record receptor activation. Proc Natl Acad Sci USA 105: 64-69.

34. Nagano S, Perentes JY, Jain RK, Boucher Y (2008) Cancer cell death enhances the penetration and efficacy of oncolytic herpes simplex virus in tumors. Cancer Res 68: 3795-3802.

35. Shashkova EV, Kuppuswamy MN, Wold WS, Doronin K (2008) Anticancer activity of oncolytic adenovirus vector armed with IFN-alpha and ADP is enhanced by pharmacologically controlled expression of trail. Cancer Gene Ther 15: 61-72.

36. Tannous BA, Christensen AP, Pike L, Wurdinger T, Perry KF, et al. (2009) Mutant sodium channel for tumor therapy. Mol Ther 17: 810-819.

37. Jiang J, Wang Z, Zhang J, Wang C, Xu X, et al. (2010) Early exposure of high-dose interleukin-4 to tumor stroma reverses myeloid cell-mediated T-cell suppression. Gene Ther 17: 991-999.

38. Puca R, Nardinocchi L, Bossi G, Sacchi A, Rechavi G, et al. (2009) Restoring wtp53 activity in HIPK2 depleted MCF7 cells by modulating metallothionein and zinc. Exp Cell Res 315: 67-75.

39. Murai R, Yoshida Y, Muraguchi T, Nishimoto E, Morioka Y, et al. (2010) A novel screen using the Reck tumor suppressor gene promoter detects both conventional and metastasis-suppressing anticancer drugs. Oncotarget 1: 252- 264.

40. Seo GM, Rachakatly RS, Balivada S, Pyle M, Shrestha TB, et al. (2011) A selfcontained enzyme activating prodrug cytotherapy for preclinical melanoma. Mol Biol Rep, May 13 [Epub ahead of print]

41. Manfredsson FP, Burger C, Rising AC, Zuobi-Hasona K, Sullivan LF, et al. (2009) Tight long-term dynamic doxycycline responsive nigrostriatal GDNF using a single rAAV vector. Mol Ther 17: 1857-1867.

42. Boy J, Schmidt T, Wolburg H, Mack A, Nuber S, et al. (2009) Reversibility of symptoms in a conditional mouse model of spinocerebellar ataxia type 3. Hum Mol Genet 18: 4282-4295.

43. Bockstael O, Chtarto A, Wakkinen J, Yang X, Melas C, et al. (2008) Differential transgene expression profiles in rat brain, using rAAV2/1 vectors with tetracycline-inducible and cytomegalovirus promoters. Hum Gene Ther 19: 1293-1305.

44. Chtarto A, Bender HU, Hanemann CO, Kemp T, Lehtonen E, et al. (2003) Tetracycline-inducible transgene expression mediated by a single AAV vector. Gene Ther 10: 84-94.
45. Chtarto A, Yang X, Bockstael O, Melas C, Blum D, et al. (2007) Controlled delivery of glial cell line-derived neurotrophic factor by a single tetracyclineinducible AAV vector. Exp Neurol 204: 387-399.
46. Liu Y, Okada T, Shimazaki K, Sheykholeslami K, Nomoto T, et al. (2008) Protection against aminoglycoside-induced ototoxicity by regulated AAV vector-mediated GDNF gene transfer into the cochlea. Mol Ther 16: 474-480.
47. Aguilar S, Scotton CJ, McNulty K, Nye E, Stamp G, et al. (2009) Bone marrow stem cells expressing keratinocyte growth factor via an inducible lentivirus protects against bleomycin-induced pulmonary fibrosis. PLoS One 4: e8013.
48. Chen M, Song K, Rao N, Huang M, Huang Z, et al. (2011) Roles of exogenously regulated bFGF expression in angiogenesis and bone regeneration in rat calvarial defects. Int J Mol Med 27: 545-553.
49. Wuebbenhorst D, Dumler K, Wagner B, Wexel G, Imhoff A, et al. (2010) Tetracycline-regulated bone morphogenetic protein 2 gene expression in lentivirally transduced primary rabbit chondrocytes for treatment of cartilage defects. Arthritis Rheum 62: 2037-2046.
50. Wolf G, Aumann N, Michalska M, Bast A, Sonnemann J, et al. (2010) Peroxiredoxin III protects pancreatic β cells from apoptosis. J Endocrinol 207: 163-175.
51. Chen KH, Wu CH, Tseng CC, Shiau JM, Lee CT, et al. (2008) Intrathecal coelectrotransfer of a tetracycline-inducible, three-plasmid-based system to achieve tightly regulated antinociceptive gene therapy for mononeuropathic rats. J Gene Med 10: 208-216.
52. Gossen M, Freundlieb S, Bender G, Mueller G, Hillen W, et al. (1995) Transcriptional activation by tetracyclines in mammalian cells. Science 268: 1766-1769.
53. Xu Y, Tian XB, An K, Yang H, Tian YK (2008) Lumbar transplantation of immortalized enkephalin-expressing astrocytes attenuates chronic neuropathic pain. Eur J Pain 12: 525-533.
54. Adler ED, Chen VC, Bystrup A, Kaplan AD, Giovannone S, et al. (2010) The cardiomyocyte lineage is critical for optimization of stem cell therapy in a mouse model of myocardial infarction. FASEB J 24:

1073-1081.

55. Chen VC, Stull R, Joo D, Cheng X, Keller G (2008) Notch signaling respecifies the hemangioblast to a cardiac fate. Nat Biotechnol 26: 1169-1178.
56. Yang J, Zhang W, Jiang W, Sun X, Han Y, et al. (2009) P21cip-overexpression in the mouse beta cells leads to the improved recovery from streptozotocininduced diabetes. PLoS One 4: e8344.
57. Thorel F, Népote V, Avril I, Kohno K, Desgraz R, et al. (2010) Conversion of adult pancreatic alpha-cells to beta-cells after extreme beta-cell loss. Nature 464: 1149-1154.
58. Wernig M, Lengner CJ, Hanna J, Lodato MA, Steine E, et al. (2008) A druginducible genetic system for direct reprogramming of multiple somatic cell types. Nat Biotechnol 26: 916-924.
59. Brambrink T, Foreman R, Welstead GG, Lengner CJ, Wernig M, et al. (2008) Sequential expression of pluripotency markers during direct reprogramming of mouse somatic cells. Cell Stem Cell 2: 151-159.
60. Vierbuchen T, Ostermeier A, Pang ZP, Kokubu Y, Südhof TC, et al. (2010) Direct conversion of fibroblasts to functional neurons by defined factors. Nature 463: 1035-1042.
61. Kaji K, Norrby K, Paca A, Mileikovsky M, Mohseni P, et al. (2009) Virus-free induction of pluripotency and subsequent excision of reprogramming factors. Nature 458: 771-776.
62. Nishiyama A, Xin L, Sharov AA, Thomas M, Mowrer G, et al. (2009) Uncovering early response of gene regulatory networks in ESCs by systematic induction of transcription factors. Cell Stem Cell 5: 420-433.
63. Matsui S, Shimosato D, Toyooka Y, Yagi R, Takahashi K, et al. (2005) An efficient system to establish multiple embryonic stem cell lines carrying an inducible expression unit. Nucleic Acids Res 33: e43.
64. Traykova-Brauch M, Schönig K, Greiner O, Miloud T, Jauch A, et al. (2008) An efficient and versatile system for acute and chronic modulation of renal tubular function in transgenic mice. Nat Med 14: 979-984.
65. Audero E, Coppi E, Mlinar B, Rossetti T, Caprioli A, et al. (2008) Sporadic autonomic dysregulation and death associated with excessive serotonin autoinhibition. Science 321: 130-133.
66. Gonzalez-Aparicio M, Mauleon I, Alzuguren P, Bunuales M,

GonzalezAseguinolaza G, et al. (2011) Self-inactivating helper virus for the production of high-capacity adenoviral vectors. Gene Ther, April 28 [Epub ahead of print]

67. Morris SJ, Fraley DC, Leppard KN (2010) Generation of cell lines to complement adenovirus vectors using recombination-mediated cassette exchange. BMC Biotechnol 10: 92.
68. Stewart HJ, Leroux-Carlucci MA, Sion CJ, Mitrophanous KA, Radcliffe PA (2009) Development of inducible EIAV-based lentiviral vector packaging and producer cell lines. Gene Ther 16: 805-814.
69. Mättö J, Maukonen J, Alakomi HL, Suikho ML, Saarela M. (2008) Influence of oral doxycycline therapy on the diversity and antibiotic susceptibility of human intestinal bifidobacterial population. J Appl Microbiol 105: 279-289.
70. Suhr ST, Gil EB, Senut MC, Gage FH (1998) High level transactivation by a modified Bombyx ecdysone receptor in mammalian cells without exogenous retinoid X receptor. Proc Natl Acad Sci USA 95: 7999-8004.
71. Hoppe UC, Marban E, Johns DC (2000) Adenovirus-mediated inducible gene expression in vivo by a hybrid ecdysone receptor. Mol Ther 1: 159-164.
72. Palli SR, Kapitskaya MZ, Kumar MB, Cress DE (2003) Improved ecdysone receptor-based inducible gene regulation system. Eur J Biochem 270: 1308- 1315.
73. Magnenat L, Schwimmer LJ, Barbas CF 3rd (2008) Drug-inducible and simultaneous regulation of endogenous genes by single-chain nuclear receptorbased zinc-finger transcription factor gene switches. Gene Ther 15: 1223-1232.
74. Lapenna S, Dinan L, Friz J, Hopfinger AJ, Liu J, et al. (2009) Semisynthetic ecdysteroids as gene-switch actuators: synthesis, structure-activity relationships, and prospective ADME properties. ChemMedChem 4: 55-68.
75. Wallbaum S, Grau N, Schmid A, Frick K, Neeb A, et al. (2009) Cell cycle quiescence can suppress transcription from an ecdysone receptor-based promoter in mammalian cells. Biotechniques 46: 433-440.
76. Taylor JL, Rohatgi P, Spencer HT, Doyle DF, Azizi B. (2010) Characterization of a molecular switch system that regulates gene expression in mammalian cells through a small molecule. BMC Biotechnol 10: 15.

77. Komita H, Zhao X, Katakam AK, Kumar P, Kawabe M, et al. (2009) Conditional interleukin-12 gene therapy promotes safe and effective antitumor immunity. Cancer Gene Ther 16: 883-891.
78. Herbermann RB, Bi M, Moreno M, Butterfield L, Buffo MJ, et al. (2011) Local and systemic anti-tumor immunity is induced by Rheoswitch regulated IL-12 production after intratumoral injection of adenovirus vector as well as vectortransduced dendritic cells. 2011 Annual Meeting of the Amercian Society of Gene and Cell therapy (abstract No. 899).
79. Schwartzentruber DJ, Kirkwood JM, Guarino MJ, Richards JM, Hamid O, et al. (2011) Immunotherapy of advanced melanoma by intratumoral injections of autologous, purified dendritic cells transduced with gene construct of interleukin-12, with dose-dependent expression under the control of an oral activator ligand. J Clin Oncol 29 (suppl; abstr 2540).
80. Salisbury BG, Mukhopadhyay G, Kostich M, Laz TM, Norris ED (2008) Inducible expression and pharmacological characterization of the mouse metabotropic glutamate 5b receptor. Eur J Pharmacol 579: 34-39.
81. Weiss A, Roscic A, Paganetti P (2009) Inducible mutant huntingtin expression in HN10 cells reproduces Huntington's disease-like neuronal dysfunction. Mol Neurodegener 4: 11.
82. Sowa G, Westrick E, Pacek C, Coelho P, Patel D, et al. (2011) In vitro and in vivo testing of a novel regulatory system for gene therapy for intervertebral disc degeneration. Spin (Phila Pa 1976): 36: E623-628.
83. Shea CM, Tzertzinis G (2010) Controlled expression of functional miR-122 with a ligand inducible expression system. BMC Biotechnol 10: 76.
84. Vegeto E, Allan GF, Schrader, WT, Tsai, MJ, McDonnell DP, et al. (1992) The mechanism of RU486 antagonism is dependent on the conformation of the carboxy-terminal tail of the human progesterone receptor. Cell 69: 703-713.
85. Wang Y, Xu J, Pierson T, O'Malley BW, Tsai SY (1997) Positive and negative regulation of gene expression in eukaryotic cells with an inducible transcriptional regulator. Gene Ther 4: 432-441.
86. Burcin MM, Schiedner G, Kochanek S, Tsai SY, O'Malley BW (1999) Adenovirus-mediated regulable target gene expression in vivo. Proc Natl Acad Sci USA 96: 355-360.

87. Abruzzese RV, Godin D, Mehta V, Perrard JL, French M, et al. (2000) Liganddependent regulation of vascular endothelial growth factor and erythropoietin expression by a plasmid-based, autoinducible GeneSwitch system. Mol Ther 2: 276-287.
88. Nordstrom JL (2002) Antiprogestin-controllable transgene regulation in vivo. Curr Opin Biotechnol 13: 453-458.
89. Pierson TM, Wang Y, DeMayo FJ, Matzuk, MM, Tsai, SY, et al. (2000) Regulable expression of inhibin A in wild-type and inhibin alpha null mice. Mol Endocrinol 14: 1075-1085.
90. Xie J, Nair A, Hermiston TW (2008) A comparative study examining the cytotoxicity of inducible gene expression system ligands in different cell types. Toxicol In Vitro 22: 261-266.
91. Wang, Y, DeMayo, FJ, Tsai, SY, O'Malley, BW (1997) Ligand-inducible and liver-specific target gene expression in transgenic mice. Nat Biotechnol 15: 239-243.
92. Matsumoto T, Kiguchi K, Jiang J, Carbajal S, Ruffino L, et al. (2004) Development of transgenic mice that inducibly express an active form of c-Src in the epidermis. Mol Carcinog 40: 189-200.
93. Zhao B, Chua SS, Burcin MM, Reynolds SD, Stripp BR, et al. (2001) Phenotypic consequences of lung-specific inducible expression of FGF-3. Proc Natl Acad Sci USA, 98: 5898-5903.
94. Lutucuta S, Tsybouleva N, Ishiyama M, Defreitas G, Wei L, et al. (2004) Induction and reversal of cardiac phenotype of human hypertrophic cardiomyopathy mutation cardiac troponin T-Q92 in switch on-switch off bigenic mice. J Am Coll Cardiol 44: 2221-2230.
95. Babij P, Psaltis G, Song D, Kulik J, Mollova N, et al. (2003) "Blue heart": characterization of a mifepristone-dependent system for conditional gene expression in genetically modified animals. Biochim Biophys Acta 1627: 15-25.
96. Bo J, Yu W, Zhang YM, Demayo, FJ, Wei, L (2005) Cardiac-specific and ligandinducible target gene expression in transgenic mice. J Mol Cell Cardiol 38: 685-691.
97. Ngan ES, Ma ZQ, Chua SS, DeMayo FJ, Tsai, SY (2002) Inducible expression of FGF-3 in mouse mammary gland. Proc Natl Acad Sci USA 99: 11187-11192.
98. Oligino T, Poliani PL, Wang Y, Tsai SY, O'Malley, BW, et al. (1998) Drug inducible transgene expression in brain using a herpes simplex

virus vector. Gene Ther 5: 491-496.

99. Wang L, Hernandez-Alcoceba R, Shankar V, Zabala M, Kochanek S, et al. (2004) Prolonged and inducible transgene expression in the liver using gutless adenovirus: a potential therapy for liver cancer. Gastroenterology 126: 278- 289.
100. Schillinger KJ, Tsai SY, Taffet GE, Reddy AK, Marian, AJ et al. (2005) Regulatable atrial natriuretic peptide gene therapy for hypertension. Proc Natl Acad Sci USA 102: 13789-13794.
101. Szymanski P, Kretschmer PJ, Bauzon M, Jin F, Qian HS, et al. (2007) Development and validation of a robust and versatile one-plasmid regulated gene expression system. Mol Ther 15: 1340-1347.
102. Harkins RN, Szymanski P, Petry H, Brooks A, Qian HS, et al. (2008) Regulated expression of the interferon-beta gene in mice. Gene Ther 15: 1-11.
103. Levitsky K, Szymanski P, Jin F, Meurer-Ogden JA, Harkins RN (2008) Development and validation of an improved inducer-regulator protein complex in the pBRES-regulated expression system. Hum Gene Ther 19: 1273-1282.
104. Emelyanov A, Parinov S (2008) Mifepristone-inducible LexPR system to drive and control gene expression in transgenic zebrafish. Dev Biol. 320: 113-121.
105. Dent CL, Lau G, Drake EA, Yoon A, Case CC et al. (2007) Regulation of endogenous gene expression using small molecule-controlled engineered zinc-finger protein transcription factors. Gene Ther 14: 1362-1369.
106. Dong A, Hu J, Zhao L, Xu H, Liu X (2007) Regulation and pharmacokinetics of inducible recombinant TRAIL expression. Cancer Biol Ther 6: 1978-1985.
107. Sivanandam VG, Stephen SL, Hernandez-Alcoceba R, Alzuguren P, Zabala M, et al. (2008) Lethality in an anti-angiogenic tumor gene therapy model upon constitutive but not inducible expression of the soluble vascular endothelial growth factor receptor 1. J Gene Med 10: 1083-1091.
108. Reboredo M, Zabala M, Mauleon I, De Las Rivas J, Kreppel F, et al. (2008) Interleukin-12 inhibits liver-specific drug-inducible systems in vivo. Gene Ther 15: 277-288.
109. Gonzalez-Aparicio M, Alzuguren P, Mauleon I, Medina-Echeverz J,

HervasStubbs S, et al. (2011) Oxaliplatin in combination with liver-specific expression of interleukin 12 reduces the immunosuppressive microenvironment of tumours and eradicates metastatic colorectal cancer in mice. Gut 60: 341-349.

110. Crettaz J, Otano I, Ochoa L, Benito A, Paneda A, et al. (2009) Treatment of chronic viral hepatitis in woodchucks by prolonged intrahepatic expression of interleukin-12. J Virol 83: 2663-2674.

111. Unniappan S, Wideman RD, Donald C, Gunn V, Wall JL, et al. (2009) Treatment of diabetes by transplantation of drug-inducible insulin-producing gut cells. J Mol Med 87: 703-712.

112. Oosman SN, Lam AW, Harb G, Unniappan S, Lam NT, et al. (2008) Treatment of obesity and diabetes in mice by transplant of gut cells engineered to produce leptin. Mol Ther 16: 1138-1145.

113. Shinoda Y, Hieda K, Koyanagi Y, Suzuki Y (2009) Efficient transduction of cytotoxic and anti-HIV-1 genes by a gene-regulatable lentiviral vector. Virus Genes 39: 165-175.

114. Roscilli, G, Rinaudo, CD, Cimino, M, Sporeno, E, Lamartina, S, et al. (2002) Long-term and tight control of gene expression in mouse skeletal muscle by a new hybrid human transcription factor. Mol Ther 6: 653-663.

115. Gallinari P, Lahm A, Koch U, Paolini C, Nardi MC, et al. (2005) A functionally orthogonal estrogen receptor-based transcription switch specifically induced by a nonsteroid synthetic ligand. Chem Biol. 12: 883-893.

116. Saunders TL (2011) Inducible transgenic mouse models. Methods Mol Biol 693: 103-115.

117. Matsuda T, Cepko CL. (2007) Controlled expression of transgenes introduced by in vivo electroporation. Proc Natl Acad Sci USA 104: 1027-1032.

118. Bosenberg M, Muthusamy V, Curley DP, Wang Z, Hobbs C, et al. (2006) Characterization of melanocyte-specific inducible Cre recombinase transgenic mice. Genesis 44: 262-267.

119. Liang CC, You LR, Chang JL, Tsai TF, Chen CM (2009) Transgenic mice exhibiting inducible and spontaneous Cre activities driven by a bovine keratin 5 promoter that can be used for the conditional analysis of basal epithelial cells in multiple organs. J Biomed Sci. 16: 2.

120. Baek WY, de Crombrugghe B, Kim JE (2010) Postnatally induced

inactivation of Osterix in osteoblasts results in the reduction of bone formation and maintenance. Bone 46: 920-928.

121. Yokoi H, Kasahara M, Mukoyama M, Mori K, Kuwahara K, et al. (2010) Podocyte-specific expression of tamoxifen-inducible Cre recombinase in mice. Nephrol Dial Transplant 25: 2120-2124.

122. Kellendonk C, Tronche F, Monaghan AP, Angrand PO, Stewart F, et al. (1996) Regulation of Cre recombinase activity by the synthetic steroid RU 486. Nucleic Acids Res 24: 1404–1411.

123. Sharma N, Moldt B, Dalsgaard T, Jensen TG, Mikkelsen JG (2008) Regulated gene insertion by steroid-induced PhiC31 integrase. Nucleic Acids Res 36: e67.

124. Malkoski SP, Cleaver TG, Lu SL, Lighthall JG, Wang XJ (2010) Keratin promoter based gene manipulation in the murine conducting airway. Int J Biol Sci 6: 68-79.

125. Chen MR, Liu SW, Wu TC, Kao VY, Yu HC, et al. (2010) RU486-inducible recombination in the salivary glands of lactoferrin promoter-driven green fluorescent Cre transgenic mice. Genesis 48: 585-595.

126. Shah VR, Koster MI, Roop DR, Spencer DM, Wei L, Li Q, et al. (2007) Doubleinducible gene activation system for caspase 3 and 9 in epidermis. Genesis 45: 194-199.

127. Rivera, VM, Clackson T, Natesan S, Pollock R, Amara JF, et al. (1996) A humanized system for pharmacologic control of gene expression. Nat Med 2: 1028-1032.

128. Pollock R, Issner R, Zoller K, Natesan S, Rivera VM, et al. (2000) Delivery of a stringent, dimerizer-regulated gene expression system in a single retroviral vector. Proc Natl Acad Sci USA 97: 13221-13226.

129. Clemons PA, Gladstone BG, Seth A, Chao ED, Foley MA, et al. (2002) Synthesis of calcineurin-resistant derivatives of FK506 and selection of compensatory receptors. Chem Biol 9: 49-61.

130. Vogel R, Mammeri H, Mallet J (2008) Lentiviral vectors mediate nonimmunosuppressive rapamycin analog-induced production of secreted therapeutic factors in the brain: regulation at the level of transcription and exocytosis. Hum Gene Ther 19: 167-178.

131. Zennou V, Serguera C, Sarkis C, Colin P, Perret E, et al. (2001) The HIV-1 DNA flap stimulates HIV vector mediated cell transduction in the brain. Nat Biotechnol 19: 446-450.

132. Rivera VM, Wang X, Wardwell S, Courage NL, Volchuk A, et al.

(2000) Regulation of protein secretion through controlled aggregation in the endoplasmic reticulum. Science 287: 826-830.
133. Song W, Dong Z, Jin T, Mantellini MG, Nunez G, et al. (2008) Cancer gene therapy with iCaspase-9 transcriptionally targeted to tumor endothelial cells. Cancer Gene Ther 15: 667-675.
134. Noer JE, Hu Y, Song W, Spencer DM, Nunez G (2002) Ablation of microvessels in vivo upon dimerization of iCaspase-9. Gene Ther 9: 444-451.
135. Ng Y, Ramm G, Lopez JA, James DE (2008) Rapid activation of Akt2 is sufficient to stimulate GLUT4 translocation in 3T3-L1 adipocytes. Cell Metab 7: 348-356.
136. Tian J, Lei P, Laychock SG, Andreadis ST (2008) Regulated insulin delivery from human epidermal cells reverses hyperglycemia. Mol Ther 16: 1146-1153.
137. Varma D, Dawn A, Ghosh-Roy A, Weil SJ, Ori-McKenney KM, et al. (2010) Development and application of in vivo molecular traps reveals that dynein light chain occupancy differentially affected dynein-mediated processes. Proc Natl Acad Sci USA 107: 3493-3498.
138. Deans TL, Cantor CR, Collins JJ (2007) A tunable genetic switch based on RNAi and repressor proteins for regulating gene expression in mammalian cells. Cell 130: 363-372.
139. Greber D, El-Baba MD, Fussenegger M (2008) Intronically encoded siRNAs improve dynamic range of mammalian gene regulation systems and toggle switch. Nucleic Acids Res 36: e101.
140. Tigges M, Dénervaud N, Greber D, Stelling J, Fussenegger M (2010) A synthetic low-frequency mammalian oscillator. Nucleic Acids Res 38: 2702- 2711.
141. Sakurai F, Katayama K, Mizuguchi H (2011) MicroRNA-regulated transgene expression systems for gene therapy and virotherapy. Front Biosci 17: 2389- 2401.
142. Brown BD, Gentner B, Cantore A, Colleoni S, Amendola M, et al. (2007) Endogenous microRNA can be broadly exploited to regulate transgene expression according to tissue, lineage and differentiation state. Nat Biotechnol 25: 1457-1467.
143. Wu C, Lin J, Hong M, Choudhury Y, Balani P, et al. (2009) Combinatorial control of suicide gene expression by tissue-specific promoter and microRNA regulation for cancer therapy. Mol Ther 17: 2058-2066.

144. Geisler A, Jungmann A, Kurreck J, Poller W, Katus HA, et al. (2011) microRNA122-regulated transgene expression increases specificity of cardiac gene transfer upon intravenous delivery of AAV9 vectors. Gene Ther 18: 199- 209.
145. Xie J, Xie Q, Zhang H, Ameres SL, Hung JH, et al. (2011) MicroRNA-regulated, systemically delivered rAAV9: a step closer to CNS-restricted transgene expression. Mol Ther 19: 526-35.
146. Cao L, Lin EJ, Cahill MC, Wang C, Liu X, et al. (2009) Molecular therapy of obesity and diabetes by a physiological autoregulatory approach. Nat Med 15: 447-454.
147. Voellmy R, Ahmed A, Schiller P, Bromley P, Rungger D (1985) Isolation and functional analysis of a human 70,000 dalton heat shock protein gene segment. Proc Natl Acad Sci USA 82: 4949-4953.
148. Schiller P, Amin J, Ananthan J, Brown ME, Scott WA, et al. (1988) Cis-acting elements involved in the regulated expression of a human hsp70 gene. J Mol Biol 203: 97-105.
149. Dreano M, Brochot J, Myers A, Cheng-Meyer C, Rungger D, et al. (1986) High-level, heat-regulated synthesis of proteins in eukaryotic cells. Gene 49: 1-8.
150. Christians ES, Benjamin IJ (2006) Heat shock response: lessons from mouse knockouts. Handb Exp Pharmacol 172: 139-152.
151. Guilhon E, Voisin P, de Zwart JA, Quesson B, Salomir R, et al. (2003) Spatial and temporal control of transgene expression in vivo using a heat-sensitive promoter and MRI-guided focused ultrasound. J Gene Med 5: 333-342.
152. Plathow C, Lohr F, Divkovic G, Rademaker G, Farhan N, et al. (2005) Focal gene induction in the liver of rats by a heat-inducible promoter using focused ultrasound hyperthermia: preliminary results. Invest Radiol 40: 729-735.
153. Vekris A, Maurange C, Moonen C, Mazurier F, De Verneuil H, et al. (2000) Control of transgene expression using local hyperthermia in combination with a heat-sensitive promoter. J Gene Med 2: 89-96.
154. Huang Q, Hu JL, Lohr F, Zhang L, Braun R, et al. (2000) Heat-induced gene expression as a novel targeted cancer gene therapy strategy. Cancer Res 60: 3435-3439.
155. Braiden V, Ohtsuru A, Kawashita Y, Miki F, Sawada T (2000) Eradication of breast cancer xenografts by hyperthermic suicide gene

therapy under the control of the heat shock protein promoter. Hum Gene Ther 11: 2453-2463.

156. Brade AM, Szmitko P, Ngo D, Liu FF, Klamut HJ (2003) Heat-directed tumor cell fusion. Hum Gene Ther 14: 447-461.
157. Fogar P, Navaglia F, Basso D, Zambon CF, Moserle L, et al. (2010) Heatinduced transcription of diphtheria toxin A or its variants, CRM176 and CRM197: implications for pancreatic cancer gene therapy. Cancer Gene Ther 17: 58-68.
158. Rohmer S, Mainka A, Knippertz I, Hesse A, Nettelbeck DM (2008) Insulated hsp70B' promoter: stringent heat-inducible activity in replication-deficient, but not replication-competent adenoviruses. J Gene Med 10: 340-354.
159. Tang QS, Zhang DS, Cong XM, Wan ML, Jin LQ (2008) Using thermal energy produced by irradiation of Mn-Zn ferrite magnetic nanoparticles (MZF-NPs) for heat-inducible gene expression. Biomaterials 29: 2673-2679.
160. Che J, Doubrovin M, Serganova I, Ageyeva L, Beresten T, et al. (2007) HSP70-inducible hNIS-IRES-eGFP reporter imaging: response to heat shock. Mol Imaging 6: 404-416.
161. Parry JJ, Sharma V, Andrews R, Moros EG, Piwnica-Worms D, et al. (2009) PET imaging of heat-inducible suicide gene expression in mice bearing head and neck squamous cell carcinoma xenografts. Cancer Gene Ther 16: 161- 170.
162. Deckers R, Quesson B, Arsaut J, Eimer S, Couillaud F, et al. (2009) Imageguided, noninvasive, spatiotemporal control of gene expression. Proc Natl Acad Sci USA 106:1175-1180.
163. Mehta HB, Popovich BK, Dillmann WH (1988) Ischemia induces changes in the level of mRNAs coding for stress protein 71 and creatine kinase M. Circ Res 63: 512-517.
164. Benjamin IJ, Kroger B, Williams RS (1990) Activation of the heat shock transcription factor by hypoxia in mammalian cells. Proc Natl Acad Sci USA 87: 6263-6267.
165. Locke M, Noble EG, Tanguay RM, Field MR, Ianuzzo SE, et al. (1995) Activation of heat-shock factor in rat heart after heat shock and exercise. Am J Physiol 268: C1387-C1394.
166. Venkataseshan VS, Marquet E (1996) Heat shock protein 72/73 in normal and diseased kidneys. Nephron 73: 442-449.

167. Salminen WF Jr, Voellmy R, Roberts, SM (1997) Differential heat shock protein induction by acetaminophen and a nonhepatotoxic regioisomer, 3'-hydroxyacetanilide, in mouse liver. J Pharmacol Exp Ther 282: 1533-1540.

168. Moseley PL (1998) Heat shock proteins and the inflammatory response. Ann NY Acad Sci 856: 206-213.

169. Bajramovic JJ, Bsibsi M, Geutskens SB, Hassankhan R, Verhulst KC, et al. (2000) Differential expression of stress proteins in human adult astrocytes in response to cytokines. J Neuroimmunol 106: 14-22.

170. Shastry S, Toft DO, Joyner, MJ (2002) Hsp70 and Hsp90 expression in leucocytes after exercise in moderately trained humans. Acta Physiol Scand 175: 138-146.

171. Zou J, Salminen WF, Roberts SM, Voellmy R (1998) Correlation between glutathione oxidation and trimerization of heat shock factor 1, an early step in the stress induction of the Hsp response. Cell Stress & Chaperones 3: 130- 141.

172. Bramson JL, Hitt M, Gauldie J, Graham, FL (1997) Pre-existing immunity to adenovirus does not prevent tumor regression following intratumoral administration of a vector expressing IL-12 but inhibits virus dissemination. Gene Ther 4: 1069-1076.

173. Vilaboa N, Fenna M, Munson J, Roberts, SM, Voellmy R (2005) Novel gene switches for targeted and timed expression of proteins of interest. Mol Ther 12: 290-298.

174. Martin-Saavedra FM, Bore A, Voellmy R, Vilaboa N (2009) Heat-Activated, Rapamycin-Dependent Gene Switches for Tight Control of Transgene Expression. Hum Gene Ther 20: 1060-1061.

175. Weichselbaum RR, Kufe D (2009) Translation of the radio- and chemoinducible TNFerade vector to the treatment of human cancers. Cancer Gene Ther 16: 609-619.

176. Gitzinger M, Kemmer C, El-Baba MD, Weber W, Fussenegger M (2009) Controlling transgene expression in subcutaneous implants using a skin lotion containing the apple metabolite phloretin. Proc Natl Acad Sci USA 106: 10638- 10643.

177. Hartenbach S, Daoud-El Baba M, Weber W, Fussenegger M (2007) An engineered L-arginine sensor of Chlamydia pneumoniae enables arginineadjustable transcription control in mammalian cells and mice. Nucleic Acids Res 35: e136.

178. Kemmer C, Gitzinger M, Daoud-El Baba M, Djonov V, Stelling J, et al. (2010) Self-sufficient control of urate homeostasis in mice by a synthetic circuit. Nat Biotechnol 28: 355-360.

179. Ye H, Daoud-El Baba M, Peng RW, Fussenegger M (2011) A synthetic optogenetic transcription device enhances blood-glucose homeostasis in mice. Science 332: 1565-1568.

6

Simultaneous Administration of ADSCs-Based Therapy and Gene Therapy Using Ad-huPA Reduces Experimental Liver Fibrosis

Alejandra Meza-Ríos,[1] Leonel García-Benavides,[2] Jesus García-Bañuelos,[1] Adriana Salazar-Montes,[1] Juan Armendáriz-Borunda,[1,3,] and Ana Sandoval-Rodríguez[1]

[1] Institute for Molecular Biology in Medicine, Department of Molecular Biology and Genomics, Health Sciences University Center, University of Guadalajara, Guadalajara, Jalisco, Mexico
[2] Unit of Cardiovascular Investigation, Department of Physiology, Health Sciences University Center, University of Guadalajara, Guadalajara, Jalisco, Mexico
[3] INNOVARE, Guadalajara, Jalisco, Mexico

ABSTRACT

Background and Aims

hADSCs transplantation in cirrhosis models improves liver function and reduces fibrosis. In addition, Ad-huPA gene therapy diminished fibrosis and

Citation: Meza-Ríos A, García-Benavides L, García-Bañuelos J, Salazar-Montes A, Armendáriz-Borunda J, Sandoval-Rodríguez A. Simultaneous Administration of ADSCs-Based Therapy and Gene Therapy Using Ad-huPA Reduces Experimental Liver Fibrosis. PLoS One. 2016;11(12):e0166849. Published 2016 Dec 16. doi:10.1371/journal.pone.0166849.

increased hepatocyte regeneration. In this study, we evaluate the combination of these therapies in an advanced liver fibrosis experimental model.

Methods

hADSCs were expanded and characterized before transplantation. Ad-huPA was simultaneously administrated via the ileac vein. Animals were immunosuppressed by CsA 24 h before treatment and until sacrifice at 10 days post-treatment. huPA liver expression and hADSCs biodistribution were evaluated, as well as the percentage of fibrotic tissue, hepatic mRNA levels of Col-αI, TGF-β1, CTGF, α-SMA, PAI-I, MMP2 and serum levels of ALT, AST and albumin.

Results

hADSCs homed mainly in liver, whereas huPA expression was similar in Ad-huPA and hADSCs/Ad-huPA groups. hADSCs, Ad-huPA and hADSCs/Ad-huPA treatment improves albumin levels, reduces liver fibrosis and diminishes Collagen α1, CTGF and α-SMA mRNA liver levels. ALT and AST serum levels showed a significant decrease exclusively in the hADSCs group.

Conclusions

These results showed that combinatorial effect of cell and gene-therapy does not improve the antifibrogenic effects of individual treatments, whereas hADSCs transplantation seems to reduce liver fibrosis in a greater proportion.

INTRODUCTION

Advanced liver fibrosis (ALF) is a major health worldwide problem and liver transplantation is the ultimate treatment for ALF, although it is infrequent due to low availability of organs, high cost and the risk of transplant rejection. Alternative therapies for liver fibrosis have been developed including stem cell therapy and gene therapy. Mesenchymal stem cells (MSCs) reside in various tissues, such as bone marrow, umbilical cord blood, placenta, liver, adipose tissue and others. MSCs undergo self-renewal but can differentiate into multiple cell lineages; and have demonstrated immunomodulation, inflammation suppression and antifibrogenic effects [1–5]. MSCs isolated from adipose tissue are called adipose derived stromal cells (ADSCs).

MSCs, particularly ADSCs, are capable of differentiating into several tissue lineages, including hepatocyte-like cells [6–8]. However, it is unlikely that MSC-derived hepatocyte-like cells could provide sufficient replacement of lost liver function in any given clinical scenario. It has been postulated that cytokine production may be the major mode of therapeutic action of MSCs in advanced liver fibrosis. ADSCs secretome include multiple growth factors, such as VEGF, HGF, GM-CSF, bFGF, BDNF and IGF-1 and interleukins, such as IL-6, IL-7, IL-8, IL-11, IL-10 [8,9]; some of which are suitable inducers of liver regeneration. ADSCs-derived IL-10 and TNFα inhibit the proliferation of hepatic stellate cells and MSC-derived HGF was shown to induce the apoptosis of hepatic stellate cells in a co-culture system [10].

On the other hand, our research group has demonstrated that urokinase Plasminogen Activator (uPA)-gene therapy is powerful resource for activating latent metalloproteinases and single chain-HGF, promoting extracellular matrix degradation and hepatic regeneration [11,12]. Ad-huPA administration reduced fibrotic tissue by 50–60% and induced brisk hepatocyte regeneration in CCl_4-induced cirrhosis [12]. Adenoviruses have been shown to be efficient vectors in cirrhosis models, introducing exogenous cDNA into the liver [13]. The aim of this study was to evaluate the combination of cell therapy with ADSCs and gene therapy using Ad-uPA in the resolution of fibrosis and comparing the results with the effects of individual therapies in an experimental model of liver fibrosis.

MATERIALS AND METHODS

hADSCs Isolation and Culture

Human adipose tissue was obtained from abdominal fat of females subjected to cosmetic liposuction who signed a written consent to donate the discarded fat tissue. The protocol was approved by the Research and Ethical Committee of the CUCS, Universidad de Guadalajara (approval number C.I. 067–2012) which review the fat obtainment procedure. Tissue was digested by a 0.075% collagenase type II solution (Invitrogen, Grand Island, NY) for 1 hour at 37°C with gentle shaking. Digestion product was filtered using a 100 um nylon mesh and centrifuged at 1200 g for 8 min. Pellet was washed with PBS once. Cells were collected and plated on plastic dishes in DMEM (Invitrogen, Grand Island, NY) supplemented with 10% fetal bovine serum (Invitrogen, Grand Island, NY) and 1% antibiotic (Invitrogen, Grand Island,

NY). The medium was changed after 48 h. Cells were harvested and seeded until passage three for achieving greater expansion. Cell characterization and transplantation was made using a unique batch of cells from a sole isolation.

ADSCs Characterization And *In Vitro* Differentiation

Surface markers for ADSCs were evaluated using a mini Guava EasyCyte flow cytometer. $5x10^4$ cells were incubated with fluorescein isothiocyanate (FITC)- or phycoerythrin (PE)-conjugated antibodies, anti-CD105, anti-CD34, anti-STRO, anti-CD73, anti-CD45, anti-HLA-ABC, anti-HLA-DR (Invitrogen, Frederick, MD) for 30 min at 4°C in PBS and washed afterwards. Cell autofluorescence in channel F1 or F2 was subtracted for obtaining the neat signal of each marker. $5x10^4$ cells were plated for differentiation tests. Adipogenic differentiation was performed in StemPro adipogenesis conditioned media (Invitrogen, Grand Island, NY) for one week. For validating differentiation, lipid droplets were stained using Oil red staining. StemPro osteogenesis conditioned media (Invitrogen, Grand Island, NY) was used for osteogenic differentiation for 11 days, replacing media every 48 hours. Extracellular calcium deposits were stained by Von Kossa. Hepatogenic differentiation started with 2 days of pre-culture in DMEM enriched with EGF and bFGF (Invitrogen, Grand Island, NY), followed by 7 days cultured in HGF, bFGF and nicotinamide enriched media. Subsequently, cells were cultured 12 days in media containing OSM (Recombinat human Oncostatin M) (Invitrogen, Grand Island, NY), dexametasone (Bio Vision, Milpitas, CA) and ITS (mixture of recombinant human insulin, human transferrin, and sodium selenite) (Sigma-Aldrich, St. Louis, MO). Media was changed 2 times per week. Albumin and α-fetoprotein expression were determined by anti-albumin (Dako Denmark A/S, Glostrup, Denmark) and anti-α-fetoprotein fluoresce-conjugated antibodies (Thermo Scientific, Rockford, IL).

Ad-huPA Production

Ad-huPA is a first generation recombinant human Ad-serotype 5 with huPA cDNA driven by CMV promoter. Ad-huPA was produced by a recombination of pMH4-huPA and pJM17 ΔE1 Ad-v backbone, resulting in a 41.4 kb Ad-vector (Ad-huPA). Ad-huPA was amplified in HEK293 cells -DMEM 10% FBS, 37°C and 5% CO_2 atmosphere. Cells were collected after 48 h of transduction disrupting cells in three cycles of freeze/thawing for initiating rAd purification. Cellular debris was removed by centrifugation and supernatant containing rAd was concentrated by two ultracentrifugation rounds at 141,000 RCF in a CsCl density gradient. Ad-huPA particles per milliliter were determined by optical densitometry (OD), and infection units using end-point titration [14].

Animal Model of Advanced Liver Fibrosis and Therapy Administration

Male Wistar rats (n = 10/group, ~120 g) were intoxicated i.p. with CCl_4 for 8 weeks. Non-manipulated rats served as healthy control (n = 10). Since ADSCs come from a xenogenic source, cirrhotic together with healthy animals were immunosuppressed using 10mg/kg/day of Cyclosporine A (orogastrically) until sacrifice. $2x10^6$ hADSCs and/or $3x10^{11}$ vp of Ad-huPA were administrated via iliac vein 48 h after first CsA administration. Animals were sacrificed 10 days posttreatment; hepatic tissue and blood were collected. A sub-group of cirrhotic rats were transplanted with DAPI-stained hADSCs for determining biodistribution. Animals were obtained from the Animal facility of the Health Sciences University Center of the University of Guadalajara and housed in a maximum of 4 animals/cage. Rats received care according to the Mexican Official Norm NOM-062-ZOO-1999 and guidelines of the Animal facility of the Health Sciences University Center of the University of Guadalajara using 12h light/dark cycles. Rats were fed *ad libitum* and had free access to water. Health of the animals was monitored daily. When animals were identified in pain during CCl_4 intoxication, had ear infection, showed slow or not movement, had brittle hair and eye dehidratation; animals were euthanized using excess of anesthesia. No animals died prior to the experiment endpoint. The protocol was approved by the Research and Ethical Committees of CUCS, Universidad de Guadalajara (approval number C.I. 067–2012) which reviewed and approved the animal mortality aspects of the protocol.

Analysis of Liver Tissue Specimens

Representative tissue from the five liver lobules was removed after sacrifice and fixed by immersion in 10% paraformaldehyde solution, dehydrated in ethylic alcohol, and embedded in paraffin. Sections 5µm thick were stained with Masson´s trichrome and Hematoxylin & Eosin. Percentage of fibrotic tissue was determined in 30 microphotographs using a computer-assisted image analyzer (Image-ProPlus 6.0, Media Cybernetics, Inc., Bethesda, MD). Immunohistochemistry was performed in liver sections incubated with monoclonal antibody against α-SMA (Dako Denmark A/S, Glostrup, Denmark) diluted 1:100 in PBS. Incubation with a biotinylated secondary antibody (Vector, Burlingame, CA) was revealed with 3,30-diaminobenzidine (Sigma Aldrich, St. Louis, MO). Percentage of α-SMA reactivity was determined in 15 microphotographs using a computer-assisted image analyzer (Image-ProPlus 6.0, Media Cybernetics, Inc., Bethesda, MD). Necroinflammation index and fibrosis staging were measured by a hepato-pathologist blinded to the study using the HAI index [15]. Liver tissue RNA isolation was performed according to the Chomczynski and Sacchi modified method [16]. Briefly, liver tissue was homogenized in the presence of a Trizol reagent (Invitrogen, Carlsbad, CA). Chloroform was added and the aqueous phase was isolated. RNA was precipitated with isopropanol. RNA quantity and quality were determined in NanoDrop equipment (Thermo Scientific, USA). For retrotranscription, 2 μg of total RNA were used with 240 ng Oligo dT, 0.5 mM dNTPs mix, 10mM DTT, 2 U of RNAse inhibitor and 200 U MMLV(Invitrogen, Carlsbad, CA). Incubation was performed for 10 min at 25°C, 50 min at 37°C, 15 min at 70°C and 5 min in ice. The samples were stored at -70°C until used. Using a LightCycler 96 instrument (Roche Diagnostics, Indianapolis, IN) the qPCR was performed as follows: 1 cycle at 50°C for 2 min, 1 cycle at 95°C for 5 min, and 30–40 cycles at 95°C for 30 s and 60°C for 40 s. The total reaction volume was 10 μL containing 2 μL of cDNA, 1X Universal PCR Master Mix (Roche, Branchburg, NJ) and 1X TaqMan primers/probe (Applied Biosystems, Foster City, CA). Levels of mRNA of Col1a2, Tgfb1, Ctgf, Mmp2, Acta2 and PAI1A were normalized using glyceraldehyde phosphate dehydrogenase (Gapdh) as housekeeping gene. See S1 File for catalog number and sequence of TaqMan Probe.

Proteins were extracted from 100–150 mg of liver minced in a buffer solution (pH 7.4) containing 65 mmol/L Tris, 310 mmol/L KCl, 1% Nonidet P40, 0.1% SDS, sodium orthovanadate 200 mmol/L, 1 mol/L NaF and 1X complete protease inhibitor cocktail tablets (Roche Diagnostics, Indianapolis, IN). After centrifugation at 21200 RCF /4°C for 1 hour, supernatant was isolated and stored at -80°C until required. The total protein quantification was performed using a Bradford assay [17]. huPA was quantified using a ELISA (Abcam Plc, Cambridge, UK) assay according to the manufacturer instructions.

Blood Sample Determinations

Blood was collected when the animals were sacrificed and serum was obtained. Serum levels of alanine aminotransferase (ALT), aspartate aminotransferase (AST) and albumin were measured in an automated Vitros DT 60 equipment (Johnson and Johnson, New Jersey, NY).

Statistical Analysis

Values are expressed as median, or as mean ± SEM. Groups were compared with the Kruskal Wallis test. Differences between two groups were analyzed with the Mann Whitney U test. Values of $p < 0.05$ were considered statistically significant. Analysis was run with SPSS 18.0. Detailed data are available in S2 File.

RESULTS

Characterization of hADSCs Expanded in Vitro

After expansion to passage 3 in supplemented-DMEN, ADSCs exhibited a characteristic elongated fibroblast-like morphology. Cell surface marker analyses by flow cytometry revealed that CD105, CD73 and HLA-ABC were positive in expression and that CD34, CD45 and HLA-DR were negative in expression "Fig 1A". The ability of these cells to differentiate into osteoblasts, adipocytes and hepatocyte-like cells was confirmed. Fig 1B represents immunocytochemical evidence of ADSCs differentiation into hepatocyte-like cells showing fluorescence with anti-albumin or anti-AFP. Adipogenic differentiated cells accumulated intracellular lipid droplets stained with oil red, whereas extracellular calcium phosphate deposits were observed in osteoblast differentiated cells using Von Kossa staining.

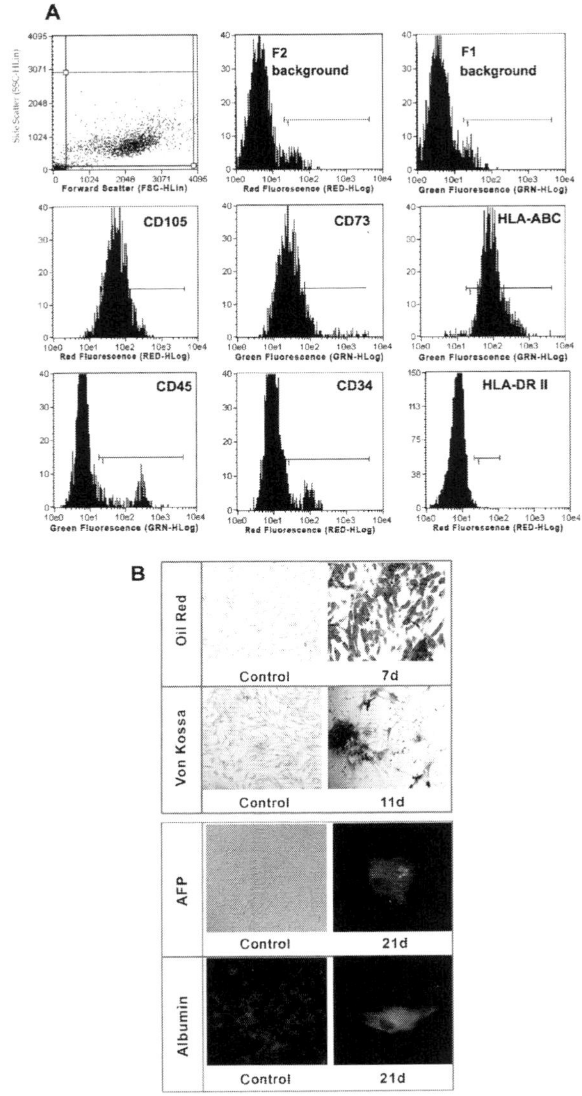

Figure 1. Characterization of hADSCs. **(A)** Phenotypic characterization of ADSCs showed cell markers CD105, CD73 and HLA-ABC positive in the population. CD45, CD34 and HLA-DR resulted negative in selected population. **(B)**Functional characterization of ADSCs into adipogenic, ostegenic and hepatic linage. Photographs (40X) showed non-differentiated cells (control) used as controls of staining and IF and cells positive for lipid droplets (oil red), extracellular calcium deposition (Von Kossa) and IF for AFP and Albumin (FITC).

In Vivo Tracking of hADSCs in Fibrotic Rats

To investigate biodistribution of hADSCs in cirrhotic animals after systemic transplantation, rats were sacrificed 10 days after DAPI-stained cell transplantation. Liver, lung, heart, spleen, kidney, brain and testis were analyzed. Detection of the DAPI fluorescence signal revealed that hADSCs migrated to the liver and some of them were arrested in lung and spleen, as seen in Fig 2A and 2B.

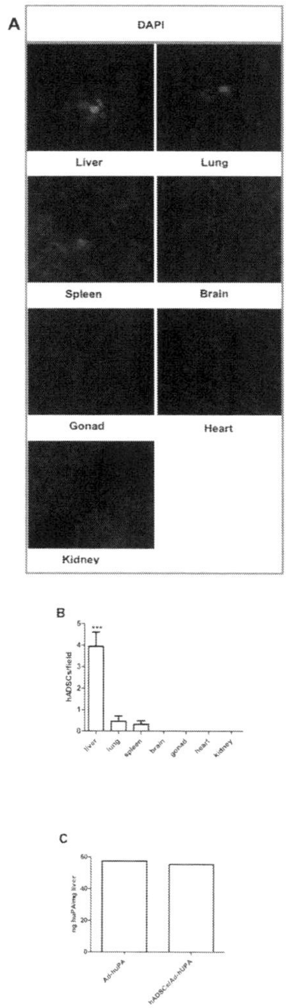

Figure 2. Biodistribution of hADSCs in cirrhotic rats. **(A)** DAPI-stained cells transplanted via iliac vein were detected mainly in liver, and in a very low frequency in lung and spleen. **(B)** Quantification of DAPI-positive cells per

field of view in collected organs (***P<0.001). **(C)** huPA protein levels in liver homogenates of cirrhotic rats treated with Ad-huPA or the combination of hAD-SCs and Ad-huPA. Data represent the median for each group.

Expression of huPA Protein in Cirrhotic Liver of Animals Treated with Ad-huPA or hADSCs/Ad-huPA

huPA protein was quantified by ELISA assay for confirming the transduction of Ad-huPA into the liver of cirrhotic animals and evaluating possible differences in transgene expression between treatments. Human uPA protein was detected in liver homogenate of animals treated with gene therapy or the combination of gene and cell therapy. The average levels of expression in liver homogenates for Ad-huPA group was 66.44± 18.50 ng/mg and 65.87± 16.69 ng/mg of tissue in the group receiving combined therapies. No statistical difference between groups was achieved, which demonstrates that transgene expression was similar between groups, even when they received either gene therapy alone or combined with hADSCs "Fig 2C".

Individual Therapies and Their Combination Diminished Liver Fibrosis, Improved Inflammation and Fibrosis Score and Reduced Hepatic Expression of A-Sma Protein and Sirius Red Collagen Staining

Therapies were systemically infused into cirrhotic animals for addressing the anti-fibrogenic effect of hADSCs, Ad-huPA or their combination. The percentage of fibrotic hepatic tissue was calculated in Masson stained liver sections using a computer-assisted morphometric analysis. In cirrhotic controls, characteristic fibrotic septa forming bridges between portal tracts (HAI score 5.57±0.20) and central veins were observed. In treated animals, fibrosis regression was induced, compared to cirrhotic controls as observed in Fig 3A. When examined by a pathologist blind to the study, the livers of treated animals were determined to have reduced bridging fibrosis (HAI score 2.60±0.22 hADSCs, 2.75±0.25 Ad-huPA and 2.62±0.26 hADSCs/Ad-huPA). Slight but statistical significant differences were found between treatments. hADSCs therapy induced fibrosis regression by 78.9±1.9%, Ad-huPA treatment induced a 65.2±2.1% fibrotic tissue regression. Interestingly enough, the combination of cell and gene therapy achieved a 72.0±1.4% fibrotic tissue regression. "Fig 3B". This piece of evidence deserves further consideration.

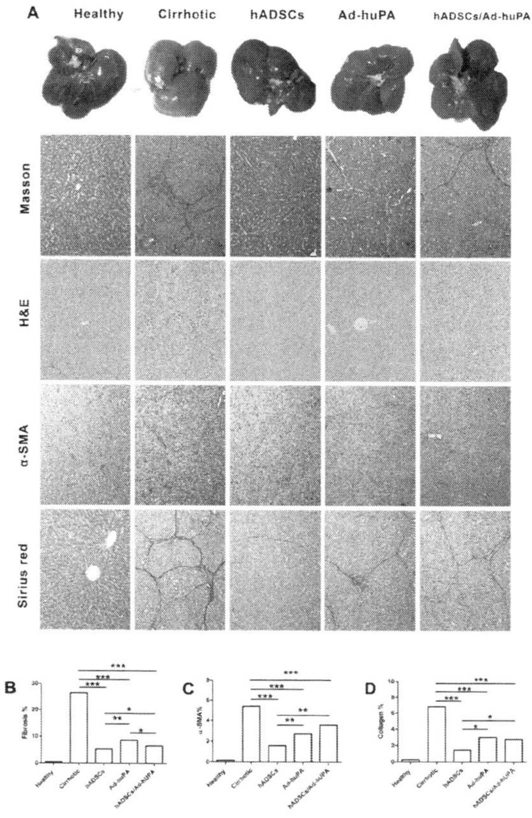

Figure 3. Assessment of liver fibrosis, inflammation, α-SMA and collagen staining in cirrhotic animals treated with cell and gene therapy. **(A)** Macroscopic appearance of the liver and Masson staining showed characteristic morphological liver alterations in cirrhotic rats. Animals treated with any of the therapies showed reduction of fibrotic tissue and decrease in inflammatory cell infiltrate in H&E-stained liver. IHC for **α**-SMA revealed decrease in staining in treated animals. Sirius red staining indicates a reduction in liver collagen content in animals that received treatment. **(B)** Fibrosis percentage in liver tissue was measured in control and treated rats **(C)** Quantification of **α**-SMA immunoreactivity in liver tissue of treated animals and controls. Data represent the median for each group (*P<0.05, **P<0.01, ***P<0.001). **(D)** Morphometric quantification of collagen staining using Sirius Red in liver tissue. Data represent the median for each group (*P<0.05, **P<0.01, ***P<0.001).

Liver samples were stained with H&E for evaluating histopathologic changes "Fig 3A". H&E staining of liver tissue from cirrhotic controls demonstrated cellular damage and centrilobular necrosis with intense neutrophilic infiltration of inflammatory cells [Histopathological Activity Index (HAI) 7.71 ± 0.28]. Necroinflammation index was improved by all therapies as shown in Table 1. HAI score was reduced by Ad-huPA to 6.62 ± 0.26, by ADSC transplantation to 5.60 ± 0.22 and by the combination of therapies to 5.87 ± 0.29.

Table 1. Necroinflammation index and fibrosis staging.

Group	Necroinflammation (0–18)	Fibrosis (0–6)
Healthy	0.37 ± 0.18	0
Cirrhotic	7.71 ± 0.28	5.57 ± 0.20
hADSCs	5.60 ± 0.22 [a***, b**]	2.60 ± 0.22 [a***]
Ad-huPA	6.62 ± 0.26 [a*]	2.75 ± 0.25 [a***]
hADSCs/Ad-huPA	5.87 ± 0.29 [a**]	2.62 ± 0.26 [a***]

[a] Statistical significance compared to cirrhotic controls

[b] statistical significance compared to the Ad-huPA group. Data represent the mean ±SEM for each group

(*P<0.05, **P<0.01, ***P<0.001).

Immunohistochemistry assays were performed to evaluate α-SMA protein expression in liver sections. α-SMA detection reveled an intense staining within the fibrotic septum in cirrhotic controls. Administration of Ad-huPA, hADSCs and hADSCs/Ad-huPA decreased α-SMA staining as shown in microphotographs “Fig 3A”. α-SMA stained area was quantified using Image-ProPlus 6.0 software. hADSCs transplantation reduced α-SMA staining 3.7 fold (P>0.001), Ad-huPA treatment 2.2 fold (P>0.001) and hADSCs/Ad-huPA 1.6 fold (P>0.001), compared to cirrhotic controls “Fig 3A”. This reduction in α-SMA staining correlates with a decrease in α-SMA mRNA levels detected in liver homogenates “Fig 4C”, and it can be speculated that this reduction may be related to a diminution in the quantity of activated HSC, a molecular mechanism that has been described for hADSCs and Ad-huPA treatments.

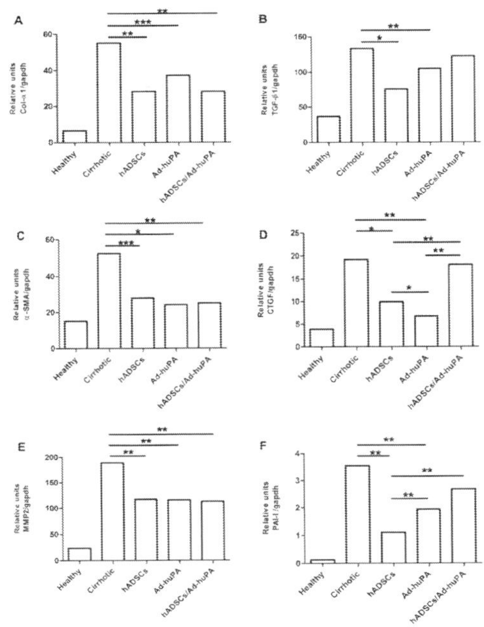

Figure 4. Analysis of profibrogenic gene expression. Mean relative expression levels of **α**-SMA, Col-**α**1, CTGF, MMP2, PAI-I and TGF-β1 in liver tissue. Data are normalized to GAPDH expression levels and represent the median for each group (*P<0.05, **P<0.01, ***P<0.001).

Sirius red staining of the liver was performed to assess collagen distribution "Fig 3A". Cirrhotic controls revealed extensive collagen bundles deposition and portal-central and central-central bridges were evident, resulting in a distorted lobular architecture. Morphometric quantification revealed a significant reduction (P>0.001) of the 1.55±0.50% stained area in the hADSCs treated animals, 2.77±0.82% in Ad-huPA animals and 2.45±0.71% for the group combining therapies compared to cirrhotic controls "Fig 3D".

Functional Recovery From Liver Damage By Hadscs Transplantation, Ad-Hupa Transduction And The Combined Therapies

We evaluated the effect of therapies on the extent of liver function. Serum level of albumin, which is an indicator of liver function, increased in treated groups to 2.7±0.26 hADSCs, 2.8±0.31 Ad-huPA and 2.7± 0.22 g/dl hADSCs/Ad-huPA (P>0.01; "Fig 5A") compared to cirrhotic controls (2.4±0.15g/

dl). Serum transaminase levels, which increase when the liver is damaged (AST 291.2±54.2 U/L, ALT 94.8±15.4 U/L) were attenuated by hADSCs treatment (AST 173.5±50.6 U/L, ALT 76.5±12.8 U/L; P>0.05). Whereas AST showed a slight diminution in Ad-huPA (AST 239.5±53.0 U/L) and hADSCs/Ad-huPA groups (AST 235.8±92.5 U/L) "Fig 5B".

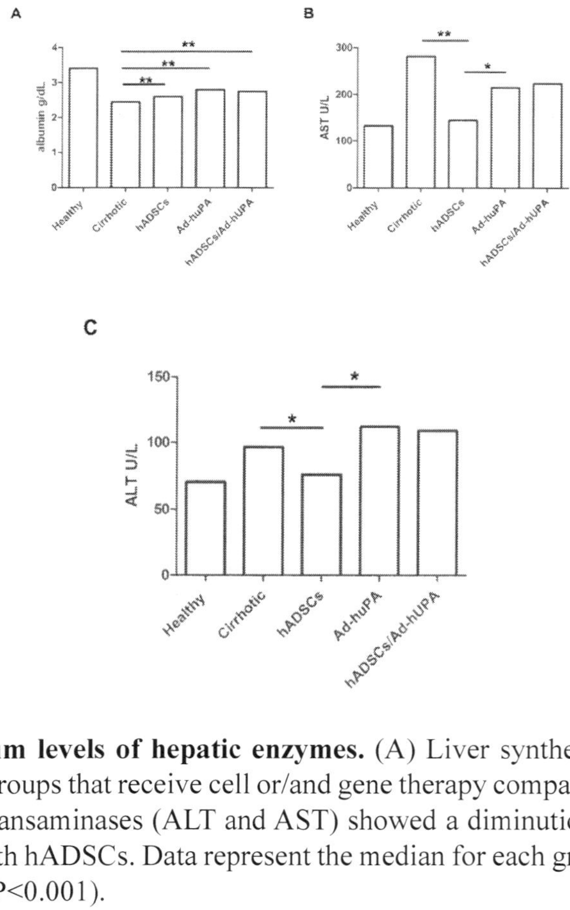

Figure 5. Serum levels of hepatic enzymes. (A) Liver synthesized albumin augmented in groups that receive cell or/and gene therapy compared to cirrhotic group. **(B,C)**Transaminases (ALT and AST) showed a diminution only in animals treated with hADSCs. Data represent the median for each group (*P<0.05, **P<0.01, ***P<0.001).

Expression of Profibrogenic Molecules Decreased By Cell and Gene Therapy and Their Combination

We tried to elucidate the mechanistic basis of the anti-fibrogenic effect of individual and combined therapies. Given that hADSCs were reported to downregulate TGF-β1 and Col-1, we assessed the expression of TGF-β1, Col-1, CTGF, and other molecules previously reported to be modified in Ad-huPA treated animals, such as MMP2, α-SMA and PAI-I.

Expressions of TGF-β1, Col-1, CTGF, MMP2, α-SMA and PAI-I were down-regulated by hADSCs, Ad-huPA and hADSCs/Ad-huPA treatments compared to cirrhotic controls. For Col-1, α-SMA and MMP2, no statistical difference between treatments was observed. "Fig 4". Lowest mRNA levels of PAI-I were found in the hADSCs group, whereas CTGF decreased mainly in the Ad-huPA group. Decrease in hepatic mRNA levels of Col-1 after treatments correlates with the diminution observed in collagen-sirius red staining; this data together with the diminish in TGF-β1 mRNA suggested ECM remodeling following hepatic hADSCs homing and Ad-uPA transduction.

DISCUSSION

In this study we show that systemic application of hADSCs, Ad-huPA, or hADSCs/Ad-huPA results in the amelioration of liver fibrosis, reducing the amount of fibrotic tissue. Furthermore, expression of profibrogenic molecules decrease, whereas hepatic cell function improves.

Fibrosis regression achieved by gene therapy is comparable to that obtained in previous publications using Ad-huPA[12]. As it is known, uPA protein cleaves pro-HGF to its active form inducing ECM degradation releasing growth factors and MMPs retained in ECM [18,19]. In this study we evaluated necroinflammation index and α-SMA staining; both parameters are also reduced in this group of animals, indicating extensive improvement in hepatic tissue due to treatment.

Regarding the application of hADSCs into fibrotic rats, our data are similar to those studies performing systemic hADSCs transplantation (in our case, iliac vein) resulting in a substantial fibrotic tissue reduction. In addition, hADSCs cell therapy in our hands, improves serum hepatic enzymes, reduces liver α-SMA staining area and mRNA, and decreases Col-α1, TGF-β1, CTGF, PAI-I and MMP2 mRNA levels. These results indicate a clear beneficial effect of hADSC transplantation in liver damage. We believe that hADSCs have clear advantages over hBMSCs as a cell transplantation source. However, their transplantation in advanced liver fibrosis models have shown some discrepancies. Kamada et al. reported that spleen administration of $1x10^5$ GFP-positive mADSCs in CCl_4-injured mice increased the area of liver fibrosis and Col-Iα1 mRNA levels, but reduced ALT and increased Albumin 4 weeks after transplantation. When the same cells are pre-treated with bFGF prior to transplantation, anti-fibrogenic effects were shown [2]. In contrast, other authors like Harn et al. reported a 1.9 fold fibrotic tissue reduction and hepatic function improvement after $1x10^6$ hADSCs administration, 14 days post-treatment in TAA cirrhotic rats [3]. Lee et al. transplanted $1X10^6$ hADSCs into DMN-cirrhotic rats via tail vein; 7 day after treatment fibrotic area reduced 3.7 fold [4]. The administration route also seems to influence the antifibrotic effect. Wang et al. proved portal vein or tail vein administration of $2x10^6$ rADSCs in CCl_4-cirrhotic animals. Six weeks after treatment only the portal vein group reduced fibrotic tissue 2.3 fold [5]. As can be observed, cell origin, administrations routes, hepatotoxic and even animal strain employed varies in reported studies, but most of them agree in a helpful effect for liver fibrosis cell-based therapies. Mechanisms proposed for the ADSC antifibrotic effect include the secretion of IL-10 and HGF, which have a paracrine influence on activated hepatic stellate cells (HSC) [10]. This assumption correlates with our results for hepatic α-SMA staining and mRNA levels that diminish, probably due to a decrease in HSC activation. To our knowledge, few studies have reported cell tracking in cirrhotic animals. Liver, spleen and lung are the organs that usually retain transplanted cells [20]. Outstandingly, we were able to detect DAPI-labeled hADSCs in fibrotic liver. We consider that biodistribution observed in this study supports a safe and efficient use of hADSCs for liver diseases, when administered in a similar vein, since cells "homing", principally in liver tissue. hADSCs are known to be chemoattracted to damaged tissue, due to the expression of SDF-1α, PDGF-AB, TGF-β1 and TNF-α; molecules broadly overexpressed in fibrotic liver [21]. However, the aim of this study was to search for an enhanced antifibrogenic effect when cell therapy and gene therapy were simultaneously administrated. Contrary to our expectations, the combination of therapies does not substantially increase beneficial properties of individual therapies; even when a slight increase in fibrosis reversion is demonstrated when compared with gene therapy alone ($p<0.05$). Since the expression of the transgene (huPA) is similar in Ad-uPA and hADSCs/Ad-uPA groups, we assume that cell therapy does not interfere with liver transduction of Ad-huPA. Conversely, it seems that Ad-huPA transduction interferes with hADSCs liver therapy since cell therapy alone, shows enhanced antifibrogenic outcome than the combination of cell and gene therapy. As previously reported, immune response is provoked after adenovirus systemic administration [22]. Besides, hADSCs are known to display anti-inflammatory and immunoregulatory properties [23,24]. Hence, ADSCs therapeutic effect could be attenuated trying to neutralize adenovector-induced transient inflammation and reducing fibrosis and necroinflammation due to CCl_4 intoxication. Our results show that liver inflammation after hADSCs transplantation was also reduced, probably due to PGE2 and IL-10, cytokines also known to be secreted by ADSCs [10,24]. However, more experiments should be run to corroborate this hypothesis. When MSCs are transduced prior to transplantation with antifibrogenic transgenes, they have exerted an improved antifibrogenic effect compared to un-modified MSC [25,26,20], with no apparent interference due to transgene expression. Therefore, this option could be explored with Ad-uPA.

CONCLUSIONS

In conclusion, cell therapy using ADSCs gene therapy with Ad-huPA and their simultaneous administration have shown substantial antifibrogenic results in a rat cirrhosis model, but conditions (time of administration, route and dose) for optimizing the conjunction of therapies must be evaluated.

FUNDING STATEMENT

The funders (CONACYT Grant 103772 to ASR and 183039 to JAB) had no role in study design, data collection and analysis, decision to publish, or preparation of the manuscript. Also, the funder (Innovare) provided support in the form of salaries for author JAB, but did not have any additional role in the study design, data collection and analysis, decision to publish, or preparation of the manuscript. The specific roles of this author are articulated in the 'author contributions' section.

DATA AVAILABILITY

All relevant data are within the paper and its Supporting Information files.

REFERENCES

1. Lin H, Xu R, Zhang Z, Chen L, Shi M, Wang FS. Implications of the immunoregulatory functions of mesenchymal stem cells in the treatment of human liver diseases. Cell Mol Immunol. 2011; 8 (1): 19–22. 10.1038/cmi.2010.57
2. Kamada Y, Yoshida Y, Saji Y, Fukushima J, Tamura S, Kiso S, et al. Transplantation of basic fibroblast growth factor-pretreated adipose tissue-derived stromal cells enhances regression of liver fibrosis in mice. Am J Physiol Gastrointest Liver Physiol. 2009; 296(2): G157–67 10.1152/ajpgi.90463.2008
3. Harn HJ, Lin SZ, Hung SH, Subeg YM, Li YS, Syu WS, et al. Adipose-derived stem cells can abrogate chemical-induced liver fibrosis and facilitate recovery of liver function. Cell Transplant. 2012; 21(12): 2753–64 10.3727/096368912X652959
4. Lee YG, Hwang JW, Park SB, Shin IS, Kang SK, Seo KW et al. Reduction of liver fibrosis by xenogeneic human umbilical cord blood and adipose tissue-derived multipotent stem cells without treatment of an immunosuppressant. J Tissue Eng Regen Med. 2008; 5: 613–21
5. Wang Y, Lian F, Li J, Fan W, Xu H, Yang X, et al. Adipose derived mesenchymal stem cell transplantation via portal vein improves microcirculation and ameliorates liver fibrosis induced by CCl4 in rats. J Transl Med. 2012; 10:133 10.1186/1479-5876-10-133
6. Banas A, Teratani T, Yamamoto Y, Tokuhara M, Takeshita F, Osaki M, et al. Rapid hepatic fate specification of adipose□derived stem cells and their therapeutic potential for liver failure. J Gastroenterol Hepatol. 2009; 24(1):70–7 10.1111/j.1440-1746.2008.05496.x
7. Lue J, Lin G, Ning H, Xiong A, Lin CS, Glenn JS. Transdifferentiation of adipose□derived stem cells into hepatocytes: a new approach. Liver Int. 2010; 30(6): 913–22 10.1111/j.1478-3231.2010.02231.x
8. Zuk P. Adipose-derived stem cells in tissue regeneration: a review. ISRN Stem Cells, 2013. (2013), Article ID 713959, 35 pages.
9. Salgado J, Reis RL, Sousa NJ, Gimble JM. Adipose tissue derived stem cells secretome: soluble factors and their roles in regenerative medicine. Curr Stem Cell Res Ther. 2010; 5(2): 103–10
10. Parekkadan B, van Poll D, Megeed Z, Kobayashi N, Tilles AW, Berthiaume F, et al. Immunomodulation of activated hepatic stellate

cells by mesenchymal stem cells. Biochem Biophys Res Commun. 2007; 363(2): 247–52 10.1016/j.bbrc.2007.05.150

11. Bueno M, Salgado S, Beas-Zárate C, Armendariz-Borunda J. Urokinase-type plasminogen activator gene therapy in liver cirrhosis is mediated by collagens gene expression down-regulation and up-regulation of MMPs, HGF and VEGF. J Gene Med. 2006; 8(11):1291–9 10.1002/jgm.961
12. Salgado S, Garcia J, Vera J, Siller F, Bueno M, Miranda A, et al. Liver cirrhosis is reverted by urokinase-type plasminogen activator gene therapy. Mol Ther. 2000; 2(6): 545–51 10.1006/mthe.2000.0210
13. Garcia-Bañuelos J, Siller-Lopez F, Miranda A, Aguilar LK, Aguilar-Cordova E, Armendariz-Borunda J. Cirrhotic rat livers with extensive fibrosis can be safely transduced with clinical-grade adenoviral vectors. Evidence of cirrhosis reversion. Gene Ther. 2002; 9(2):127–34 10.1038/sj.gt.3301647
14. Armendariz-Borunda J, Bastidas-Ramírez BE, Sandoval-Rodríguez A, González-Cuevas J, Gómez-Meda B, García-Bañuelos J. Production of first generation adenoviral vectors for preclinical protocols: amplification, purification and functional titration. J Biosci Bioeng. 2011; 112(5):415–21 10.1016/j.jbiosc.2011.07.018
15. Ishak K, Baptista A, Bianchi L, Callea G, De Groote J, Gudat F, et al. Histological grading and staging of chronic hepatitis. J Hepatol. 1995; 22(6): 696–9
16. Chomczynski P, Sacchi N. Single-step method of RNA isolation by acid guanidinium thiocyanate-phenol-chloroform extraction. Anal Biochem. 1987; 162(1):156–9 10.1006/abio.1987.9999
17. Bradford MM. A rapid and sensitive method for the quantitation of microgram quantities of protein utilizing the principle of protein-dye binding. Anal Biochem. 1976; 72:248–54
18. Naldini L, Tamagnone L, Vigna E, Sachs M, Hartmann G, Birchmeier W, et al. Extracellular proteolytic cleavage by urokinase is required for activating hepatocyte growth factor/scatter factor. EMBO J. 1992;11(13):4825–33
19. Carmeliet P, Moons L, Lijnen R, Baes M, Lemaitre V, Tipping P, et al. Urokinase-generated plasmin activates matrix metalloproteinases during aneurysm formation. Nat Genet. 1997; 17(4): 439–44 10.1038/ng1297-439

20. Fiore EJ, Bayo JM, Garcia MG, Malvicini M, Lloyd R, Piccioni F, et al. Mesenchymal Stromal Cells Engineered to Produce IGF-I by Recombinant Adenovirus Ameliorate Liver Fibrosis in Mice. Stem Cells Dev. 2015; 24(6): 791–801 10.1089/scd.2014.0174
21. Baek SJ, Kang SK, Ra JC. In vitro migration capacity of human adipose tissue-derived mesenchymal stem cells reflects their expression of receptors for chemokines and growth factors. Exp Mol Med. 2011; 43(10): 596–603 10.3858/emm.2011.43.10.069
22. Muruve DA, Barnes MJ, Stillman IE, Libermann TA. Adenoviral gene therapy leads to rapid induction of multiple chemokines and acute neutrophil-dependent hepatic injury in vivo. Hum Gene Ther.1999; 10(6): 965–76 10.1089/10430349950018364
23. Mohammadzadeh A, Pourfathollah AA, Shahrokhi S, Hashemi SM, Moradi SL, Soleimani M. Immunomodulatory effects of adipose-derived mesenchymal stem cells on the gene expression of major transcription factors of T cell subsets. Int Immnopharmacol. 2014; 20(2): 316–21
24. Cui L, Yin S, Liu W, Li N, Zhang W, Cao Y. Expanded adipose-derived stem cells suppress mixed lymphocyte reaction by secretion of prostaglandin E2. Tissue Eng. 2007;13(6): 1185–95 10.1089/ten.2006.0315
25. Yu Y, Yao AH, Chen N, Pu LY, Fan Y, Lv L, et al. Mesenchymal stem cells over-expressing hepatocyte growth factor improve small-for-size liver grafts regeneration. Mol Ther. 2007; 15(7): 1382–9 10.1038/sj.mt.6300202
26. Li H, Zhang B, Lu Y, Jorgensen M, Petersen B, Song S. Adipose tissue-derived mesenchymal stem cell-based liver gene delivery. J Hepatol. 2011; 54(5): 930–8 10.1016/j.jhep.2010.07.051

7

Gene Therapy: Advances, Challenges and Perspectives

Giulliana Augusta Rangel Gonçalves, Raquel de Melo Alves Paiva

Hospital Israelita Albert Einstein, São Paulo, SP, Brazil.

ABSTRACT

The ability to make site-specific modifications to the human genome has been an objective in medicine since the recognition of the gene as the basic unit of heredity. Thus, gene therapy is understood as the ability of genetic improvement through the correction of altered (mutated) genes or site-specific modifications that target therapeutic treatment. This therapy became possible through the advances of genetics and bioengineering that enabled manipulating vectors for delivery of extrachromosomal material to target cells. One of the main focuses of this technique is the optimization of delivery vehicles (vectors) that are mostly plasmids, nanostructured or viruses. The viruses are more often investigated due to their excellence of invading cells and inserting their genetic material. However, there

Citation: Gonçalves, Giulliana Augusta Rangel, & Paiva, Raquel de Melo Alves. (2017). Gene therapy: advances, challenges and perspectives. Einstein (São Paulo), 15(3), 369-375. https://doi.org/10.1590/s1679-45082017rb4024.

is great concern regarding exacerbated immune responses and genome manipulation, especially in germ line cells. *In vivo* studies in in somatic cell showed satisfactory results with approved protocols in clinical trials. These trials have been conducted in the United States, Europe, Australia and China. Recent biotechnological advances, such as induced pluripotent stem cells in patients with liver diseases, chimeric antigen receptor T-cell immunotherapy, and genomic editing by CRISPR/Cas9, are addressed in this review.

Keywords: Gene therapy; Genetic Vectors; Gene transfer, horizontal; CRISPR-Cas9; CAR-T cell; Genetic therapy; Clustered regularly inter-spaced short palindromic repeats

INTRODUCTION

In 1991, James Watson declared that "many people say they are worried about the changes in our genetic instructions. But these (genetic instructions) are merely a product of evolution, shaped so we can adapt to certain conditions which might no longer exist. We all know how imperfect we are. Why not become a little better apt to survive?".[1] Since the beginning, humans understand that the peculiar characteristics of the parents can be transmitted to their descendents. The first speculation originated from the ancient Greek students, and some of these theories continued for many centuries. Genetic-scientific studies initiated in the early 1850s, when the Austrian monk, Gregor Mendel, in a series of experiments with green peas, described the inheritance pattern by observing the traces that were inherited as separate units, which we know today as genes. Up until 1950, little was known as to the physical nature of genes, which was when the American biochemist, James Watson, and the British biophysicist, Francis Crick, developed the revolutionary model of the double strand DNA. In 1970, researchers discovered a series of enzymes that enabled the separation of the genes in predetermined sites along the DNA molecule and their reinsertion in a reproducible manner. These genetic advances prepared the scenario for the emergence of genetic engineering with the production of new drugs and antibodies, and as of 1980, gene therapy has been incorporated by scientists.[2,3]

In this review, we cover gene therapy, the different methodologies of genetic engineering used for this technique, its limitations, applications, and perspectives.

Gene Therapy

The ability to make local modificiations in the human genome has been the objective of Medicine since the knowledge of DNA as the basic unit of heredity. Gene therapy is understood as the capacity for gene improvement by means of the correction of altered (mutated) genes or site-specific modifications that have therapeutic treatment as target. Further on, diffrent strategies are described, which are often used for this purpose.(4)

Currently, gene therapy is an area that exists predominantly in research laboratories, and its application is still experimental.(5) Most trials are conducted in the United States, Europe, and Australia. The approach is broad, with potential treatment of diseases caused by recessive gene disorders (cystic fibrosis, hemophilia, muscular dystrophy, and sickle cell anemia), acquired genetic diseases such as cancer, and certain viral infections, such as AIDS.(3,6)

One of the most often used techniques consists of recombinant DNA technology, in which the gene of interest or healthy gene is inserted into a vector, which can be a plasmidial, nanoestrutured, or viral; the latter is the most often used due to its efficiency in invading cells and introducing its genetic material. On table 1, a few gene therapy protocols are summarized, approved and published for clinical use, exemplifying the disease, the target, and the type of vector used.(3)

Table 1. Gene therapy protocols

Disease	Objective	Stem cells	Release mode	Countries with the protocol
Adenosine deaminase deficiency	Substitution of the adenosine deaminase deficiency	Blood	Retrovirus	Italy, Holland, and the United States
α 1-antitrypsin deficiency	Substitution of α 1-antitrypsin	Respiratory epithelium	Liposome	United States
AIDS	Inactivation of the HIV-presenting antigen	Blood and bone marrow	Retrovirus	United States
Cancer	Improvement of immune function	Blood, bone marrow, and tumor	Retrovirus, liposome, electroporation, and cell-mediated transfer	Austria, China, France, Germany, Italy, Holland [Netherlands], and the United States

Cancer	Tumor removal	Tumor	Retrovirus, non-complexed DNA, cell-mediated transfer	United States
Cancer	Chemoprotection	Blood and bone marrow	Retrovirus	United States
Cancer	Stem cell marking	Blood, bone marrow, and tumor	Retrovirus	Canada, France, Sweden and United States
Cystic fibrosis	Enzymatic substitution	Respiratory epithelium	Adenovirus and liposome	England and the United States
Familial hypercholesterolemia	Substitution of low-density lipoprotein receptors	Liver	Retrovirus	United States
Fanconi anemia	Complement C gene release	Blood and bone marrow	Retrovirus	United States
Gaucher Disease	Glucocerebrosidase substitution	Blood and bone marrow	Retrovirus	United States
Hemophilia B	Factor IX substitution	Skin fibroblasts	Retrovirus	China
Rheumatoid arthritis	Cytokine release	Synovial membrane	Retrovirus	United States

Source: Adapted from Misra S. Human gene therapy: a brief overview of the genetic revolution. J Assoc Physicians India. 2013;61(2):127-33. Review.[3]

Although several protocols have been successful, the gene therapy process remains complex, and many techniques need new developments. The specific body cells that need treatment should be identified and accessible. A way to effectively distribute the gene copies to the cells must be available, and the diseases and their strict genetic bonds need to be completely understood.[3] There is also the important issue of the target cell type of gene therapy that currently is subdivided into two large groups: gene therapy of the germline[7] and gene therapy of somatic cells.[8] In germline gene therapy, the stem cells, *e.g.*, with the sperm and egg, are modified by the introduction of functional genes, which are integrated into the genome. The modifications are hereditary and pass on to subsequent generations. In theory, this approach should be highly effective in the fight against genetic and hereditary diseases. Somatic cell gene therapy is when therapeutic genes are transferred to a patient's somatic cells. Any modification and any effects are restricted only to that patient and are not inherited by future generations.

Gene Therapy Process: Release of the Gene

In gene therapy, a normal gene is inserted into the genome to replace an abnormal gene responsible for causing a certain disease. Of the various challenges involved in the process, one of the most significant is the difficulty in releasing the gene into the stem cell. Thus, a molecular carrier called a "vector" is used to release the gene, which needs to be very specific, display efficiency in the release of one or more genes of the sizes necessary for clinical applications, not be recognized by the immune system, and be purified in large quantities and high concentrations so that it can be produced and made available on a large scale. Once the vector is inserted into the patient, it cannot induce allergic reactions or inflammatory process; it should increase the normal functions, correct deficiencies, or inhibit deleterious activities. Furthermore, it should be safe not only for the patient, but also for the environment and for the professionals who manipulate it. Finally, the vector should be capable to express the gene, in general, for the patient's entire life.(3,9)

Although the efficacy of viral vectors is confirmed, recently some studies demonstrated that the use of these carriers presented with several limitations. The presence of viral genetic material in the plasmid is a strong aggravating factor, since it can induce an acute immune response, besides a possible oncogenic transformation. Currently, there are two main approaches for genetic modifications of the cells, namely: virus-mediated (Table 2) and via physical mechanisms, from preparations obtained by advanced nanotechnology techniques.(5) Within this context, included are polymers that form networks that capture a gene and release its load when they penetrate the cells, such as DNA microinjections,(10) cationic polymers,(11) cationic liposomes,(12,13) and particle bombardment.(14)

Table 2. Viral vectors for gene therapy

	Retrovirus	Lentivirus	Herpes virus	Adenovirus	Adenoassoci-ated	Plasmid
Provirus	RNA	RNA	RNA	DNA	DNA	DNA
Capacity	~9 kB	~10 kB	>30 kB	~30 kB	~4.6 kB	Unlimited
Integration into the recipient genome	Yes	Yes	Yes	No	Extremely rare	No
Duration of transgene expression	Long	Long	Transient	Transient	Long in post-mitotic cells	Transient

Preexisting immunity in the recipient	No	No	Yes	Yes	Yes	No
Adverse effects	Insertional mutagenesis	Insertional mutagenesis	Inflammatory response	Inflammatory response	Mild inflammatory response	No
Germline transmission	May occur	Yes	No	No	May occur	No

Source: Modified from Linden R. Gene therapy: what it is, what it is not, and what it will be. Estud Av. 2010;24(70):31-69.[5]

Each exogenous material introduction technique differs from the other and depends on the type of application proposed. Some are more efficient, others more apt to carry large genes (>10kB) and integrate with the genome, allowing a permanent expression.[1]

Gene Therapy and Hematopoietic Stem Cells

Hematopoietic stem cells have become ideal targets for gene transfer due to the high potential for longevity and the capacity for self-renovation. One example of this combination of gene therapy and stem cells would be the production of gene transfer vectors for the creation of induced pluripotent stem cells (iPS), in order to generate the differentiation of the iPS and afford an additional phenotype from this differentiated derived cell. Patients with chronic liver disease and infection by the hepatitis virus (*e.g.*, hepatitis B virus and hepatitis C virus), which require a liver transplant, may be likely to undergo the hepatic transplantation of mature hepatocytes or those derived from iPS.[15] Not only the transfer of genes might be needed to convert stem cells into hepatocytes; since the transplanted cells are susceptible to reinfection by the hepatitis virus, the transfer of a vector that encodes a short hairpin RNA directed against the virus would provide the transferred cells with resistance or 'immunity' to reinfection. Resistant cells can repopulate the liver over time and restore normal hepatic function (Figure 1).[15]

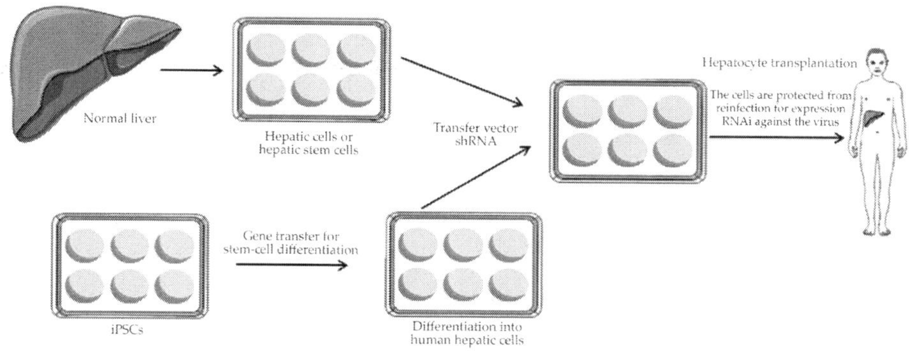

Source: Adapted from Kay MA. State-of-the-art gene-based therapies: the road ahead. Nat Rev Genet. 2011;12(5):316-28. Review.[15]

shRA: short hairpin RNA; iPS: induced pluripotent stem cells.

Figure 1. Combination of stem cells and gene therapy.

Therapy with T cells of the Recipient of a Chimeric Antigen

Chimeric antigen recipient T (CAR-T) cell therapy is a type of immunotherapy that involves manipulation/reprogramming of immune cells (T lymphocytes) of the patients themselves, in order to recognize and attack the tumor T cells. Initial advancement in the design of the first CAR generation, by Eshhar et al.,[16] was marked by the fusion of a single chain fragment variable (scFv) to a transmembrane domain and an intracellular signaling unit: chain CD3 zeta.[17,18] This design combined the active element of a well-characterized monoclonal antibody with a signaling domain, increasing the recognition of the tumor-specific epitope and the activation of T cells, without depending on molecules from the histocompatibility complex.

An improvement in the first generation of CAR was made by means of integrating co-stimulating molecules necessary for signal transduction. The stimulatory recipient most commonly used in this CAR generation is CD28. This recipient acts as a second activating event of the route, enabling a marked proliferation of T cells along with an increased expression of cytokines.[19]

The most recent generation of CAR incorporated the addition of a co-stimulatory domain addition to increase the CAR function. Co-stimulatory molecules as recipients of the tumor necrosis factor (CD134 or CD137) are required for this methodology. In summary, the most recent forms of CAR include scFv, the initial chain of CD3-ζ, along with the stimulatory chains of CD28 and CD134 or CD137.(20)

With the third CAR generation, Zhong et al., demonstrated an improvement in T cell activation of the Akt route (protein kinase B), which regulates the cell cycle. According to other studies, this last generation shows greater persistence of the T cells in comparison with the second generation of CAR.(21)

The most critical point of the adverse effects of CAR-T therapy is the identification of non-tumor cells that express the target epitope by CAR. Tumor antigens are molecules highly expressed in the tumor cells, but are not exclusive of these cells. For example, the CD19 antigen can be found in normal or malignant B cells, and the CAR design for the CD19 target in not capable of distinguishing them.(20,22) Other common toxicity for CAR-T therapy (and many other types of immunotherapy for cancer) is the cytokine release syndrome (CRS). Activation of the immune system after CAR-T infusion can induce a rapid increase in the levels of inflammatory cytokines.(20,23)

New developments in the design of vectors and trials with CAR-T provide balance and reinforcement in safety for amplification of the clinical application. The progressive improvement in the CAR trials has already advanced, as was observed from the first to the third generation. Knowledge and experience acquired in the assessment of CAR-T toxicity will increase the success of the progressive improvements for future trials.

CRISPR-Cas9

During the 1980's, in the genome of *Escherichia coli*, a region was identified with an uncommon pattern, in which a highly variable sequence was intercalated by a repeated sequence with no known function. In 2005, it was assumed that the variable sequences were of extra-chromosomal origin, acting as an immune memory against phages and plasmids, starting the then unknown CRISPR system (Clustered Regularly Interspaced Short Palindromic Repeats*)* and Cas (Associated Proteins), that shines since 2012 as one of the primary biotechnological tools for gene edition.(24) Originating in the immune-adaptive system of procaryontes, this mechanism recognizes the invading genetic mateiral, cleaves it into small fragments, and integrates it into its own DNA. In a second infection by the same agent, the following sequence occurs: transcription of the CRISPR locus, RNAm processing, and creation of small fragments of RNA (crRNAs) that form complexes with the Cas proteins, and these recognize the alien nucleic acids and finally destroy them.[24]

Based on this natural mechanism, the CRIPSR technique was developed enabling editing of the target-specific DNA sequences of the genome of any organism by means of basically three molecules: nuclease (Cas9), responsible for cleavage of the double-strand DNA; an RNA guide, which guides the complex to the target; and the target DNA, as is shown in figure 2.[25,26]

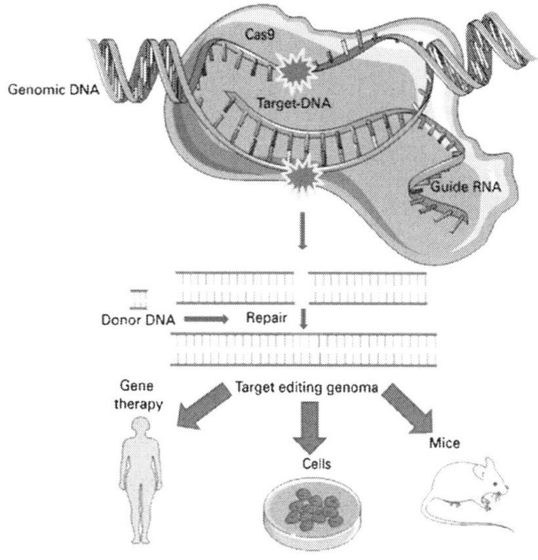

Source: Modified from Vieira GV, Cecílio NT, Arruda LM, Sales KU. Visão geral do mecanismo básico de ação. In: Pereira TC, organizador. Introdução à técnica de CRISPR. Ribeirão Preto: Cubo; 2016. Cap. 2. p. 54.[27]

Figure 2. CRISPR Cas-9 system. The technique involves basically three molecules: one nuclease (generally wild type Cas-9 of *Streptococcus pyogenes*), an RNA guide (known as single guide RNA), and the target (frequently the DNA).

Due to its simplicity and its precision when compared to other techniques (Zinc-Finger Nucleases, TALENs, and Gene Targeting), the CRISPR system arrives as a versitile tool that promotes the genetic editing by means of inactivation (knockout gene – KO), integration of exogenous sequences (knock-in), and allele substitution, among others.[27,28]

The guide RNA hybridizes with the target DNA. Cas-9 recognizes this complex and should mediate cleavage of the DNA double strand and reparation in the presence of a (homologous) donor DNA. The result of this process is the integration of an exogenous sequence into the genome (knock-in) or allele substitution.

The rapid advancement of this new technology allowed the performance of translational trials in human somatic cells, using genetic editing by CRISPR. The first applications with a therapeutic focus already stood out in describing even the optimization steps of the delivery systems and specificity for the safety and efectiveness of the system.[28,29]

Researchers from the University of California and of Utah recently were successful in correcting the mutation of the hemoglobin gene, which originates sickle cell anemia. CD34+ cells from patients who are carriers of sickle cell anemia were isolated, edited by CRISPR-Cas9, and after 16 weeks, the results showed a reduction in the expression levels of the mutated gene and an increased gene expression of the wild type.[29]

The technology referred to is in use mainly in monogenic genetic pathologies, which, despite being rare, can reach about 10 thousand diseases already described.[4] Phase 1 clinical trials are foreseen for 2017, as well as the appearance of companies geared toward the clinical use of this system.

Ethical Issues

The possibility of genetically modifying germlines has been the object of heated discussion in the field of science for a long time. Bioethics is always present when new techniques are created, in order to assess the risks of the procedure and the moral implications involved.

A large part of the scientific community approves genetic therapy in somatic cells, especially in cases of severe disorders, such as cystic fibrosis and Duchenne muscular dystrophy.

In 2015, however, Chinese researchers went beyond the moral issues and announced, for the first time, the genetic modification of embryonic cells using the CRISPR-Cas9 technique. Next, another Chinese group also reported the conduction of the same process done with the intention of conferring resistance to HIV by insertion of the CCR5 gene mutation. The genetic analysis showed that 4 of the 26 embryos were successfully modified. The result clearly reveals the need for improving the technique, alerting that, possibly, such trials could be previously tested in animal models.[4,30]

These recent publications rekindled the debate regarding genetic editing. On one side, the Japanese Ethics Committee declared that the manner in which the experiment was conducted was correct, since there had been approval by the local Ethics Committee for the study conducted, as well as the consent of the egg donors. In the United Kingdom, the first project for healthy human embryo editing was approved. On the other hand, American research groups remained conservative, reiterating their position of not supporting this type of experiment and declaring that they await improvement in the techniques and of the definitions of ethical issues.[30]

COMMENT

Since the declaration of James Watson in 1991, in reference to the likely optimization of human genetics, gene therapy has advanced throughout the decades, whether by optimization of the types of vectors, by the introduction of new techniques, such as induced pluripotent stem cells in combination with current models of genetic editing (CRISPR-Cas9), and even by trials in germ cells, bringing with it the contradictory ethical and moral aspects that accompany the technique.

Local successes have already solidified the viability of treatments using gene therapy in clinical practice, as an alternative form for patients with congenital diseases or monogenic disorders and cancer, especially when the pharmacological or surgical interventions do not show good results.

The design of new experimental vectors, the increase in efficiency, the specificity of the delivery systems, and the greater understanding of the inflammatory response induction may balance the improvement of safety with the expansion of techniques in clinical applications. Yet the knowledge and experience acquired with the careful assessment of toxicity of these technologies also allow significant advances in the application of these methods.

Therefore, historically, gene therapy and the discovery of antibiotics and chemotherapy agents, or any new technology, need more clarifying preclinical studies. In the future, there is the promise of applying these techniques in several fields of Medicine and a greater percentage of clinical trials.

REFERENCES

1. Mukherjee S. Genetic therapies: posthuman gene therapy. In: Mukherjee S. The gene: an intimate history. Nova York: Scribner; 2016. Chap. 34. p. 415.
2. Friedmann T. A brief history of gene therapy. Nat Genet. 1992;2(2):93-8. Review.
3. Misra S. Human gene therapy: a brief overview of the genetic revolution. J Assoc Physicians India. 2013;61(2):127-33. Review.
4. Tebas P, Stein D, Tang WW, Frank I, Wang SQ, Lee G, et al. Gene editing of CCR5 in autologous CD4 T cells of persons infected with HIV. N Engl J Med. 2014;370(10):901-10.
5. Linden R. Gene therapy: what it is, what it is not and what it will be. Estud Av. 2010;24(70):31-69.
6. Ginter EK. [Gene therapy of hereditary diseases]. Vopr Med Khim. 2000; 46(3):265-78. Review. Russian.
7. Mathews QL, Curiel DT. Gene therapy: human germline genetics modifications-assessing the scientific, socioethical, and religious issues. South Med J. 2007; 100(1):98-100.
8. Bank A. Human somatic cell gene therapy. Bioessays. 1996;18(12):999-1007. Review.
9. Gardlík R, Pálffy R, Hodosy J, Lukács J, Turna J, Celec P. Vectors and delivery systems in gene therapy. Med Sci Monit. 2005;11(4):RA110-21. Review.
10. McDonnell WM, Askari FK. DNA vaccines. N Engl J Med.1996;334(1):42-5. Review.
11. Plank C, Tang MX, Wolfe AR, Szoka FC Jr. Branched cationic peptides for gene delivery: role of type and number of cationic residues in formation and in vitro activity of DNA polyplexes. Hum Gene Ther. 1999;10(2):319-32. Erratum in: Hum Gene Ther. 1999;10(13):2272.
12. Caplen NJ, Kinrade E, Sorgi F, Gao X, Gruenert D, Geddes D, et al. In vitro liposome-mediated DNA transfection of epithelial cell lines using the cationic liposome DC-Chol/DOPE. Gene Ther. 1995;2(9):603-13.
13. Nabel GJ, Chang AE, Nabel EG, Plautz GE, Ensminger W, Fox BA, et al. Immunotherapy for cancer by direct gene transfer into tumors. Hum Gene Ther. 1994;5(1):57-77.

14. Yang Y, Nunes FA, Berencsi K, Furth EE, Gönczöl E, Wilson JM. Cellular immunity to viral antigens limits E1-deleted adenoviruses for gene therapy. Proc Natl Acad Sci U S A. 1994;91(10):4407-11.
15. Kay MA. State-of-the-art gene-based therapies: the road ahead. Nat Rev Genet. 2011;12(5):316-28. Review.
16. Eshhar Z, Waks T, Gross G, Schindler DG. Specific activation and targeting of cytotoxic lymphocytes through chimeric single chains consisting of antibody-binding domains and the gamma or zeta subunits of the immunoglobulin and T-cell receptors. Proc Natl Acad Sci U S A. 1993;90(2):720-4.
17. Gross G, Waks T, Eshhar Z. Expression of immunoglobulin-T-cell receptor chimeric molecules as functional receptors with antibody-type specificity. Proc Natl Acad Sci U S A. 1989;86(24):10024-8.
18. Parham P. Antigen recognition by T lymphocytes. In: Parham P. The immune system. 3rd edition. Nova York: Garland Science; 2009. Chap. 5. p. 113.
19. Chambers CA, Allison JP. Co-stimulation in T cell responses. Curr Opin Immunol. 1997;9(3):396-404. Review.
20. Wilkins O, Keeler AM, Flotte TR. CAR T-Cell Therapy: Progress and Prospects. Hum Gene Ther Methods. 2017;28(2):61-6. Review.
21. Zhong XS, Matsushita M, Plotkin J, Riviere I, Sadelain M. Chimeric antigen receptors combining 4-1BB and CD28 signaling domains augment PI3kinase/AKT/Bcl-XL activation and CD8+ T cell-mediated tumor eradication. Mol Ther. 2010;18(2):413-20.
22. Maude SL, Frey N, Shaw PA, Aplenc R, Barrett DM, Bunin NJ, et al. Chimeric antigen receptor T cells for sustained remissions in leukemia. N Engl J Med. 2014;371(16):1507-17.
23. Lee DW, Gardner R, Porter DL, Louis CU, Ahmed N, Jensen M, et al. Current concepts in the diagnosis and management of cytokine release syndrome. Blood. 2014;124(2):188-95. Erratum in: Blood. 2015;126(8):1048. Dosage error in article text.
24. Marraffini LA, Sontheimer EJ. CRISPR interference: RNA-directed adaptive immunity in bacteria and archaea. Nat Rev Genet. 2010;11(3):181-90. Review.
25. Jinek M, Chylinski K, Fonfara I, Hauer M, Doudna JA, Charpentier E. A programmable dual-RNA-guided DNA endonuclease in adaptive bacterial immunity. Science. 2012;337(6096):816-21.

26. Cong L, Ran FA, Cox D, Lin S, Barretto R, Habib N, et al. Multiplex genome engineering using CRISPR/Cas systems. Science. 2013;339(6121):819-23.
27. Vieira GV, Cecílio NT, Arruda LM, Sales KU. Visão geral do mecanismo básico de ação. In: Pereira TC, organizador. Introdução à técnica de CRISPR. Ribeirão Preto: Cubo; 2016. Cap. 2. p. 54.
28. Gori JL, Hsu PD, Maeder ML, Shen S, Welstead GG, Bumcrot D. Delivery and Specificity of CRISPR/Cas9 Genome Editing Technologies for Human Gene Therapy. Hum Gene Ther. 2015;26(7):443-51. Review.
29. De Witt MA, Magis W, Bray NL, Wang T, Berman JR, Urbinati F, et al. Selection-free genome editing of the sickle mutation in human adult hematopoietic stem/progenitor cells. Sci Transl Med. 2016;8(360):360ra134.
30. Callaway E. Second Chinese team reports gene editing in human embryos. Nature. 2016; 08 April 2016. doi:10.1038/nature.2016.19718.

8

Recent Advances in Mitochondria-Targeted Gene Delivery

Yoon-ha Jang [1] and Kwang-il Lim [1,2,3]

[1]Department of Chemical and Biological Engineering, Sookmyung Women's University, Seoul 04310, Korea
[2]Institute of Advanced Materials and Systems, Sookmyung Women's University, Seoul 04310, Korea
[3]Research Institute of ICT Convergence, Sookmyung Women's University, Seoul 04310, Korea

ABSTRACT

Mitochondria are the energy-producing organelles of cells. Mitochondrial dysfunctions link to various syndromes and diseases including myoclonic epilepsy and ragged-red fiber disease (MERRF), Leigh syndrome (LS), and Leber hereditary optic neuropathy (LHON). Primary mitochondrial diseases often result from mutations of mitochondrial genomes and nuclear genes

Citation: Jang, Y.-H.; Lim, K.-I. Recent Advances in Mitochondria-Targeted Gene Delivery. Molecules 2018, 23, 2316. https://doi.org/10.3390/molecules23092316.

that encode the mitochondrial components. However, complete intracellular correction of the mutated genetic parts relevant to mitochondrial structures and functions is technically challenging. Instead, there have been diverse attempts to provide corrected genetic materials with cells. In this review, we discuss recent novel physical, chemical and biological strategies, and methods to introduce genetic cargos into mitochondria of eukaryotic cells. Effective mitochondria-targeting gene delivery systems can reverse multiple mitochondrial disorders by enabling cells to produce functional mitochondrial components.

KEYWORDS: mitochondrial gene delivery; mitochondrial disease; mitochondrial genomes; heteroplasmy; mitochondrial matrix

INTRODUCTION

Mitochondria are intracellular organelles that produce adenosine triphosphate (ATP), which is a usable form of energy for cells. In eukaryotes, cells convert nutrients to ATP through oxidative phosphorylation that is carried out by the electron transport chain complex in the membranes of mitochondria. In addition, mitochondria take part in the regulation of cell growth and synthesis of multiple biological molecules such as pyrimidine, heme, neurotransmitters, and hormones. Mitochondrial proteins play key roles in these cellular processes and they are encoded by mitochondrial DNA (mtDNA) or nuclear DNA (nDNA). In human cells, mitochondria are present in a wide range of numbers from a few hundred to about one thousand depending on their tissue origins and one mitochondrion contains two to eight copies of mtDNA in the mitochondrial matrix [1,2,3]. mtDNA encodes 13 mRNAs that are used to produce multiple protein components for the electron transport chain complex (Figure 1). nDNA encodes genes for not only the remaining components of the complex but also proteins involved in various mitochondrial functions.

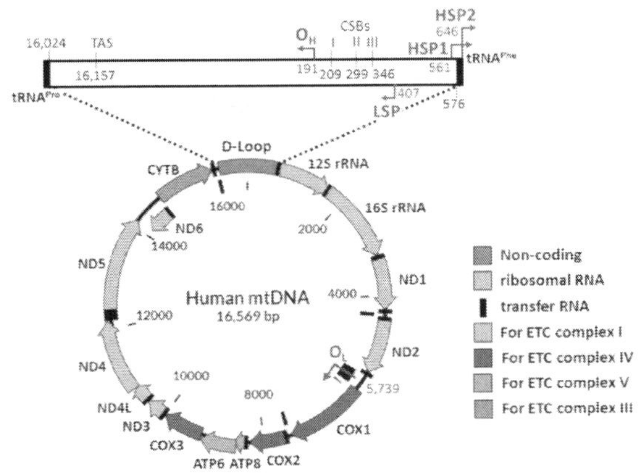

Figure 1. Schematic diagram of human mitochondrial genomic DNA. Mitochondrial DNA is 16,569 bp long. mtDNA encodes 13 mRNAs required to produce components for the electron transport chain complex and 2 rRNAs and 22 tRNAs for mitochondrial translation. D-Loop is the major non-coding region that contains the promoter components called HSP1, HSP2, and LSP. The replication origins of heavy and light strands known as O_H and O_L control the mtDNA replications. Compared with O_H, locating within the D-Loop, O_L is located between the region encoding three tRNAs and the region encoding two other tRNAs. The names of the protein-coding genes are shown in the diagram. LSP = light strand promoter. HSP1 and HSP2 = heavy strand promoters. CSB = conserved sequence block. TAS = termination-associated sequence.

Mutations in mtDNA or parts of nDNA alter the sequences of mitochondrial proteins, which ultimately causes mitochondrial dysfunction. Neurodegeneration, seizures, myopathy, cardiomyopathy, deafness, optic atrophy, and developmental delays are common symptoms of mitochondrial disorders. Multiple therapeutic strategies to treat mitochondrial dysfunctions that were caused by mtDNA mutations, in particular, have been based on the consideration of mtDNA heteroplasmy. Mitochondria normally contain more than one type of genome including possibly the wild type mtDNA and mutant variants in some cases. Therefore, a small percentage of mtDNA with mutations around multiple copies of wild type mtDNA is not sufficient to lead to a disease phenotype. However, if the percentage is over the threshold (typically >80%), mitochondrial disease occurs [4]. Based on this heteroplasmic feature of mtDNA, it might be possible to recover mitochondrial functions by supplementing the impaired genes above the clinical threshold with the corresponding normal genes in gene therapy.

Development of gene delivery systems that can pass through the mitochondrial double membranes is a key to the success of mitochondrial gene therapy. Mitochondria are roughly structured by two phospholipid bilayers including the mitochondrial outer membrane and the inner membrane, which are separated by the intermembrane space and enclose the mitochondrial matrix. Small molecules such as ions, ATP, and proteins smaller than 10 kDa can freely diffuse across the channels of the outer membrane. The mitochondrial inner membrane, however, forms a barrier that selectively transports molecules into the mitochondrial matrix. The mitochondrial inner membrane is composed of cardiolipin, which has four alkyl tails, which render the membrane impermeable to hydrophilic molecules. Outflow or inflow of molecules related to mitochondrial functions is only enabled with the help of various proteins such as transporters and translocases.

Delivery of DNA molecules encoding the corrected versions of genes into mitochondria is a promising strategy to treat mitochondrial disease. Due to the impermeability of the mitochondrial inner membrane to hydrophilic molecules, however, passage of DNA through the mitochondrial double membrane remains a challenge. In this review, we summarize various recent physical, chemical, and biological approaches to transfer DNA into mitochondria and discuss the advantages and limitations for these approaches (Table 1 and Figure 2). Methods for enhancing the expression of the cargo DNA are also introduced. At last, we highlight the applications of these strategies to treat mitochondrial diseases.

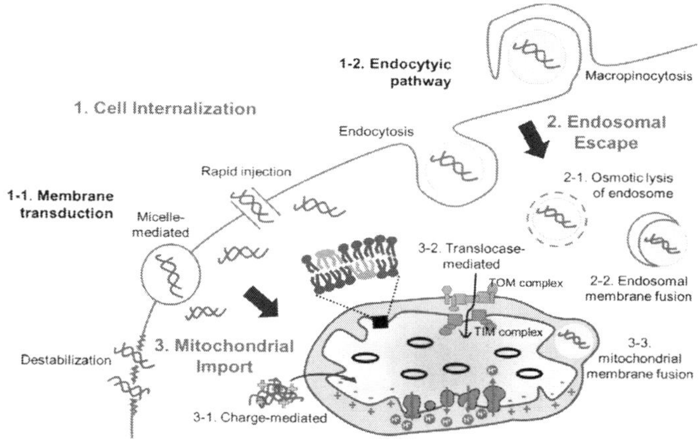

Figure 2. Barriers against delivery of exogenous DNA into mitochondrial matrix. There are three main barriers (words in blue) that need to be overcome

to deliver DNA into mitochondria: the cell membrane (against cell internalization, (1), the endosomes (against endosome escape, (2), and mitochondrial double membranes (against mitochondrial import, (3). The hydrophilic feature of DNA molecules links to their low cell membrane permeability. Various strategies have been applied to enhance the passage of DNA molecules through the cell membrane. Physical, chemical, and biological methods enhance the DNA molecule internalization into cells by generating pores in the plasma membrane (via membrane transduction) or depending on the cellular endocytic pathways. In the latter case, cargo DNA needs to escape from endosomes before being degraded by hydrolytic enzymes in the late endosomes or lysosomes. Multiple carrier molecules can facilitate endosomal escape of DNA molecules often via osmotic lysis of endosomes. Liposome-based carriers have fusogenic activity with an endosomal membrane to enhance the release of DNA molecules from endosomes to the cytoplasm. DNA or carrier molecule-DNA complex in the cytoplasm needs to pass through mitochondrial double membranes to reach the mitochondrial matrix, which is the final destination. Commonly, cationic molecules are used as carrier molecules to facilitate the transfer of DNA molecules into the matrix because the matrix is negatively charged by the actions of an electron transport chain complex. In addition, mitochondrial targeting signal peptides that can be recognized by receptors of translocases in the mitochondrial membranes are often linked to carrier molecules. Like fusion with endosomal membranes, liposome-based carrier molecules can be imported into the mitochondrial matrix through membrane fusion when their lipid composition is properly designed.

Table 1. Summary of approaches for mitochondrial gene delivery.

Classification	**Key Acting Component**	**Delivery-Target Systems**	**Strategy**	**Advantages**	**Limitations**	**Ref.**
Physical	Hydrodynamic injection	Rat, Mice	Cell penetration by hydrodynamic force	Simplicity	No mitochondria-targeting	[5,6,7]
	Biolistics	Yeast	Cell penetration by bombardment	Cell type-independent	Potential cell damage, No mitochondria-targeting	[8]
Chemical	Gemini surfactants	HeLa cell	Formation of cationic micelle-like structure with DNA	Works in small dose	Weak specificity to mitochondria	[9]

	Rhodamine 123	Normal human dermal fibroblast (NHDF) adult donor cell	Precipitation of DNA using lipophilic molecule with delocalized positive charge	Can be traced due to fluorescence	Expression from transferred DNA has not been confirmed	[**10**,**11**]
	DQAsomes	BT 20 cell	Transport of DNA by ampiphilic and cationic lipid-based vesicle	High specificity to mitochondria	Low transfection efficiency, Cytotoxicity	[**12**,**13**,**14**]
	DQA80s	Primary human dermal fibroblasts, HeLa cell	Lipid incorporation into DQAsomes	Improved transfection efficiency	Delivery of DNA around mitochondria not into the matrix	[**15**]
	MITO-Porter	HeLa cell, Rat	Membrane fusion by lipid-based nano carrier	Easy surface modification		[**16**,**17**, **18**,**19**]
	R8-MITO-Porter	Rat	Enhancing cellular uptake by functionalization with cationic peptide	High fusogenic activity with mitochondrial outer membrane	Low fusogenic activity with mitochondrial inner membrane, Moderate cytotoxicity	[**16**,**17**]
	KALA-MITO-Porter	HeLa cell, Mice	Enhancing cellular uptake by functionalization with membrane destabilizing peptide	Improved transfection efficiency	High cytotoxicity	[**19**,**20**]
	STPP-liposome	4T1 cell, Mice	Conjugation of stearyl residue to lipophilic and cationic TPP for ampiphilic property	Selective accumulation in mitochondria	Cytotoxicity	[**21**]
	TPP-PEG-PE liposome	HeLa, 4T1 cell, Mice	Substitution of stearyl moiety with biocompatible PEG-PE polymer	Decreased cytotoxicity	Transgene expression not confirmed	[**22**]

	TPP-PAMAM dendrimer	HeLa, MCF-7, 4T1, NIH 3T3 cell	DNA condensation by high positive surface charge, endosomal escape by free tertiary amine groups	Efficient endosomal escape and high serum resistance	Transgene expression not confirmed	[**23**]
Biological	MTS-PNA	Myoblasts, Fibroblasts, NT 2, IMR 32, HeLa, HepG2, C2C12 cell	MTS-guided localization of DNA hybridized with PNA to mitochondria	High specificity to mitochondria via actions of translocase	Only can transfer short nucleic acids	[**24**,**25**]
	MTS-KH peptide	HEK 293 cell	Mitochondrial localization of MTS-conjugated DNA-binding peptide and exogenous DNA complex	Can transfer large DNA with high specificity to mitochondria		[**26**]
	MTS-AAV	Neuronal G11778A NT 2 cybrid, HEK 293T cell, Mice	Mitochondrial localization of DNA by inserting MTS into the AAV capsid	Proven effects of transgene expression	Inability to carry large DNA	[**27**,**28**]

PHYSICAL APPROACHES

Physical gene delivery systems can provide a simple and direct way to transfer exogenous genes into cells because they penetrate the cell membrane without the need for carrier molecules to enhance accessibility to the cells or intracellular organelles. One such physical method known as microinjection uses a fine glass micropipette or needle to directly introduce liquids containing genetic materials into cells. Another example known as particle bombardment uses helium gas flow at a high velocity to propel DNA-coated micropellets into target cells. In addition, pulse electric fields and ultrasonic sound waves have been used in electroporation and sonoporation, respectively, to penetrate the cell membrane. Recently, these physical approaches have been applied for exogenous gene delivery specifically into the mitochondria.

Using hydrodynamic limb vein injection, Yasuzaki et al. demonstrated that naked plasmid DNA could be delivered into rat liver mitochondria in vivo [5]. Hydrodynamic injection uses hydrodynamic pressure generated by injecting a large volume of fluid into a blood vessel to permeabilize the capillary endothelium and induce pores in the plasma membranes of surrounding cells [29]. They injected a solution containing plasmid DNA into the distal vein of limbs in anesthetized rats [5,6]. As a result, they verified mitochondrial delivery of the plasmid DNA by quantifying it in the mitochondria-enriched fraction based on quantitative polymerase chain reaction (PCR).

Biolistic technology, which is another physical method, introduces DNA into the plasma membrane of target cells by coating it onto heavy metal particles and moving these complexes at high speed. Gene delivery to mitochondria using biolistic technology was carried out in *Saccharomyces cerevisiae* [8]. Since mitochondrial genetic systems of single-cell eukaryotes are well known, methods to introduce DNA into mitochondria in single-cell eukaryotes can provide a useful reference for multi-cell eukaryote systems. First, DNA was precipitated onto tungsten particles (<1 μm), which were then penetrated into the plasma membrane by high-pressure helium gas-driven acceleration. After bombardment, the DNA was randomly imported into mitochondria and cells with transformed mitochondria were subsequently selected. Biolistic gene delivery is advantageous in that it can be used regardless of the cell type and non-toxic metal particles are available. However, to date, there have been few reports of its use for mitochondrial gene delivery in mammalian cells.

In contrast to chemical and biological methods, physical methods do not require carrier molecules (Table 1). Therefore, they do not introduce any toxicity associated with these molecules. However, physically imported DNA is evenly distributed throughout the cytoplasm and it enters the mitochondrial matrix randomly, which makes it difficult to deliver DNA specifically to the mitochondria. Another limitation of physical methods is that the target cells can be damaged in the process of penetrating the cell membrane (Table 1).

CHEMICAL APPROACHES

Most methods developed for mitochondrial gene delivery are chemical-based. While physical methods enhance the efficiency of gene delivery by increasing cellular uptake of DNA with physical forces, chemical methods are often based on specific chemical interactions with mitochondria. Mitochondrial membranes have hydrophobic characteristics and are also negatively charged. Therefore, it is necessary to enclose negatively charged DNA with carrier molecules that have cationic and amphiphilic properties for efficient delivery. In chemical strategies, carrier molecules encounter several challenges before reaching the mitochondria (Figure 2). In most cases, they must first cross the cellular membrane via endocytosis or translocation. If carrier molecules follow the endocytic pathway, they are not accessible to mitochondria until they escape from the endosome. Accordingly, it is important to find molecules that facilitate the passage of these carriers from the endosome. In addition, there should be strategies to direct the floating DNA-carrier complex in the cytoplasm to mitochondria. Thus, mitochondria-targeting molecules are conjugated with DNA-binding motifs or presented on the surface of DNA-enclosing vesicles composed of dendrimers, surfactants, or liposomes. Many carrier molecules can be easily improved by altering the incorporated mitochondria-targeting motifs.

Cardoso et al. used cationic surfactants with two hydrophilic head groups and two hydrophobic groups as DNA carriers for mitochondrial delivery [9,30]. Even at a low concentration, the cationic gemini surfactants could form micelle-like structures surrounding DNA strands with a negative charge in aqueous solution. This occurred by hydrophobic and electrostatic attractive interactions between the surfactant molecules and DNA [31]. Based on the interactions of gemini-surfactant-DNA complexes with vesicles mimicking the mitochondrial membrane structure, they hypothesized that four modified gemini surfactant-DNA complexes would differentially cross the cellular membrane through either endolysosomal routes or membrane translocation. The former releases plasmid DNA into the mitochondrial surroundings and the latter directly interacts with the mitochondrial membrane. Although they did not identify the precise mechanism underlying the delivery, confirmation of expression of the gene transferred to the mitochondria demonstrated successful entry of the carrier plasmid into the mitochondria [30]. Gemini surfactants usually deliver exogenous DNA into mitochondria passively in that the DNA is delivered to the mitochondrial surroundings instead of the interior. However, specificity for mitochondria can be improved through chemical modifications. Moreover, gemini surfactant-mediated gene delivery can reduce potential cytotoxicity because only small amounts of chemicals are required to transfer DNA.

To improve targeting to mitochondria, rhodamine 123, which is a biocompatible dye with a delocalized positive charge and green fluorescence, was incorporated into nanoparticles carrying plasmid DNA [11]. The lipophilicity combined with the delocalized positive charge of this dye specifically allows nanoparticles to pass through both the plasma and mitochondrial membranes and to ultimately accumulate in the mitochondrial matrix, which has a negative charge. The rhodamine-plasmid DNA nanoparticle complex can be easily prepared by co-precipitation with calcium carbonate. An addition of cellulose increased the stability of the complex in aqueous solution and reduced its size to facilitate penetration of the plasma and mitochondrial membranes [11]. In a subsequent study, successful delivery of plasmid DNA into mitochondria using a rhodamine-based carrier was confirmed by fluorescence imaging of the rhodamine [10].

Liposome-based vesicles can be used to encapsulate plasmid DNA and deliver the complex into mitochondria across the plasma membrane and cytoplasm. For example, the dequalinium (1,1′-(1,10-decamethylene-*bis*-[aminoquinaldinium])-chloride) vesicle (DQAsome) is an amphiphilic and cationic lipid-based vesicle made by self-assembly of dequalinium [12]. The intrinsic ability of dequalinium to accumulate in the mitochondrial matrix was confirmed using a cardiolipin-rich liposome, which is a model of the mitochondrial membrane [14]. When the DQAsome/DNA complex (DQAplex) came into contact with the cardiolipin-rich liposome, cargo DNA was released. In a subsequent study with a mammalian cell system, it was demonstrated that DQAplexes could be released from the endosome and subsequent dissociation of the DQAplex occurred selectively at the mitochondria [12]. Although DQAsomes have high specificity for mitochondria, their use has several limitations such as low transfection efficiency, cytotoxicity, and difficulty in generating homogenous DQAsome solutions. To overcome these limitations, Bae et al. devised the DQA80, which incorporated lipid molecules 1,2-dioleoyl-3-trimethylammonium-propane (DOTAP) and 1,2-dioleoyl-*sn*-glycerol-3-phosphoethanolamine (DOPE) to improve transfection efficiency and increase endosomal escape, respectively [15]. Mixtures of dequalinium and DOTAP/DOPE in different proportions were assessed for their ability to form a carrier for mitochondrial gene delivery. Complexes of plasmid DNA and DQA80s with 10% of DOTAP and DOPE showed enhanced stability and effective mitochondrial targeting.

Using such functionalization methods, existing carriers for mitochondrial gene delivery have been greatly improved. Harashima's group developed the MITO-Porter, which is a liposome-based nanocarrier that targets mitochondria, and they have continuously improved the MITO-Porter in several studies [16,17,18,19]. The major advantage of the MITO-Porter is that it can be easily multi-functionalized by adding ligands. To enhance cellular uptake, which is the first step in delivery of exogenous DNA into mitochondria, the MITO-Porter surface was modified with octa-arginine (R8) peptides mimicking the transactivator of the transcription (TAT) peptide. The TAT peptide fuses to proteins and is commonly used for the delivery of biological molecules crossing the cell membranes [17,18,32,33]. The cationic property of TAT peptide facilitates interactions with negatively charged phospholipid and carbohydrate components of the cell membrane [32].

R8-modified liposomes were found to have improved fusion with mitochondrial membranes and could introduce nucleic acids, peptides, proteins, sugars, or chemicals into mitochondria [16]. MITO-Porter surface-modified with R8 was constructed by multi-layering methods including encapsulation of nanoparticles with a mitochondria-fusogenic inner envelope and endosome-fusogenic outer envelope [18]. R8 peptides in the outer layer allowed efficient internalization of carriers and those in the inner layer fused with the mitochondrial membrane [18]. Lastly, the MITO-Porter, which has an inner envelope composed of sphingomyelin, was found to have both a high mitochondrial membrane fusogenic activity and low cytotoxicity [17].

Another attempt was made to enhance the mitochondrial targeting activity of MITO-Porter by functionalizing it with cationic amphipathic cell-penetrating peptides called KALA-peptides, which are composed of repeated lysine-leucine-alanine residues instead of R8 peptides [19,20]. The resultant KALA-MITO-Porter showed comparable cellular uptake with Lipofectamine 2000, which is a well-known transfection reagent, and improved cellular uptake when compared to the R8-MITO-Porter. Moreover, expression of cargo DNA delivered by the KALA-MITO-Porter was the highest among the three carriers. However, in the BCA protein assay of cellular protein concentration in MITO-Porter-treated cells, KALA-MITO-Porter-treated cells showed lower levels of proteins when compared to R8-MITO-Porter-treated cells, which indicates higher cytotoxicity. It was hypothesized that this cytotoxicity was caused by the destabilization of the mitochondrial membrane by the KALA peptide.

A new approach to constructing carrier systems for mitochondrial gene delivery is the conjugation of the triphenylphosphonium (TPP) cation to liposome-containing DNA as a ligand for mitochondrial targeting [21]. TPP is a lipophilic cation with a delocalized positive charge, which facilitates accumulation into the mitochondrial matrix due to the membrane potential across the mitochondrial inner membrane [23]. Boddapati et al. incorporated TPP into the lipid bilayer of liposomes by conjugating a stearyl residue to form stearyl triphenyl phosphonium (STPP), which is an amphiphilic molecule. As a result, the liposome surface-modified with STPP easily targeted mitochondria and successfully delivered cargo molecules. However, cytotoxic effects of STPP-modified liposomes were observed and, accordingly, the stearyl moiety was substituted with a biocompatible polyethylene glycol (PEG)-phosphatidyl ethanolamine (PE) polymer to incorporate TPP in the liposomes [22]. Novel surface-modified liposomes with TPP-PEG-PE delivered cargo into mitochondria without cytotoxicity.

Biswas et al. conjugated TPP to the surface of polyamidoamine (PAMAM) dendrimers, which can encapsulate biomolecules in their core due to the multi-valency effect of dendrimer-facilitated conjugation of many ligands [23]. Dendrimers are classified by the generation number, which refers to the size and number of repeated functional groups, and generation 5 has been used specifically as a carrier in gene or drug delivery applications. Due to their size, generation 5 PAMAM dendrimers can solubilize multiple hydrophobic entities and simultaneously diffuse through the tissue [34,35,36]. Generation 5 PAMAM dendrimers also have a high positive surface charge, which means they can condense negatively charged DNA by electrostatic interaction and subsequently form PAMAM-DNA complexes [23,37]. The resultant complex has a net positive charge that allows efficient endosomal escape. Subsequently, TPP-conjugated PAMAM dendrimers were localized to mitochondria with reduced cytotoxicity, which was often caused by nonspecific targeting. The latter was achieved by conjugating an acetyl group, which neutralizes the highly cationic dendrimer and allows a high level of association with the anionic cell membrane [23].

The chemical approaches for mitochondrial gene delivery discussed here require chemically synthesized carrier molecules to load DNA, pass through intracellular barriers, and ultimately release DNA into the mitochondria (Table 1 and Figure 2). Their common advantage is that the carrier molecules can be easily and widely modified to overcome limitations such as low efficiency of mitochondrial targeting or high cytotoxicity to target cells. Studies of carrier molecules have focused on their ability to bypass intracellular barriers and pass through the mitochondrial double membrane mostly via membrane fusion. These characteristics can be improved by linking ligands that target specific mitochondrial molecules such as receptors rather than a broad feature of the membrane. Since the ultimate goal of exogenous DNA delivery into the mitochondrial matrix is to supplement the genes that produce defective mitochondrial proteins, it is necessary to confirm whether transferred DNA is properly released and transgene expression is possible.

BIOLOGICAL APPROACHES

Mitochondrial proteins encoded by nDNA account for more than 99% of total mitochondrial proteins [38]. Mitochondrial proteins expressed in the cytosol as mitochondrial precursor proteins enter the mitochondria via mitochondrial targeting signal peptide (MTS)-mediated translocation. Most MTSs are located at the amino-terminus of precursor proteins and are cleaved by proteolysis upon import into mitochondria. The MTS presequences form an amphipathic α-helix that has hydrophobic residues on one side and positively charged residues on the other side. MTS-conjugated proteins enter the mitochondrial matrix through the translocase of the mitochondrial outer membrane (TOM) and that of the mitochondrial inner membrane (TIM). The MTS at the *N*-terminus of a precursor protein is preferentially recognized by the TOM20-TOM22 receptor subcomplex, which contains clusters of negatively charged residues. The protein passes through the import pore formed by TOM40 [39,40]. Consequently, precursor proteins enter the channel formed by TIM17 and TIM23 across the inner membrane in the presence of a membrane potential. The *N*-terminal domain of TIM23 encodes an MTS receptor that has negatively charged residues, which interact with the positively charged amphipathic MTS. Through this interaction, MTS conjugates can pass through the mitochondrial double membranes. Accordingly, conjugation of MTS to carrier molecules that have a DNA-binding function has been a strategy to enable specific mitochondrial gene delivery. For successful MTS-mediated gene delivery, the selection of an effective MTS is required. However, highly hydrophobic proteins and macromolecules are not suitable for MTS-mediated gene delivery [41].

In an initial study, a peptide nucleic acid (PNA) was used as a DNA carrier molecule conjugated with MTS [35]. PNA, which is a DNA oligomer with a polyamide bond instead of a deoxyribose phosphate backbone, was conjugated to an MTS peptide to retain base-specific hybridization [24,42]. PNA and MTS-PNA were biotinylated to evaluate intracellular distribution and introduced into different cell lines including human fibroblasts, HeLa cells, HepG2 cells, and SY5Y cells. Treated cells were subsequently labeled with streptavidin-fluorescein. While PNA without MTS was localized to the nucleus, MTS-conjugated PNA was imported into the mitochondria to some degree. The level of mitochondrial targeting was low because nuclear localization tendency of PNA counteracted mitochondrial targeting. As expected, MTS-mediated PNA entered mitochondria through the TOM and TIM complexes [25]. Transferred oligonucleotides can selectively inhibit the replication of mutated mtDNA if the oligonucleotide is designed to be the anti-sense of mutated mtDNA. Even though this approach cannot transfer large genes, it is a promising early-stage application of MTS in mitochondrial gene delivery.

Similarly, Chuah et al. linked a nonspecific DNA-binding peptide to an MTS to form a mitochondrial gene carrier [26]. A KH peptide comprising alternating lysine and histidine residues that are positively charged was designed to bind to DNA that has the negatively charged phosphate backbone. This peptide enabled cargo DNA to easily pass through the cellular membrane by condensing the DNA. Histidines in the KH peptide are known to facilitate endosomal escape by the effect of a proton sponge, which increases osmolality of the endosome and causes swelling due to accumulation of histidines in the endosome [43]. They conjugated the pre-sequences of yeast cytochrome c oxidase subunit IV and human hepatic enzyme ornithine transcarbamylase to the KH peptide and tested their applicability as mitochondrial gene carriers. As a result, MTS-KH peptides and naked plasmid DNA were electrostatically self-assembled and the resultant complex successfully entered mitochondria, which was followed by the expression of the exogenous plasmid DNA [26].

Viruses have been extensively used in gene therapy as vehicles due to their innate ability to insert their genome into the host cell. Among the various viral vectors used for gene delivery to the nucleus, the adeno-associated virus (AAV) vector has also been used for gene delivery to mitochondria [27]. AAV has an ssDNA genome consisting of the *rep* gene encoding non-structural proteins (REP78, REP68, REP52, and REP40) and the *cap* gene encoding capsid proteins (VP1, VP2, and VP3). The two genes are flanked by inverted terminal repeat (ITR) sequences, which provide *cis*-acting sequences required for replication, packaging, and integration of the viral genome. For application in gene therapy, the promoter and transgene are inserted between the two ITR sequences in the viral vector and both the *rep* and *cap* genes are provided in trans for packaging the vector particles [44].

With respect to the AAV entry process, capsid proteins participate in internalization into the host cell, intracellular transport, and localization. To redirect AAV to mitochondria, Yu et al. constructed MTS-AAV by inserting the cytochrome oxidase subunit VIII (COX8) pre-sequence into VP2, which is one of the capsid proteins in AAV serotype 2 [27]. Among the three capsid genes, it was feasible to insert signal peptides into VP2 because, even though both VP1 and VP2 are involved in nuclear translocation, VP1 affects viral infectivity [45]. In addition, the *N*-terminus of VP2 can tolerate insertion of large peptides [46]. In Leber's hereditary optic neuropathy (LHON), the NADH ubiquinone oxidoreductase subunit 4 (ND4) gene is defective. The wild type human or rodent mitochondrial genes were loaded into the MTS-AAV genome for delivery to the mitochondria of human cells with ND4 mutations or the cells of LHON rodent models, respectively. Since the ND4 gene encodes for a subunit of electron transport chain complex I, LHON cells lack ATP. When the MTS-AAV-mediated ND4 gene was introduced into LHON disease cell lines, rescue of ATP synthesis was observed. In LHON rodents, visual loss and optic atrophy as well as phenotypic symptoms of LHON were also ameliorated by MTS-AAV-mediated ND4 gene therapy. The exact entry mechanism of the MTS-AAV system was not identified. However, in a subsequent study, the research group demonstrated through next generation sequencing that the transferred genes remained episomal [47]. The major advantage of the MTS-AAV-mediated gene delivery system is that it can be applied in vitro and eventually in vivo once clinical safety is established. Additionally, it can deliver hydrophobic proteins and macromolecules, which is a limitation of MTS-peptide conjugates [**28**]. However, the size of the cargo gene that can be delivered by AAV is still limited. Thus, the use of retroviral vectors or lentiviral vectors that can accommodate larger sizes of genes may be required.

COMBINATORIAL APPROACHES

We have classified the various methods of gene delivery to mitochondria into physical, chemical, and biological ones (Table 1). Since each strategy has advantages and disadvantages, individual methods can be combined to improve the import of genes into mitochondria. We discuss several examples where methods have been successfully combined.

When DNA was hydrodynamically injected into mouse liver, condensed DNA rather than naked DNA was used [6]. Plasmid DNA was biologically condensed with protamine, which is a positively charged protein molecule. During injection, the condensed plasmid DNA was internalized into cells by using hydrodynamic force. It was hypothesized that condensed plasmid DNA has a small aggregated form with a positive charge, which enhances mitochondrial association [6]. This strategy helps to overcome the limitation of non-selective hydrodynamic delivery of naked DNA into mitochondria.

Similarly, rhodamine 123, as described above, formed nanoparticles with plasmid DNA by co-precipitation [11]. Incorporation of polymers such as cellulose and gelatin, which are derived from biological components, into nanoparticles condensed the DNA-nanoparticle complex since cellulose and gelatin bind Ca^{2+}, which reduces electrostatic repulsion between hydroxyl and carboxyl groups in the complex [48]. Additionally, cellulose and gelatin have properties such as biocompatibility and stability in aqueous solution, which enhanced the transfection efficiency of the rhodamine 123-based mitochondrial gene delivery system.

Yamada et al. linked MTS to MITO-Porter to integrate the mitochondria-specific targeting activity of MTS with the MITO-Porter delivery system [49]. MTS was conjugated with DOPE (MTS-DOPE) using 4-(*N*-maleimidomethyl)-cyclohexane-1-carboxylic acid *N*-hydroxysuccinimide ester (SMCC) as a bifunctional cross-linker for displaying MTS on the surface of liposomes. MTS-DOPE was added to a lipid solution for preparing liposomes. The resultant MTS-modified MITO-Porter enhanced mitochondrial selectivity and, thereby, more frequently fused with the mitochondrial membrane as compared with non-modified liposomes. These findings suggest the possibility that MTS-modified MITO-Porter could be an efficient carrier for gene delivery through mitochondrial membrane fusion [49].

Unlike existing approaches based on the properties of mitochondrial structure, a novel strategy based on the biological function of mitochondria was suggested. A study showed that CdSe/ZnS core/shell quantum dot (QD) nanoparticles coated with glutathione (GSH) were selectively accumulated in the mitochondrial matrix despite the negative charge of GSH [50]. GSH is a key in maintaining the appropriate mitochondrial redox environment. It is synthesized in the cytosol and easily passes through the mitochondrial outer membrane through porin channels. The anionic property of GSH is unfavorable for entry against the electrochemical gradient present in the mitochondrial inner membrane. However, GSH transporters anchored to the mitochondrial inner membrane including the dicarboxylate carrier (DCc) and the 2-oxoglutarate carrier (OGc) make the entry possible [51]. Zheng et al. observed that GSH-QD clusters were localized to mitochondria by TEM imaging and organelle staining. The specific mitochondrial targeting mechanism of GSH was not identified, but it was shown that GSH can act as a surface ligand for nanocarriers by targeting mitochondria [50].

CARGO DNAS

The main objective in developing exogenous gene delivery systems targeted to mitochondria is to compensate for defective proteins caused by mtDNA mutations. Therefore, successful mitochondrial gene delivery requires not only introduction of exogenous DNA into the mitochondrial matrix but also expression of functional proteins from introduced DNA. Lyrawati et al. designed and generated an artificial mini-mitochondrial genome that contained regulatory domains present in the endogenous mitochondrial genome and a reporter gene to demonstrate transgene expression in mitochondria [13].

The mitochondrial genome is made of circular and double-stranded DNA that is composed of the heavy (H) strand and the light (L) strand (Figure 1). Most of the mitochondrial genome comprises coding regions [52]. These consist of the L-strand replication origin (O_L) and the displacement loop (D-loop) (Figure 1) [53]. The D-loop includes many regulatory domains such as the major H-strand promoter (HSP2), the L-strand promoter (LSP), the H strand replication origin (O_H), the conserved sequence blocks (CSBs), and the termination-associated sequence (TAS) [54,55]. These non-coding regions play roles in replication, transcription, and translation of mtDNA. Both ends of the D-loop are flanked by genes encoding $tRNA^{Phe}$ on the H-strand and genes encoding $tRNA^{Pro}$ on the L-strand [56,57]. These regions are known to be a major site for the control of gene expression (Figure 1).

The newly constructed mini-mitochondrial genome possessed its own genetic expression system similar to that of the mitochondrial genome [13]. For stable expression of transgenes in mitochondria, the mini-mitochondrial genome was constructed to have a D-loop region with flanking sequences that encoded $tRNA^{Phe}$ and $tRNA^{Pro}$ and an O_L region with flanking sequences that encoded $tRNA^{Asn}$, $tRNA^{Cys}$, and $tRNA^{Tyr}$. The transgene encoding the reporter green fluorescent protein (GFP) was designed to be located between the D-loop region and the O_L region. To ensure the reporter gene was expressed in mitochondria, the codon for tryptophan (TGG) in the interior of the gene was changed to the stop codon TGA, which is read as tryptophan in vertebrate mitochondria [56]. Accordingly, translation in the cytosol of the resultant reporter gene with the modified genetic code (mtGFP) was terminated prematurely and could not produce green fluorescence. Translation only in the mitochondria can produce functional GFP with full length. The resultant mini-mitochondrial genome was delivered into the mitochondria of the RAW 264.7 macrophage cell line via DQAsome-mediated gene delivery.

After transfection, a small fraction of transfected cells showed intense green fluorescence within the cytoplasm but not in the nucleus. The fraction of cells producing the modified reporter protein encoded by mtGFP was increased over 14 days post-transfection and the presence of the mini-mitochondrial genome in mitochondria was confirmed by PCR 30 days after transfection. The sustained expression and stability of the mini-mitochondrial genome as an artificial mitochondrial genome are advantageous features for mitochondrial therapeutics [13].

Based on the artificial mini-mitochondrial genome, researchers have constructed cargo DNA molecules with promoters originally encoded in the mitochondrial genome and reporter genes or mitochondrial endogenous genes optimized to the mitochondrial codon system [7,27]. HSP2, which drives transcription of 12 of the 13 protein-coding genes in mitochondria, has been used to mediate transgene expression [53,58] (Figure 1). Yamada et al. showed for the first time that the cytomegalovirus (CMV) promoter, which has been widely used to achieve a high level of nuclear transgene expression in mammalian cells, could trigger the transcription of mitochondrial genes efficiently [19,59]. An HSP-driven reporter gene, which consists of a luciferase gene adjusted to the mitochondrial codon system (mtLuc), produced no luminescence when the corresponding DNA construct was delivered to HeLa cells by Lipofectamine 2000. The novel pCMV-mtLuc vector was constructed by substitution of HSP with the CMV promoter and resulted in a high level of luciferase activity under the same delivery approach. Moreover, the luciferase activity of transferred CMV-mtLuc was similar before and after treatment with α-amanitin, which is a specific inhibitor of RNA polymerase II. This result implies the transcription of CMV-mtLuc in mitochondria. These findings indicate the presence of new mechanisms involved in the process of transcription in mitochondria in addition to the well-known mitochondrial transcriptional system. This strategy may allow the production of functional proteins required for normal mitochondrial activity in mitochondrial disease when various approaches are applied for mitochondrial gene delivery.

In contrast to the above-mentioned approaches that aimed to provide cells with corrected versions of mitochondrial genetic materials, several research groups applied new approaches to eliminate mitochondrial genomes with mutations. They targeted disease-causing point mutations within mitochondrial genomes including T8993G that links to neuropathy ataxia, retinitis pigmentosa (NARP) syndrome, and Leigh syndrome, A8344G associated with MERRF, and A3243G, which may lead to mitochondrial encephalopathy, lactic acidosis, and stroke-like episodes (MELAS) [60,61,62,63,64]. To specifically destroy mitochondrial genomes carrying the targeted mutations, endonucleases such as *Sma* I were first employed. When the gene that encoded the MTS-tagged *Sma* I was introduced into cells, the fraction of mitochondrial genomes with the mutation of T8993G was greatly decreased via cutting of the mutated sequence by the action of the endonuclease [63]. Given a set of existing endonucleases with defined recognition sites, target DNA sequences are limited. To more easily target other mitochondrial DNA mutations, zinc fingers and transcription activator-like effectors (TALEs) have been combined to DNA nucleases [60,61,62,64]. These two types of DNA-binding proteins can be engineered to have specificities for new DNA sequences of interest. For the same goal of easy targeting various mitochondrial mutations, clustered regularly interspaced short palindromic repeats (CRISPR)/Cas9 proteins can be also introduced into cells through the addition of the corresponding encoding genes. If the supply of corrected genetic materials and elimination of mutated mitochondrial genomes are applied at the same time, more effective treatment of mitochondrial malfunctions might be possible.

APPLICATIONS AS DISEASE THERAPIES

Genetic mutations in mtDNA cause a number of diseases and the major sites of mutations are well known [65]. For example, mitochondrial diseases such as MERRF disease are caused by the A8344G mutation in the $tRNA^{Phe}$-coding gene. Neurogenic muscle weakness, ataxia, retinitis pigmentosa (NARP), and Leigh syndrome are caused by the T8993G mutation in the ATP6 gene. Other genetic mutations have been linked to Kearns-Sayre syndrome, Ataxia neuropathy syndrome, and Alpers-Huttenlocher [66].

LHON is a well-studied mitochondrial disease. It involves retinal ganglion cell death and degeneration of optic nerves. Over 90% of LHON patients carry one of three point mutations in their mtDNA: G3460A in the ND1 gene, G11778A in the ND4 gene, and T14484C in the ND6 gene. Among these mutations, G11778A accounts for more than 70% of the cases of LHON [67,68]. Yu et al. successfully achieved delivery of the wild type ND4 gene and ND4 expression using MTS-AAV in an LHON cell line harboring G11778A [27]. The treated cells showed restoration of ATP synthesis, which is deficient in the LHON cell line. Additionally, the MTS-AAV-mediated ND4 gene delivery was demonstrated to be effective in in vivo testing. In a mouse model of LHON, mutant mice injected with MTS-AAV carrying the wild type ND4 gene showed improved visual function after one year post injection when compared with control groups injected with either AAV not flanked by MTS or MTS-AAV carrying only GFP gene. Histological cross-sections of the optic nerves confirmed that the mutant mice injected with MTS-AAV carrying the wild type ND4 gene had thicker optic nerves, which indicated that optic atrophy was relieved by the wild type ND4 allele [27].

MELAS are other well-known symptoms of the mitochondrial disease caused by various point mutations in the NADH dehydrogenase gene or mitochondrial tRNA-coding genes [69]. Gene delivery using MITO-Porter functionalized with KALA peptides (KALA-MITO-Porter) was conducted with cells obtained from a patient suffering from MELAS-like symptoms including epilepsy, hearing loss, and an elevated lactate level. KALA-MITO-Porter-mediated delivery of the luciferase gene and wild type ND4 gene under the CMV promoter were tested in fibroblasts harboring the G625A mutation in the mitochondrial $tRNA^{Phe}$-coding gene. In this study, the biocompatibility and efficiency of both the KALA-MITO-Porter-mediated gene delivery system and the CMV-driven ND4 expression were investigated. As a result, cell viability was not affected by KALA-MITO-Porter regardless of cargo DNA and cell type, but there was no detectable luciferase activity in diseased cells. To facilitate binding of KALA-MITO-Porter to diseased cells and improve cellular uptake and mitochondrial targeting, a mitochondrial RNA aptamer was added. Using this approach, they achieved efficient mitochondrial transgene expression. However, a more practical approach would have been to include the delivery of cargo DNA that was deficient in diseased cells. In this case, it includes the gene encoding $tRNA^{Phe}$. Nevertheless, the results of this research highlight the importance of practical application of mitochondrial gene delivery systems.

Until now, there have been few attempts to directly deliver genes into the mitochondria of cells or animals that are actually under mitochondrial dysfunction. Instead, several groups have attempted to transfer genes to cells under mitochondrial disease states without mitochondria-targeting. Previously, the yeast NDI1 gene encoding NADH-quinone oxidoreductase was reported as being capable of modulating oxidative phosphorylation in 293 human kidney cells. In another study, this gene restored the respiratory NADH oxidase activity of complex I-deficient mammalian cells [70,71]. Based on these findings, the *S. cerevisiae* NDI1 gene was packaged into recombinant AAV (rAAV-NDI1) and was introduced into human C4T cells, which carry a homoplasmic frameshift mutation in the ND4 gene. The C4T cells cannot form the NADH dehydrogenase complex and, thus, do not grow in galactose medium because they lack glycolytic activity. The C4T cells transduced with rAAV-NDI1 produced NDI1 proteins in the cytosol, which were then imported into the mitochondria due to the NADH-binding site facing the mitochondrial matrix. As a result, the C4T cells highly expressing NDI1 could grow in a galactose medium [70]. To demonstrate if this mitochondrial gene delivery system is applicable to diseased cells or animals, construction of both various diseased models and efficient delivery systems is required.

CONCLUDING REMARKS

In contrast to chromosomes in the nucleus, mitochondria possess many copies of their genome and, thus, only a small percentage of mutated alleles do not cause mitochondrial disease. This feature can be exploited to treat mitochondrial diseases caused by genetic mutations in mtDNA by complementing essential genes. The major challenge of gene delivery is passage through the mitochondrial double membranes to transfer cargo DNA into the mitochondrial matrix (Figure 2). Based on the structure of mitochondria, cationic amphiphiles linked to DNA binding or encapsulating motifs have been used as carriers for transgene delivery. Another challenge is the stable expression of the transgene to produce functional proteins to restore mitochondrial activity. To date, few studies have reported successful delivery of genetic cargo and transgene expression in cells or animals with mitochondrial disease. Various approaches have been used to deliver wild type genes into the mitochondrial matrix and have been improved steadily to overcome the challenges including passage through the mitochondrial membranes, low and unstable transgene expression, or cytotoxicity derived from carrier molecules (Table 1). Most studies reviewed here provide targeting strategies to transfer genetic cargo into mitochondria with ongoing improvements. However, compared to nuclear gene delivery methods, mitochondrial gene delivery methods are currently far from being clinically applicable. We think that new approaches and trials would be required to effectively deliver parts of or the entire wild-type mitochondrial genome to cells of interest in order to achieve long-term expression of transgenes. Additionally, more disease models need to be developed to cover the wide range of genetic mutations in mitochondria.

FUNDING

This work was supported by the Medical Research Center Program (No. 2011-0030074) through the National Research Foundation of Korea (NRF) grant funded by the Korea government (MSIP).

CONFLICTS OF INTEREST

The authors declare no conflict of interest.

REFERENCES

1. Robin, E.D.; Wong, R. Mitochondrial DNA molecules and virtual number of mitochondria per cell in mammalian cells. *J. Cell. Physiol.* 1988, *136*, 507–513.
2. Legros, F.; Malka, F.; Frachon, P.; Lombes, A.; Rojo, M. Organization and dynamics of human mitochondrial DNA. *J. Cell Sci.* 2004, *117 Pt 13*, 2653–2662.
3. D'Erchia, A.M.; Atlante, A.; Gadaleta, G.; Pavesi, G.; Chiara, M.; De Virgilio, C.; Manzari, C.; Mastropasqua, F.; Prazzoli, G.M.; Picardi, E.; et al. Tissue-specific mtDNA abundance from exome data and its correlation with mitochondrial transcription, mass and respiratory activity. *Mitochondrion* 2015, *20*, 13–21.
4. Stewart, J.B.; Chinnery, P.F. The dynamics of mitochondrial DNA heteroplasmy: Implications for human health and disease. *Nat. Rev. Genet.* 2015, *16*, 530–542.
5. Yasuzaki, Y.; Yamada, Y.; Kanefuji, T.; Harashima, H. Localization of exogenous DNA to mitochondria in skeletal muscle following hydrodynamic limb vein injection. *J. Control. Release* 2013, *172*, 805–811.
6. Yasuzaki, Y.; Yamada, Y.; Fukuda, Y.; Harashima, H. Condensation of plasmid DNA enhances mitochondrial association in skeletal muscle following hydrodynamic limb vein injection. *Pharmaceuticals* 2014, *7*, 881–893.
7. Yasuzaki, Y.; Yamada, Y.; Ishikawa, T.; Harashima, H. Validation of Mitochondrial Gene Delivery in Liver and Skeletal Muscle via Hydrodynamic Injection Using an Artificial Mitochondrial Reporter DNA Vector. *Mol. Pharm.* 2015, *12*, 4311–4320.
8. Bonnefoy, N.; Fox, T.D. Directed alteration of Saccharomyces cerevisiae mitochondrial DNA by biolistic transformation and homologous recombination. *Methods Mol. Biol.* 2007, *372*, 153–166. [PubMed]
9. Cardoso, A.M.; Morais, C.M.; Cruz, A.R.; Cardoso, A.L.; Silva, S.G.; do Vale, M.L.; Marques, E.F.; de Lima, M.C.P.; Jurado, A.S. Gemini Surfactants Mediate Efficient Mitochondrial Gene Delivery and Expression. *Mol. Pharm.* 2015, *12*, 716–730.
10. Salvado, R.; Sousa, F.; Queiroz, J.; Costa, D. Development of mitochondrial targeting plasmid DNA nanoparticles: Characterization

and in vitro studies. *Colloid Surf. A* 2015, *480*, 287–295.

11. Santos, J.; Sousa, F.; Queiroz, J.; Costa, D. Rhodamine based plasmid DNA nanoparticles for mitochondrial gene therapy. *Colloids Surf. B Biointerfaces* 2014, *121*, 129–140.
12. D'Souza, G.G.M.; Rammohan, R.; Cheng, S.M.; Torchilin, V.P.; Weissig, V. DQAsome-mediated delivery of plasmid DNA toward mitochondria in living cells. *J. Control. Release* 2003, *92*, 189–197.
13. Lyrawati, D.; Trounson, A.; Cram, D. Expression of GFP in the mitochondrial compartment using DQAsome-mediated delivery of an artificial mini-mitochondrial genome. *Pharm. Res.* 2011, *28*, 2848–2862.
14. Weissig, V.; Lizano, C.; Torchilin, V.P. Selective DNA release from DQAsome/DNA complexes at mitochondria-like membranes. *Drug Deliv.* 2000, *7*, 1–5.
15. Bae, Y.; Jung, M.K.; Song, S.J.; Green, E.S.; Lee, S.; Park, H.S.; Jeong, S.H.; Han, J.; Mun, J.Y.; Ko, K.S.; et al. Functional nanosome for enhanced mitochondria-targeted gene delivery and expression. *Mitochondrion* 2017, *37*, 27–40.
16. Kogure, K.; Akita, H.; Yamada, Y.; Harashima, H. Multifunctional envelope-type nano device (MEND) as a non-viral gene delivery system. *Adv. Drug Deliv. Rev.* 2008, *60*, 559–571.
17. Yamada, Y.; Fukuda, Y.; Harashima, H. An analysis of membrane fusion between mitochondrial double membranes and MITO-Porter, mitochondrial fusogenic vesicles. *Mitochondrion* 2015, *24*, 50–55.
18. Yamada, Y.; Furukawa, R.; Yasuzaki, Y.; Harashima, H. Dual function MITO-Porter, a nano carrier integrating both efficient cytoplasmic delivery and mitochondrial macromolecule delivery. *Mol. Ther.* 2011, *19*, 1449–1456.
19. Yamada, Y.; Ishikawa, T.; Harashima, H. Validation of the use of an artificial mitochondrial reporter DNA vector containing a Cytomegalovirus promoter for mitochondrial transgene expression. *Biomaterials* 2017, *136*, 56–66.
20. Ma, X.C.; Wang, X.B.; Zhou, M.; Fei, H. A Mitochondria-Targeting Gold-Peptide Nanoassembly for Enhanced Cancer-Cell Killing. *Adv. Healthc. Mater.* 2013, *2*, 1638–1643.
21. Boddapati, S.V.; D'Souza, G.G.; Erdogan, S.; Torchilin, V.P.; Weissig, V. Organelle-targeted nanocarriers: Specific delivery of liposomal

ceramide to mitochondria enhances its cytotoxicity in vitro and in vivo. *Nano Lett.* 2008, *8*, 2559–2563.

22. Biswas, S.; Dodwadkar, N.S.; Deshpande, P.P.; Torchilin, V.P. Liposomes loaded with paclitaxel and modified with novel triphenylphosphonium-PEG-PE conjugate possess low toxicity, target mitochondria and demonstrate enhanced antitumor effects in vitro and in vivo. *J. Control. Release* 2012, *159*, 393–402.
23. Biswas, S.; Dodwadkar, N.S.; Piroyan, A.; Torchilin, V.P. Surface conjugation of triphenylphosphonium to target poly(amidoamine) dendrimers to mitochondria. *Biomaterials* 2012, *33*, 4773–4782.
24. Chinnery, P.F.; Taylor, R.W.; Diekert, K.; Lill, R.; Turnbull, D.M.; Lightowlers, R.N. Peptide nucleic acid delivery to human mitochondria. *Gene Ther.* 1999, *6*, 1919–1928.
25. Flierl, A.; Jackson, C.; Cottrell, B.; Murdock, D.; Seibel, P.; Wallace, D.C. Targeted delivery of DNA to the mitochondrial compartment via import sequence-conjugated peptide nucleic acid. *Mol. Ther.* 2003, *7*, 550–557.
26. Chuah, J.A.; Matsugami, A.; Hayashi, F.; Numata, K. Self-Assembled Peptide-Based System for Mitochondrial-Targeted Gene Delivery: Functional and Structural Insights. *Biomacromolecules* 2016, *17*, 3547–3557.
27. Yu, H.; Koilkonda, R.D.; Chou, T.H.; Porciatti, V.; Ozdemir, S.S.; Chiodo, V.; Boye, S.L.; Boye, S.E.; Hauswirth, W.W.; Lewin, A.S.; et al. Gene delivery to mitochondria by targeting modified adenoassociated virus suppresses Leber's hereditary optic neuropathy in a mouse model. *Proc. Natl. Acad. Sci. USA* 2012, *109*, E1238–E1247.
28. Yu, H.; Ozdemir, S.S.; Koilkonda, R.D.; Chou, T.H.; Porciatti, V.; Chiodo, V.; Boye, S.L.; Hauswirth, W.W.; Lewin, A.S.; Guy, J. Mutant NADH dehydrogenase subunit 4 gene delivery to mitochondria by targeting sequence- modified adeno-associated virus induces visual loss and optic atrophy in mice. *Mol. Vis.* 2012, *18*, 1668–1683.
29. Suda, T.; Liu, D. Hydrodynamic gene delivery: Its principles and applications. *Mol. Ther.* 2007, *15*, 2063–2069.
30. Wang, W.X.; He, Y.F.; Shang, Y.Z.; Liu, H.L. Interaction between the Gemini Surfactant (12-6-12) and DNA. *Acta Phys. Chim. Sin.* 2011, *27*, 156–162.

31. Cardoso, A.M.; Faneca, H.; Almeida, J.A.; Pais, A.A.; Marques, E.F.; de Lima, M.C.; Jurado, A.S. Gemini surfactant dimethylene-1,2-bis(tetradecyldimethylammonium bromide)-based gene vectors: A biophysical approach to transfection efficiency. *Biochim. Biophys. Acta* 2011, *1808*, 341–351.
32. Del Gaizo, V.; Payne, R.M. A novel TAT-mitochondrial signal sequence fusion protein is processed, stays in mitochondria, and crosses the placenta. *Mol. Ther.* 2003, *7*, 720–730.
33. Futaki, S.; Ohashi, W.; Suzuki, T.; Niwa, M.; Tanaka, S.; Ueda, K.; Harashima, H.; Sugiura, Y. Stearylated arginine-rich peptides: A new class of transfection systems. *Bioconjug. Chem.* 2001, *12*, 1005–1011.
34. Choi, S.K.; Myc, A.; Silpe, J.E.; Sumit, M.; Wong, P.T.; McCarthy, K.; Desai, A.M.; Thomas, T.P.; Kotlyar, A.; Holl, M.M.B.; et al. Dendrimer-Based Multivalent Vancomycin Nanoplatform for Targeting the Drug-Resistant Bacterial Surface. *ACS Nano* 2013, *7*, 214–228.
35. Santos, J.L.; Pandita, D.; Rodrigues, J.; Pego, A.P.; Granja, P.L.; Balian, G.; Tomas, H. Receptor-Mediated Gene Delivery Using PAMAM Dendrimers Conjugated with Peptides Recognized by Mesenchymal Stem Cells. *Mol. Pharm.* 2010, *7*, 763–774.
36. Van Dongen, M.A.; Orr, B.G.; Banaszak Holl, M.M. Diffusion NMR study of generation-five PAMAM dendrimer materials. *J. Phys. Chem. B* 2014, *118*, 7195–7202.
37. Wang, X.Y.; Shao, N.M.; Zhang, Q.; Cheng, Y.Y. Mitochondrial targeting dendrimer allows efficient and safe gene delivery. *J. Mater. Chem. B* 2014, *2*, 2546–2553.
38. Chacinska, A.; Koehler, C.M.; Milenkovic, D.; Lithgow, T.; Pfanner, N. Importing Mitochondrial Proteins: Machineries and Mechanisms. *Cell* 2009, *138*, 628–644.
39. Pfanner, N.; Meijer, M. The Tom and Tim machine. *Curr. Biol.* 1997, *7*, R100–R103.
40. Van der Laan, M.; Meinecke, M.; Dudek, J.; Hutu, D.P.; Lind, M.; Perschil, I.; Guiard, B.; Wagner, R.; Pfanner, N.; Rehling, P. Motor-free mitochondrial presequence translocase drives membrane integration of preproteins. *Nat. Cell Biol.* 2007, *9*, 1152–1159.
41. Esaki, M.; Kanamori, T.; Nishikawa, S.; Endo, T. Two distinct mechanisms drive protein translocation across the mitochondrial outer membrane in the late step of the cytochrome b(2) import pathway. *Proc.*

Natl. Acad. Sci. USA 1999, *96*, 11770–11775.

42. Nielsen, P.E.; Egholm, M.; Berg, R.H.; Buchardt, O. Sequence-selective recognition of DNA by strand displacement with a thymine-substituted polyamide. *Science* 1991, *254*, 1497–1500.
43. Midoux, P.; Pichon, C.; Yaouanc, J.J.; Jaffres, P.A. Chemical vectors for gene delivery: A current review on polymers, peptides and lipids containing histidine or imidazole as nucleic acids carriers. *Br. J. Pharmacol.* 2009, *157*, 166–178.
44. Daya, S.; Berns, K.I. Gene Therapy Using Adeno-Associated Virus Vectors. *Clin. Microbiol. Rev.* 2008, *21*, 583–593.
45. Tu, M.; Liu, F.; Chen, S.; Wang, M.; Cheng, A. Role of capsid proteins in parvoviruses infection. *Virol. J.* 2015, *12*, 114.
46. Warrington, K.H.; Gorbatyuk, O.S.; Harrison, J.K.; Opie, S.R.; Zolotukhin, S.; Muzyczka, N. Adeno-associated virus type 2 VP2 capsid protein is nonessential and can tolerate large peptide insertions at its N terminus. *J. Virol.* 2004, *78*, 6595–6609.
47. Yu, H.; Mehta, A.; Wang, G.F.; Hauswirth, W.W.; Chiodo, V.; Boye, S.L.; Guy, J. Next-generation sequencing of mitochondrial targeted AAV transfer of human ND4 in mice. *Mol. Vis.* 2013, *19*, 1482–1491.
48. Coutinho, E.; Batista, C.; Sousa, F.; Queiroz, J.; Costa, D. Mitochondrial Gene Therapy: Advances in Mitochondrial Gene Cloning, Plasmid Production, and Nanosystems Targeted to Mitochondria. *Mol. Pharm.* 2017, *14*, 626–638.
49. Yamada, Y.; Harashima, H. Enhancement in selective mitochondrial association by direct modification of a mitochondrial targeting signal peptide on a liposomal based nanocarrier. *Mitochondrion* 2013, *13*, 526–532.
50. Zheng, H.; Mortensen, L.J.; Ravichandran, S.; Bentley, K.; DeLouise, L.A. Effect of Nanoparticle Surface Coating on Cell Toxicity and Mitochondria Uptake. *J. Biomed. Nanotechnol.* 2017, *13*, 155–166.
51. Mari, M.; Morales, A.; Colell, A.; Garcia-Ruiz, C.; Fernandez-Checa, J.C. Mitochondrial glutathione, a key survival antioxidant. *Antioxid. Redox Signal.* 2009, *11*, 2685–2700.
52. Bibb, M.J.; Van Etten, R.A.; Wright, C.T.; Walberg, M.W.; Clayton, D.A. Sequence and gene organization of mouse mitochondrial DNA. *Cell* 1981, *26 Pt 2*, 167–180.

53. Bonawitz, N.D.; Clayton, D.A.; Shadel, G.S. Initiation and beyond: Multiple functions of the human mitochondril transcription machinery. *Mol. Cell* 2006, *24*, 813–825.
54. Taanman, J.W. The mitochondrial genome: Structure, transcription, translation and replication. *Biochim. Biophys. Acta* 1999, *1410*, 103–123.
55. Walberg, M.W.; Clayton, D.A. Sequence and properties of the human KB cell and mouse L cell D-loop regions of mitochondrial DNA. *Nucleic Acids Res.* 1981, *9*, 5411–5421.
56. Anderson, S.; Bankier, A.T.; Barrell, B.G.; de Bruijn, M.H.; Coulson, A.R.; Drouin, J.; Eperon, I.C.; Nierlich, D.P.; Roe, B.A.; Sanger, F.; et al. Sequence and organization of the human mitochondrial genome. *Nature* 1981, *290*, 457–465.
57. Holt, I.J.; Reyes, A. Human mitochondrial DNA replication. *Cold Spring Harb. Perspect. Biol.* 2012, *4*.
58. Zollo, O.; Tiranti, V.; Sondheimer, N. Transcriptional requirements of the distal heavy-strand promoter of mtDNA. *Proc. Natl. Acad. Sci. USA* 2012, *109*, 6508–6512.
59. Boshart, M.; Weber, F.; Jahn, G.; Dorsch-Hasler, K.; Fleckenstein, B.; Schaffner, W. A very strong enhancer is located upstream of an immediate early gene of human cytomegalovirus. *Cell* 1985, *41*, 521–530.
60. Gammage, P.A.; Gaude, E.; Van Haute, L.; Rebelo-Guiomar, P.; Jackson, C.B.; Rorbach, J.; Pekalski, M.L.; Robinson, A.J.; Charpentier, M.; Concordet, J.P.; et al. Near-complete elimination of mutant mtDNA by iterative or dynamic dose-controlled treatment with mtZFNs. *Nucleic Acids Res.* 2016, *44*, 7804–7816.
61. Hashimoto, M.; Bacman, S.R.; Peralta, S.; Falk, M.J.; Chomyn, A.; Chan, D.C.; Williams, S.L.; Moraes, C.T. MitoTALEN: A General Approach to Reduce Mutant mtDNA Loads and Restore Oxidative Phosphorylation Function in Mitochondrial Diseases. *Mol. Ther.* 2015, *23*, 1592–1599.
62. Pereira, C.V.; Bacman, S.R.; Arguello, T.; Zekonyte, U.; Williams, S.L.; Edgell, D.R.; Moraes, C.T. mitoTev-TALE: A monomeric DNA editing enzyme to reduce mutant mitochondrial DNA levels. *EMBO Mol. Med.* 2018.
63. Tanaka, M.; Borgeld, H.J.; Zhang, J.; Muramatsu, S.; Gong, J.S.; Yoneda,

M.; Maruyama, W.; Naoi, M.; Ibi, T.; Sahashi, K.; et al. Gene therapy for mitochondrial disease by delivering restriction endonuclease SmaI into mitochondria. *J. Biomed. Sci.* 2002, *9 Pt 1*, 534–541.

64. Yang, Y.; Wu, H.; Kang, X.; Liang, Y.; Lan, T.; Li, T.; Tan, T.; Peng, J.; Zhang, Q.; An, G.; et al. Targeted elimination of mutant mitochondrial DNA in MELAS-iPSCs by mitoTALENs. *Protein Cell* 2018, *9*, 283–297.
65. Tuppen, H.A.; Blakely, E.L.; Turnbull, D.M.; Taylor, R.W. Mitochondrial DNA mutations and human disease. *Biochim. Biophys. Acta* 2010, *1797*, 113–128.
66. Menezes, M.P.; Ouvrier, R.A. Peripheral neuropathy associated with mitochondrial disease in children. *Dev. Med. Child Neurol.* 2012, *54*, 407–414.
67. Cwerman-Thibault, H.; Augustin, S.; Ellouze, S.; Sahel, J.A.; Corral-Debrinski, M. Gene therapy for mitochondrial diseases: Leber Hereditary Optic Neuropathy as the first candidate for a clinical trial. *C. R. Biol.* 2014, *337*, 193–206.
68. Man, P.Y.W.; Turnbull, D.M.; Chinnery, P.F. Leber hereditary optic neuropathy. *J. Med. Genet.* 2002, *39*, 162–169.
69. Tzen, C.Y.; Thajeb, P.; Wu, T.Y.; Chen, S.C. Melas with point mutations involving tRNALeu (A3243G) and tRNAGlu(A14693g). *Muscle Nerve* 2003, *28*, 575–581.
70. Bai, Y.; Hajek, P.; Chomyn, A.; Chan, E.; Seo, B.B.; Matsuno-Yagi, A.; Yagi, T.; Attardi, G. Lack of complex I activity in human cells carrying a mutation in MtDNA-encoded ND4 subunit is corrected by the Saccharomyces cerevisiae NADH-quinone oxidoreductase (NDI1) gene. *J. Biol. Chem.* 2001, *276*, 38808–38813.
71. Seo, B.B.; Kitajima-Ihara, T.; Chan, E.K.; Scheffler, I.E.; Matsuno-Yagi, A.; Yagi, T. Molecular remedy of complex I defects: Rotenone-insensitive internal NADH-quinone oxidoreductase of Saccharomyces cerevisiae mitochondria restores the NADH oxidase activity of complex I-deficient mammalian cells. *Proc. Natl. Acad. Sci. USA* 1998, *95*, 9167–9171.

9

Gene Therapy for Cancer: Present Status and Future Perspective

Magid H Amer

Department of Medicine, St Rita's Medical Center, 825 West Market Street, Suite #203, Lima, OH 45805, USA

ABSTRACT

Advancements in human genomics over the last two decades have shown that cancer is mediated by somatic aberration in the host genome. This discovery has incited enthusiasm among cancer researchers; many now use therapeutic approaches in genetic manipulation to improve cancer regression and find a potential cure for the disease. Such gene therapy includes transferring genetic material into a host cell through viral (or bacterial) and non-viral vectors, immunomodulation of tumor cells or the host immune system, and manipulation of the tumor microenvironment, to reduce tumor vasculature or to increase tumor antigenicity for better recognition by the host immune

Citation: Amer, M.H. Gene therapy for cancer: present status and future perspective. Mol and Cell Ther 2, 27 (2014). https://doi.org/10.1186/2052-8426-2-27.

system. Overall, modest success has been achieved with relatively minimal side effects. Previous approaches to cancer treatment, such as retrovirus integration into the host genome with the risk of mutagenesis and second malignancies, immunogenicity against the virus and/or tumor, and resistance to treatment with disease relapse, have markedly decreased with the new generation of viral and non-viral vectors. Several tumor-specific antibodies and genetically modified immune cells and vaccines have been developed, yet few are presently commercially available, while many others are still ongoing in clinical trials. It is anticipated that gene therapy will play an important role in future cancer therapy as part of a multimodality treatment, in combination with, or following other forms of cancer therapy, such as surgery, radiation and chemotherapy. The type and mode of gene therapy will be determined based on an individual's genomic constituents, as well as his or her tumor specifics, genetics, and host immune status, to design a multimodality treatment that is unique to each individual's specific needs.

INTRODUCTION

Over the last two decades, cancer research and genomics have experienced considerable advancements. In 2001, two independent draft versions of the human genome sequence and the concomitant identification of approximately 30,000 genes were published [1, 2]. As expected, this was followed by extensive research in molecular genetics with advanced tools used to dissect gene function and explore the biological processes involved in causing genetic aberration and malignant transformation [3]. The recent proliferation of knowledge of cancer has led to the development of novel therapeutic approaches in cancer management, particularly gene therapy.

Gene therapy implies any procedure intended to treat or alleviate a disease by genetically modifying the cell of a patient [4]. The material to be transferred into patient cells may be genes, gene segments, or oligonucleotides. Gene transfer therapy can be conducted either as in vivo or ex vivo approaches. In the in vivo approach, targeted cells are approached directly, such as the intradermal injection of a metastatic nodule, or intravesical therapy for superficial bladder cancer. In the ex vivo approach, targeted cells from a tumor are selected, then collected, grown in culture media at a controlled microenvironment, manipulated genetically by the insertion of a new gene or protein (transgene) in the cell genome, then introduced back into the host. The ex vivo approach is much simpler to achieve as it is easier to manipulate target cells externally.

Concerning cancer, initial efforts to deactivate oncogenes and replace non-functioning tumor suppressor genes were barely successful. Subsequently, new approaches have been developed to transfer genetic materials (transgenes) directly into target cells aiming to transiently or permanently change their phenotypes [5]. Target cells may be normal cells, cancerous cells, immune mediated cells, or pluripotent stem cells. Once the transgene enters a cancer cell, it can then assists in its death or restore normal cellular functions, whereas for normal cells, the transgene can protect them from drug-induced toxicities, or activate an immune cell to get rid of the cancer cell. Gene and vector-based molecular therapies for cancer comprise a wide range of treatment modalities to modify cancer cells, normal cells, and/or a tumor microenvironment [6].

HISTORY

The history of cancer therapy dates back to the eighteenth century, when surgery was the primary treatment for early stages of cancer, and patients suffered from frequent relapses [7]. Once the disease spread, patients were treated with herbal medications, castor oil, or arsenic. In 1895, radiation therapy was discovered, but resulted in few cures [8]. At that time, several cases of spontaneous cancer regression following bacterial infection were reported [9]. In 1868 a patient with soft tissue sarcoma went into remission following an erysipelas infection, but this regression lasted for a short duration [9]. In 1943, nitrogen mustard was used in the management of patients with lymphoma [10], and in 1948 folic acid antagonists led to transient remission in childhood leukemia [11]. Since then, there has been a dramatic advancement in chemotherapy treatment for cancer [7]. Viruses were also found to be effective in controlling malignancies in animal models, and subsequently in humans in 1956 [12]. Adenoviruses in particular have been studied more intensively in humans, with the subsequent development of gene therapy [13]. In 1987, immunotherapy was introduced in the management of cancer patients with subsequent FDA approval of rituximab antibodies in the treatment of patients with lymphoma (1997) [14].

The first FDA-approved gene therapy experiment in the United States occurred in 1990 for a patient with severe combined immunodeficiency disorder [15]. Since then, many clinical trials have been conducted for patients with cancer, using different approaches in gene therapy, with successful results reported in patients with chronic lymphocytic leukemia, acute lymphocytic leukemia, brain tumors, as well as others. Several commercially approved medications for gene therapy were released, including ONYX-15 (Onyx Pharmaceuticals) for refractory head and neck cancer (2005) [16]; human papilloma virus vaccine (Gardasil) (Merck Sharp & Dohme) for the prevention of cancer cervix (2006) [17]; and modified dendritic cells, sipuleucel-T (Provenge) (Dendreon Corporation, Seattle, WA), for minimally symptomatic, castration-resistant metastatic prostate cancer (2010) [18].

METHODS OF GENE THERAPY

Gene therapy implies an approach that aims to modify, delete, or replace abnormal gene(s) at a target cell. Such target cells may be malignant primary or metastatic nodules, circulating tumor cells or dormant stem cells, and specific cells such as T-cell lymphocytes or dendritic cells. With the presence of over 20,000 active genes in human cells, exposed to numerous factors whether hereditary, environmental, infectious or spontaneous, unlimited possibilities for gene mutation, aberration, dysfunction or deletion have been expected, leading to clinical presentation of various medical disorders, including cancer. Furthermore, genomics of cancer evolve between primary and metastases. For example, estrogen receptor gene (ESR J) mutations in breast cancer were found in proportion of metastases but not in primary tumors [19, 20]. Whole-exome sequencing of metastatic samples reported among the top 17 mutated genes, only five were mutated in primary tumors [21]. The evolution from minority clone to lethal metastases follows branched evolution. Thus, tumors with high a level of intratumor heterogenicity and genomic instability could be more likely to escape from targeted therapies such as gene therapy, unless such a branched evolution is taken into consideration. Hence, gene therapy is somewhat difficult to achieve, with limited success. Presently, most approaches are for monogenic gene therapy, tackling one or more critical gene defects. Selection of the appropriate mode of gene therapy is based on the assessment of the immune status, and determination of the molecular nature of a patient's disease. With the recent increases in knowledge of molecular biology of various medical disorders, a more advanced and comprehensive gene therapy approach will ultimately become available, with anticipated improved results.

GENE TRANSFER DELIVERY SYSTEM

Several methods have been developed to facilitate the entry of genetic materials (transgenes) into target cells, using various vectors. They are broadly divided into two major categories: viral (or bacterial) and non-viral vectors [Table 1]. Viruses usually bind to target cells and introduce their genetic materials into the host cell as part of their replication process. As they enter target cells, they can carry a load of other genetic material called "transgenes". For non-viral vectors, different approaches have been utilized, using physical, chemical, as well as other modes of genetic transfer. Transferring genetic material directly into cells is referred to as "transfection", while moving them into cells carried by a viral or bacterial vector is termed "transduction". Non-viral approaches have the advantage of safety and easy modifiability, but have a lower transfection efficiency compared to viral vectors [22].

Table 1. Gene transfer and immunomodulation in cancer therapy

Predominant action	**Examples**	**Commercially available***	**Clinical trials, Phases II,III,IV ****
Gene transfer			
Non-Viral	Electroporation, nanoparticles, hydrodynamics, cationic liposomes, transposon, synthetic viruses		18,1,0
Bacterial	Escherichia coli, Salmonella, Clostridium, Listeria, CEQ508		6,0,0
Viruses			
ssDNA viruses	Adeno-Associated: Parvovirus		
dsDNA viruses	Adenoviruses: Ad5-D24, CG870, Ad5-CD/TKrep, Recombinant H103, Gutless adenovirus, OBP-301	ONYX-015	11,3,0
dsDNA viruses	Herpetic viruses: Herpes simplex-1, TVEC		42,10,0
ssRNA viruses	Lentiviruses: HIV-1, HIV-2, Simian IV, Feline IV.		8,2,0
dsRNA viruses	Reoviruses		9,1,0
Immunomodulation			

Active immunotherapy			41,3,0
	Single Tumor cell surface antigen vaccine		
	Antigen-specific plasmid-based vaccine: PSA, HER/2, Modified CEA vaccine.		
	Tumor cells, irradiated as vaccine		
	Genetically modified tumor cell vaccine: Using Poxvirus, Vaccinia virus, Recombinant fowlpox virus, Combination (TRICOM) (Prostvac-VF vaccine).		
Passive immunotherapy			219,29,2
	Antibodies against:	Rituximab	
	CD20 Protein on lymphoma cells	Rituximab	
	HER/2 receptor protein in breast cancer	Trastuzumab	
	CD52 Protein on CLL	Alemtuzumab	
	CD20 Protein on lymphoma cells	Ibritumomab	
	CD20 Protein on lymphoma cells	Tositumomab	
	EGFR Receptor on squamous CA	Cetuximab	
	EGFR Receptor on colorectal CA	Panitumumab	
	CD20 Protein on CLL	Ofatumumab	
	CD30 Protein on Hodgkin lymphoma cells	Brentuximab	
	HER/2 receptor protein in breast cancer	Pertuzumab	
	HER/2 receptor protein in breast cancer	Ado-Trastuzumab	
	CD20 Protein on CLL	Obinutzumab	
Adoptive immunotherapy			15,1,0

	Autologous activated T- lymphocytes	Sipuleu-cel-T	
	Genetically modified activated T-lymphocytes		
	Chimeric antigen receptor integrated T-lymphocytes		
	Activated dendritic cells		
	Genetically modified dendritic cells		
Immune enhancement			11,1,0
	Antibodies blocking CTLA-4 Inhibitors for malignant melanoma.	Ipilim-umab	
Microenvironment modification			
Impact on vasculature	Humanized monoclonal antibodies against VEGFR-A	Bevaci-zumab	
	Anti-angiogenic genes (against VEGFR-A): Endostatin, Angiostatin		22,4,0

Abbreviations: *CA* Cancer; *CEA* carcinoembryonic antigen; *CLL* chronic lymphocytic leukemia; *ds* double stranded; *CTLA-4* cytostatic T-lymphocyte antigen 4; *DNA* deoxy nucleic acid; *EGFR* epidermal growth factor receptor; *FDA* Food and Drug Administration in United States; *HER/2* human epidermal growth factor receptor-2; *HIV* human immunodeficiency virus; *PSA* prostatic acid phosphatase antigen; *RNA* ribonucleic acid; *ss* single stranded; *VEGF-A* vascular endothelial growth factor A receptor.

1. *Commercially approved medications by FDA US as of July 1, 2014. ONYX-015 was previously approved by FDA China.
2. **Clinical trials: Number of active clinical trials on gene therapy for cancer (Phases-II, -III, and –IV) as of July 1, 2014 (http://www.clinicaltrials.gov).

Physical Mediated Gene Transfer

DNA genetic material that is coated with nanoparticles from gold or other minerals, and with their kinetic energy supplemented by compressed air or fluid (gene gun), or using ultrasound, can force the genetic material into the target cell, followed by the release of DNA into its nucleus. They are best suited for gene delivery into tissue or in case of gene vaccination [23].

The electroporation gene therapy approach aims to achieve cellular membrane disruption with high-voltage electrical pulses, resulting in the formation of nanopores through which naked DNA, foreign genetic materials, and even chemotherapeutic agents can enter cells [23, 24]. This approach is best suited for plasmid DNA-based gene transfer therapy with the advantage of effectiveness in a vast array of cell types, ease of its administration, lack of genome integration with the risk of malignancy, as well as the low potential for unwanted immunogenicity [22]. Electroporation is presently being tested in several clinical trials, especially on patients with malignant melanoma, prostate cancer, colorectal cancer, and leukemia [22].

Chemical Mediated Gene Transfer

Cationic liposomes are microscopic vesicles of synthetic phospholipids and cholesterol that can enter into cells by endocytosis [25], with the capability of carrying a variety of molecules such as drugs, nucleotides, proteins, plasmids and large genes [23]. Their advantage is selectivity to endothelial cells, a relatively high rate of gene transfer efficiency, a broad application as carriers for many genes, and the lack of severe side effects [26]. When combined with small interfering RNA (siRNA), cationic liposomes may lead to the inhibition of tumor proliferation, inducement of apoptosis, and enhancement of radiosensitivity to tumor cells [27].

Synthetic viruses have been developed to exploit the efficiency of viral vectors and the advantage of liposomes [28]. Once they enter the target cell, DNA is released from the endosome. This method has shown promising results in preclinical studies [29–32]. Transposons can also transport genetic material inside the cell as well as into the nucleus [33].

Bacterial Mediated Gene Transfer

Some bacteria have the capability of specifically targeting tumor cells, leading to RNA interference (RNAi) and gene silencing with blockage of RNA functions, including cellular metabolism and protein synthesis. Examples include Escherichia coli, Salmonella typhimurium, Clostridium, and Listeria [34]. Bacterial vectors can deliver pro-drug-converting enzymes and cytotoxic agents into tumor cells, and can mediate the host immune response. They can be engineered to carry magnetic or fluorescent material to enhance the utility of diagnostic approaches in tumor localization, such as with magnetic resonance imaging (MRI) [35], and even in the development of cancer vaccines [36]. However, the outcome has been far less pronounced compared to other RNA interference silencing techniques. Overall, genetically engineered bacteria acting as vectors for RNA interference are relatively safe, effective, practical and cheaper to manufacture compared to viral vectors. They selectively colonize and grow within the tumor. They can also be administered orally, hence their use in the management of gastrointestinal disorders [34].

Viral Mediated Gene Transfer

Viruses are small particles that contain either ribonucleic acid (RNA) or deoxyribonucleic acid (DNA), and may be single-stranded (ss) or double-stranded (ds). The viral structure consists of a genome surrounded by a protective protein coat (viral capsid) which helps the virus attach to host cell receptors, and prevents viral destruction by cell nuclease enzymes. Some viruses may also have a lipid bilayer envelope derived from the host cell's membrane, and an outer layer of viral envelope made of glycoprotein. A complete viral particle (virion) by itself is unable to replicate. For propagation, the virus needs to insert its genetic material into a host cell, in order to acquire metabolic and biosynthetic products for viral transcription and replication.

Viruses can be modified genetically to be noninfectious. As they enter the cell, they can carry genetic material for delivery into a target cell's cytoplasm, and subsequently into the nucleus. In monogenic gene therapy, virus vectors can carry a load of 2–10 kb of relevant genes. In high complex gene therapy, other supporting molecules can also be added, such as immune-stimulatory molecules to the virus's DNA for subsequent release during viral replication. The advantage of viral vectors in gene therapy is the ease of purification into high titers, and prolonged gene expression with minimal side effects. Retroviruses including lentiviruses can integrate themselves into host cell genome at the nucleus, while adenoviruses and adeno-associated viruses predominantly persist as extrachromosomal episomes [24, 37].

RNA viruses comprise about 70% of all viruses, and vary greatly in genomic structures. They usually have a higher mutation rate with increased adaptation to attack different host cells. Single-stranded RNA viruses may have a viral reverse transcriptase enzyme in their genome, which helps in genetic transcription of the viral genome inside the host nucleus, into double-stranded pro-viral DNA. With viral integrase, the pro-viral DNA then integrates with the host DNA making the subsequent transcription of other parts of the virus possible, in order to give rise to a new retrovirus progeny. The proteins of the mature virion are then rearranged to form the new viral particles. Viral particles subsequently destroy the host cell, and release mature viruses to attack neighboring cells. Double-stranded DNA viruses enter the host cells by endocytosis through interaction between the virus and cell receptor. The virus then enters the nucleus through nuclear pores once they escape cellular endosome [38, 39]. The virus subsequently releases two gene products which bind to the retinoblastoma and p53 tumor suppressor genes, thus allowing viral replication. New viruses then cause cell lysis and the released viruses spread to attack neighboring cells [38].

Adenoviruses

Adenoviruses are double-stranded DNA viruses that usually cause mild respiratory, digestive and ocular infection in humans. In gene therapy, modified versions of adenovirus and adeno-associated viral vectors have been designed. Compared to wild-type, they are more potent in infecting cells, both dividing and non-dividing, replicate exclusively in tumor cells [40], and selectively target certain cellular receptors or molecular defects. They pose a very high transduction efficiency, which may approach 100%, with fewer tendencies for viral shedding and latent infection. They can easily be produced commercially in large quantities, and are capable of carrying pro-drug genes as well as others [41]. However, they have several pitfalls, including the tendency to develop genetic instability of carried genes. Subsequent chromosomal aberration may lead to the development of lymphoproliferative disorders. As nearly half of all humans have been exposed to these viruses during their lifetime, with the generation of neutralizing antibodies, they may lead to high immunogenicity, with shorter duration of adenoviral-mediated transgene expression [42].

Modified oncolytic adenoviruses are presently tested in different clinical trials, especially in patients with astrocytoma of the brain, in combination with radiation and/or temozolomide chemotherapy [43]. ONYX-015 (Onyx Pharmaceuticals) is a modified oncolytic adenovirus that was previously approved by the Chinese Food and Drug Administration (2005) in the management of refractory head and neck cancer in combination with cisplatinum [16]. It is presently being investigated in the management of other solid malignancies. Other oncolytic adenoviruses include Ad5-D24, recombinant H103, Ad5-CD/TKrep, CG7870, KH901, and OBP-301 (Telomelysin). The latest generation of adenoviral vectors is the gutless adenovirus; it has an impressive safety profile, less in vivo immune response and long-term sustained gene expression [24]. Most clinical trials using oncolytic adenoviruses rarely produce dramatic tumor response. However, when combined with other modalities of cancer therapy, good tumor regression has been reported [38, 42].

Adeno-Associated Virus

This represents small, single-stranded DNA viruses, which do not usually cause infection without co-infection of a helper virus, such as adenovirus, or herpes simplex virus. They have the advantage of broad host range, low level of immune response, and longer gene expression. One example is the Eukaryotic adeno-associated virus, which is a chimeric virus vector containing parvovirus and adenovirus [44]. It is capable of transfecting mitotic and quiescent cells, lacks immunogenicity and pathogenicity in humans, and integrates stably into the host DNA at a predictable location within a chromosome-19 in cell culture, but not in mammalian cells.

Herpes Simplex Virus

This is a large, enveloped double-stranded DNA virus (150 kb), naturally neurotropic (prefer nerve cells), that infects humans particularly at the oral and genital mucosa, but ultimately spreads to sensory nerves to replicate or become dormant at the sensory ganglions. Viral reactivation may lead to oral or genital ulcerations, skin rashes, or even encephalitis. Up to 80% of the population are seropositive to the virus [45, 46]. With genetic engineering, a modified oncolytic recombinant replication-selective herpes simplex virus has been developed, and has exhibited several advantages: it has broad tropism, potent in causing tumor cell lysis, it is non-integrating in targeting the cell genome (apart from nonessential genes), can evade the host immune system; and in case of toxicities, several effective antiviral therapies are presently available to control viral replication. Another advantage is its viral capability to carry a large load of transgenes, such as a pro-drug-activating gene thymidine kinase enzyme that enhances tumor lysis when ganciclovir medication is subsequently administered intravenously (suicide gene) [45]; therapeutic immunomodulatory transgenes that augment the antitumor immune response (such as talimogene laherparepvec) (OncoVEX GM-CSF) [47]; and antiangiogenic genes to suppress tumor vasculature [48]. Presently, modified oncolytic herpes simplex viruses such as Talimogene laherparepvec (TVEC) as well as others, are being tested in several clinical trials either as a monotherapy, or in association with surgery, radiation therapy or chemotherapy, particularly on patients with high-grade glioma. Currently, some success has been reported [45].

Lentivirus Vector

Lentiviruses are retroviruses that infect bovine, equine, nonhuman primates and humans [49]. One of the most destructive human pathogens is human immunodeficiency virus infection (HIV). It constitutes a class of enveloped viruses that contain a single-stranded 9.2 kb RNA genome. The lentivirus carries a reverse transcriptase enzyme that transcribes RNA into double-stranded DNA once it enters the cytoplasm. It then integrates permanently into the nuclear genome of the target cells. Examples include lentiviral vectors derived from immunodeficiency viruses such as HIV-1, HIV-2. With genetic engineering, researchers have removed the infectious parts of the virus and added other parts from different viruses such as cytomegalovirus, generating a highly modified lentivirus [50]. Another genetic modification employed by previous researches created an integration-deficient lentiviruses that did not integrate into a host genome [51, 52], though with slightly lower transduction efficiency. Such a modified lentivirus has the advantage of being relatively safe, of having variable specificity to either a particular cell or is broad enough to infect all cells, and of having efficient transduction of both dividing and non-dividing cells. Modified viruses have low antivirus immunity, low potential for genotoxicity due to insertional mutagenesis, and the capability of carrying genes inside the nucleus [49]. Major disadvantages include inadequate immune responses as well as antitumor response, the risk of viral transformation into pathogenic HIV infection, especially in immunized individuals, and insertional mutagenesis of new cancer genes into the host genome, with the risk of second malignancy [49].

Reovirus

This is an oncolytic virus that usually infects animals. In humans, it rarely causes major illness except for respiratory and gastrointestinal symptoms. Nearly 100% of human adults are seropositive to the virus [53]. It is non-enveloped, double-stranded RNA (dsRNA), and its oncolytic activities are mainly through stimulation of the immune system, particularly through bystander immune activation. The release of tumor-associated antigens following cellular lysis further stimulates innate immunity against tumor cells. The virus is considered relatively benign, with good safety records, does not require genetic alterations to become an oncolytic virus, and is less expensive to be produced commercially. Because of its relative safety, the virus is presently being used in several clinical trials, as oncolytic reovirus monotherapy, administered intratumoral, intravenously, or intraperitoneally; or as polytherapy, in combination with radiation therapy or chemotherapy [53].

GENE THERAPY IMPLEMENTATION

Once genetic materials are transferred into target cells and incorporated into nuclear genetic DNA, they may induce silencing, down-regulation, modification, or repair of the target cell genes. Depending on the intensity of the gene expression, it may lead to cell death and tumor necrosis (as with the suicide gene), or impaired cell growth with tumor regression (as with the silencing gene). Modification of the gene may improve the response from subsequent cancer therapy, such as chemotherapy, immunotherapy, or radiation. Repair of the target gene may help in preventing subsequent malignancy or cancer-related complications such as thrombosis. They may also be helpful in the future by preventing hereditary cancer syndromes.

Suicide Gene

These are transgenes that make up products that can cause a cell to kill itself through apoptosis. Such gene products are usually transcribed by various factors (promoters) leading to cell death and necrosis. One example of such promoters is the human *H19* RNA gene which is highly expressed in most fetal organs but rapidly cleared immediately after birth [54]. This gene has shown an abnormal expression in various types of cancer cells, and plays an important role in cancer cell proliferation, genetic instability, vascular angiogenesis, tumor metastases, multidrug resistance as well as cell survival despite hypoxia, with secondary tumor progression and dissemination [55]. Blocking *H19* gene function leads to marked tumor regression, cellular death and necrosis. Another important promoter is human telomerase reverse transcriptase (hTERT), which is a critical factor for cell immortalization and tumorigenesis [56]. Its blockage with agents such as OBP-301 (Telomelysin) (Oncolys BioPharma) leads to cell necrosis and tumor regression (Figure 1). Other approaches to induce tumor cell death include the use of small-molecule drugs, monoclonal antibodies, and toxin gene therapy with agents such as Corynebacterium diphtheriae toxin-A chain (DTA-H19 therapy) [57].

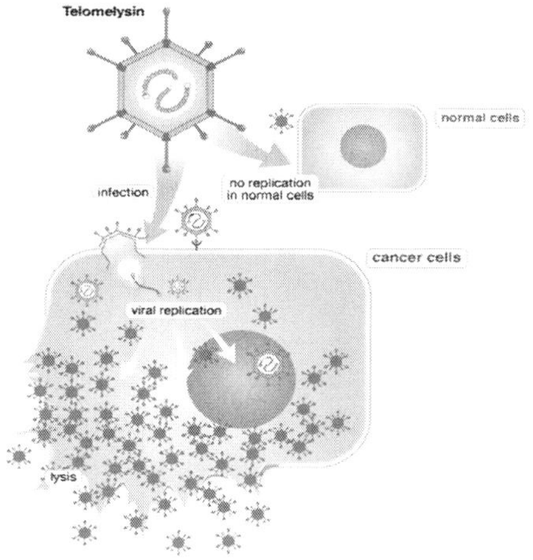

Figure 1. Genetically-modified adenovirus acting as a suicide gene. The above mode of action represents an example of a modified virus acting as a suicide gene, namely OBP-301 (Telomelysin) (Courtesy Oncolys BioPharma Company, Tokyo, Japan).

Gene Silencing

This has been achieved through specific delivery of a small interfering double-stranded RNA (siRNA) into target cells, and subsequent duplex formation of RNA-induced silencing complex (RISC) that destroys messenger-RNA (mRNA), thus leading to interference with RNA functions and protein synthesis within the target cells [58]. Through the appropriate design of siRNA, it is theoretically possible to use the technology in silencing any gene in the body, providing a greater therapeutic potential in cancer therapy [59], as well as in the management of other medical disorders such as the hepatitis B virus, human papilloma virus, hypercholesterolemia and liver cirrhosis [59, 60]. As siRNA does not interact with chromosomal DNA, it does have a lower risk of inducing target cell gene alterations and possible mutagenesis. It is highly specific against target genes, with low systemic toxicities, and does not induce multidrug resistance. Furthermore, these genes can induce potent gene silencing of many cancer-related genes, leading to tumor regression, but do not abolish abnormal genes. siRNA therapy can be administered directly into tumors; however, for systemic administration, it is somewhat difficult as a naked siRNA protein is liable for host-mediated clearance by enzymatic degradation, renal filtration, and host cellular phagocytosis. Several delivery systems for siRNA have been developed to protect them from enzymatic degradation, and facilitate their effect in silencing specific genes. Examples of siRNA systemic delivery system presently in clinical trials include CALAA-01 (Calando Pharmaceuticals) for patients with malignant melanoma [61], and ALN-VSPOI (Alnylam Pharmaceuticals) for liver cancer and solid tumors [62]. However, limited success has been achieved mainly due to relatively high toxicity and low transfection efficiency [58, 59].

Gene Modification

This may be helpful in improving cancer therapy results, such as with radiation therapy. Radiosensitizing gene therapy promotes transgene expression in tumor tissue, thus increasing tumor sensitivity to radiation with better tumor control [63–65]. In contrast, radioprotective gene therapy distributes transgenes and their products to surrounding normal tissue, thus limiting radiation induced toxicities to normal tissue [66]. The concept of combining both approaches is presently being investigated in several preclinical studies.

Gene Repair

This can be achieved using zinc finger nuclease attached to the lentiviral vector. Once the viral vector enters the nucleus, it binds to a specific location in the double-stranded DNA, breaking it at specific location, with subsequent endogenous repair mechanisms, to create a newly edited double-stranded DNA [23]. Other technological approaches include transcription activator-like effector nucleases (TALENs) [67, 68], and clustered regularly interspaced short palindromic repeats (CRISPR) [69].

Gene Therapy for Mitochondria

Gene therapy may also be directed to cytoplasmic organelles such as mitochondria. The mammalian mitochondria are responsible for metabolic functions. Nearly 300 of the known mutations causing metabolic diseases are secondary to disorders affecting the mitochondrial genome [23]. Several approaches have been used to transfer genes successfully into cell mitochondria.

IMMUNOMODULATION

It has become apparent that the immune system is a crucial element in cancer regression or progression. There are two types of immune responses: humoral immunity and cellular immunity. Furthermore, the tumor microenvironment plays an important role in host immune effects against cancer cells [Table 1].

Humoral immunity is mediated by antibodies released by B-cells with a high-binding affinity to specific tumor antigens. The Food and Drug Administration in the United States (FDA) approved several antibodies against malignant cells, which include trastuzumab for breast cancer [70], rituximab for indolent lymphoma [71], cetuximab for lung cancer [72], and bevacizumab for various solid tumors [49, 73], and many others [74] [Table 2].

Table 2. Monoclonal antibodies in cancer management

Name	**Class**	**Target**	**Approved initial indications**	**FDA Approved**
Rituximab (Rituxan)	Chimeric IgG1	CD20	Non-Hodgkin lymphoma	1997
Trastuzumab (Herceptin)	Humanized IgG1	HER2	Breast cancer	1998
Alemtuzumab (Campath)	Humanized IgG1	CD52	B-cell CLL	2001
Ibritumomab tiuxetan (Zevalin)	Murine IgG1	CD20	Non-Hodgkin lymphoma	2002
Tositumomab (Bexxar)	Murine IgG2a	CD20	Non-Hodgkin lymphoma	2003
Cetuximab (Erbitux)	Chimeric IgG1	EGFR	Squamous cancer head & neck	2004
Bevacizumab (Avastin)	Humanized IgG1	VEGF	Colorectal cancer	2004
Panitumumab (Vectibix)	Humanized IgG2	EGFR	Colorectal cancer	2006
Ofatumumab (Arzerra)	Humanized IgG1	CD20	CLL	2009
Denosumab (Xgeva)	Humanized IgG2	RANKL	Bone metastases	2010
Ipilimumab (Yervoy)	Humanized IgG1	CTLA-4	Metastatic melanoma	2011

Brentuximab vedotin (Adcetris)	Chimeric IgG1	CD30	Hodgkin lymphoma	2011
Pertuzumab (Perjeta)	Humanized IgG1	HER2	Breast cancer	2012
Trastuzumab emtansine (Kadcyla)	Humanized IgG1	HER2	Breast cancer	2013
Obinutzumab (Gazyva)	Humanized IgG1	CD20	B-cell CLL	2013
Name	**Class**	**Target**	**Present indications**	**Clinical Trials****
Amatuximab	Chimeric IgG1	mesothelin	mesothelioma	Phase-I
Elotuzumab	Humanized IgG1	CS1	Multiple myeloma	Phase-III
Farletuzumab	Humanized IgG1	FRA	Ovarian and lung cancers	Phase-III
Inotuzumab ozogamicin	Humanized IgG4	CD22	ALL, Malignant lymphoma	D/C
Moxetumomab pasudotox	Murine Fv-CD22	CD22	Hairy cell Leukemia	Phase-III
Naptumomab estafenatox	Murine Fab	5 T4	Renal and solid malignancies	Phase-II
Necitumumab	Humanized IgG1	EGFR	NSCL (Squamous cell)	Phase-III
Nivolumab	Humanized IgG4	PDI	NSCL, Melanoma, Renal	Phase-III
Ontuximab	Humanized IgG1	TEM1	Solid tumors	Phase-I/II
Onartuzumab	Humanized IgG1	c-Met	NSCL, Gastric	D/C
Racotumomab vaccine (Vaxira)	Murine	GM3	NSCL	Phase-III
Rilotumumab	Humanized IgG2	HGF/SF	Gastric, GEJ	Phase-III

Abbreviations: *5 T4* Antigen expressed on several solid tumors; *ALL* acute lymphocytic leukemia; *c-MET* MNNG HOS proto-oncogene that encodes hepatocyte growth factor receptor; *CLL* chronic lymphocytic leukemia; *CTLA-4* cytotoxic T lymphocyte inhibitors mediated by malignant cells; *CS1* human CS1 antigen glycoprotein belonging to CD2 subset of the immunoglobulin superfamily; *D/C* clinical trials that were discontinued due to the lack of efficacy or excessive toxicities; EGFR, epidermal growth factor receptor; *FDA* Food and Drug Administration in United States; *FRA* folate receptor alpha; *GM3* tumor antigen N-glycolil, a type of ganglioside present on the surface of breast and lung cancer cells; *HGF* human hepatocyte growth factor receptors; *Mesothelin* mesothelin is cell surface glycoprotein overexpressed in multiple malignancies such as mesothelioma; *NSCL* non-small cell lung cancer; *PD-1* human cell surface receptor programed death-1, results in activation of T cell mediated immune responses; *RANKL* RANK ligand protein that acts as the primary signal for bone removal, loss, or destruction; *TEM1* tumor endothelial Marker-1 and CD248 (Morphotek Inc, Exton, PA); *VEGF* vascular endothelial growth factor receptor.

1. **Active clinical trials on monoclonal antibodies in cancer management as of July 1st, 2014 (http://www.clinicaltrials.gov).

Cellular immunity is mediated by cell-to-cell contact that leads to antigen recognition and cell destruction of a target cell. Based on the intensity of tumor-associated antigens (TAAs) on the surface of tumor cells, they are recognized by the host immune system [75]. Dendritic cells are specialized in antigen recognition as well as mediation of immune responses against infectious agents or malignant cells, through direct stimulation or inhibition of immune effector cells such as T-cells, B-cells, and natural killer (NK) cells [76]. Dendritic cells are derived from the bone marrow and migrate to lymph nodes and distant tissue, looking for those foreign antigens [49].

Cancer cells can evade the immune system by secreting immunosuppressive cytokines that can downregulate major histocompatibility molecules, can recruit regulatory T-cells, and can kill reactive cytotoxic lymphocytes. Thus, the tumor microenvironment is highly immunosuppressive, which allows a tumor to grow and metastasize [77]. Several efforts have been undertaken to manipulate the tumor microenvironment in order to induce tumor regression.

INNATE IMMUNITY

Most tumor cells express antigens that can mediate antitumor immune responses [78]. Earlier studies on antigens for therapeutic targeting were based on shared antigens that are expressed on self-tissue or peripheral cells, which can lead to immunologic tolerance for the interaction between antigen peptide, major histocompatibility complex (MHC), and T-cell antigen receptor (TCR). Generated immunologic responses were restricted with low therapeutic efficacy [78]. Recently, it has been found that neoantigens generated by point mutation in normal genes, which are unique to particular tumors, can result in much more potent antitumor T-cell response. Some cancers display hundreds or even thousands of mutations in coding exons, representing a large resource of potential targets for recognition by the immune system. However, despite such a plethora of antigens, most cancers progress and evade immune-system mediated destruction [78].

Antigens recognition by dendritic cells induce a T-cell inflamed reaction consisting of infiltrating T-cells, a broad chemokine profile, and type I interferon signature indicative of innate immune activation. Presence of excessive infiltration by CD8+ cells both within the tumor and in the peritumoral stroma (high immunoscore) had a favorable prognostic significance with improved survival, even in advanced cancer stages, compared to tumors with poor or no T-cell infiltration (low immunoscore) at an earlier stage of malignancy [79–81]. Hence, a heavy presence of activated CD8+ T-cells reflects good innate immune responses against such malignancy. Recent investigations have suggested two explanations for tumor escape recognition by host immune system, based primarily on cellular and molecular characteristics of the tumor microenvironment [78].

One explanation is that tumors resist immune attacks through inhibitory effects mediated by immune system suppressive pathways. This was evident in some tumors such as melanoma with high expression of PD-LI and indoleamine-2,3-dioxygenase (IDO) [82], leading to T-cell anergy and dysfunction with subsequent immune escape detection [83]. The presence of transcription factor forkhead box 03 proteins (Fox3) in the peritumoral microenvironment leads to the inhibition of tumor-infiltrating dendritic cell stimulatory functions [84]. The US FDA's approval in 2011 of the anti-CTLA-4 monoclonal antibody ipilimumab for the treatment of patients with advanced malignant melanoma [85] represents the first-in-class strategy of uncoupling inhibitory pathways for initial antigen recognition by the host immune system [78] [Figures 2 and 3].

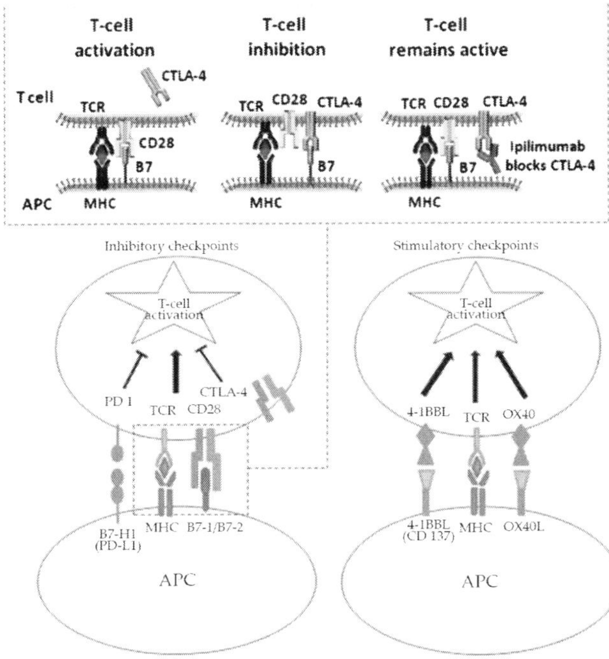

Figure 2. Mechanism of action of monoclonal antibody ipilimumab. Generation of an immune signal requires presentation of tumor antigen by major histocompatibility complex (MHC) class I or II molecules, on an antigen presenting cell (APC) such as dendritic cell. However, T-cell activation and proliferation requires a second signal, typically generated by CD28 antigen. When CD28 antigen on T-cell surface simultaneously binds to costimulatory B7-1/B7- ligand on the antigen presenting cell (APC), T-cell upregulate and translocate CTLA-4 receptor molecules to the surface, which binds B7 with a higher avidity than CD28, leading to suppressor effects, with T-cell inhibition, reduction of interleukin-2 (IL-2) secretion, and prevention of immune response against malignant cell. Ipilimumab blocks cytotoxic T-lymphocyte antigen-4 (CTLA-4) receptor, thus prevents such inhibitory effect, and allows T-cell to proliferate and mediate an immune reaction against malignant cells. Other regulatory checkpoints with the potential for modulation include the coinhibitory molecule PD-1, as well as costimulatory molecules such as OX40 and 4-1BB. Abbreviations: APC, antigen presenting cells; CTLA-4, cytotoxic T-lymphocyte antigen-4; MHC, major histocompatibility antigen; TCR, T-cell receptor. (Courtesy of Annals of New York Academy of Science and Wiley, Hoboken, New Jersey, Publisher) [86].

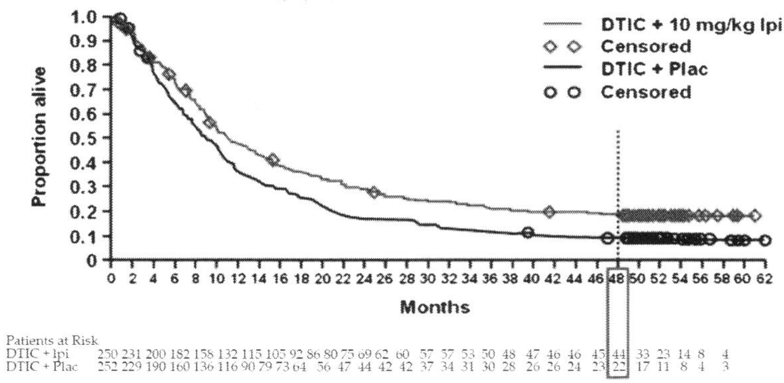

	Ipilimumab + DTIC N=250	Placebo + DTIC N=252
1-year OS rate, %	47.5	36.4
2-year OS rate, %	28.8	17.8
3-year OS rate, %	21.2	12.1
4-year OS rate, %	19.0	9.6
Median OS, months	11.2	9.1

Figure 3. Analysis of overall survival comparing monoclonal antibody ipilimumab plus dacarbazine to placebo plus dacarbazine in metastatic melanoma patients. Kaplan–Meier analysis of overall survival in the phase III study CA184-024. Survival analysis of overall survival in treatment-naive patients with advanced melanoma who received ipilimumab at 10 mg/kg plus DTIC or placebo plus DTIC in the phase III trial, CA184-024. The survival curves reach a plateau beginning at approximately three years after initiation of treatment. Continued survival follow-up of more than four years demonstrates a long-term survival benefit that is consistent with the results of other ipilimumab studies. Abbreviations: DTIC, dacarbazine; Ipi, ipilimumab, Plac, placebo (Courtesy of Annals of New York Academy of Science and Wiley, Hoboken, New Jersey, Publisher) [86].

The other mechanism is immune system exclusion or ignorance with subsequent poor or no T-cell inflammatory reaction. Such tumors appear to lack a type I interferon signature and/or chemokines for recruitment of T-cells. Microenvironment vasculature may be nonpermissive for entry by T-cells, and the stromal element may prevent trafficking and/or function of T-cells. Radiations of tumors have shown to induce productions of interferon-beta and augment function of intratumoral dendritic cells with improved accumulation of T-cells leading to tumor regression [87]. Imatinib in gastrointestinal stromal cell tumors may cause down-modulation of IDO with improved antitumor response [88]. In patients with malignant melanoma, inhibition of R-Raf enzyme activity with vemurafenib can induce a T-cell infiltration within 1–2 weeks of therapy with some tumor responses [89]. It has been suggested that combination regimens consisting of strategies to improve innate immune system activation, T-cell trafficking in the tumor microenvironment, vaccination or adoptive T-cell transfer, and blockage of immune inhibitory pathways may be necessary to achieve clinical benefit in patients with a non-inflamed tumor phenotype. Such an approach is presently being tested in clinical trials [90, 91].

IMMUNOMODULATORY APPROACHES IN CANCER THERAPY

Immunotherapy in cancer can be classified into four major categories [92]. Active immunotherapy includes strategies that directly sensitize the host immune system to tumor-specific antigens, exemplified as cancer vaccines. Passive immunotherapy utilizes humanized or chimeric antibodies to specifically target tumor antigens without direct activation of the immune system. Adoptive immunotherapy utilizes patients' immune cells, whether T-cells or dendritic cells, stimulated or manipulated ex vivo, then infused back, to better react against tumor antigens. Immune enhancement therapy aims to augment co-stimulatory molecules or block inhibitory molecules. Immune-based therapy may include one or more of the above approaches, either as distinct immunotherapy treatment, or in combination with other modalities of cancer therapy [Table 1].

Autologous Stimulated T-lymphocytes

Adoptive T-cell therapy has been shown to induce tumor regression in some patients with solid malignancies. In a recent study on patients with human papilloma virus (HPV)-induced metastatic cervical cancer who failed to respond to chemotherapy and radiation, and were selected for HPV-E6 and HPV-E7 reactivity, researchers collected tumor-infiltrating T-cell lymphocytes (TIL), and infused them back to patients. This was preceded by a non-myeloablative conditioning regimen and followed by a high-dose of bolus aldesleukin (interleukin-2). Three out of six patients with HPV reactivity achieved objective tumor responses, including two patients with metastatic disease that achieved complete tumor regression for 18 and 11 months after therapy. Side effects were minimal [93].

Autologous Activated T-Lymphocytes

Host T-cell lymphocytes have been found to be successful in controlling metastatic cancer with transient side effects. The first commercially available vaccine was modified dendritic cells, sipuleucel-T (Provenge) (Dendron Corporation), which was approved by the FDA for minimally symptomatic castration-resistant metastatic prostate cancer. CD54 T-cell lymphocytes were obtained from the patients using density gradient centrifugation, and then activated ex vivo with a prostatic specific antigen in addition to granulocyte macrophage colony stimulating factor (GM-CSF) to form sipuleucel-T. Autologous activated T-cell lymphocytes, at a dose of at least fifty million CD54 cells were infused back to the patient, intravenously over 60 minutes, every two weeks for three infusions. Premedications included acetaminophen and diphenhydramine. Side effects included transient fever, chills, fatigue, asthenia, backaches, and headaches. However, infusion-induced hypersensitivity reactions with cerebrovascular events have been reported in 3.5% of patients. Compared to a control group treated with a placebo, there were significant improvements in the survival of 20.5% versus 16.1% at four years [94].

Genetically Modified Activated T-Lymphocytes

The adoptive transfer of lymphokine-activated lymphocytes can mediate the cellular immune response against cancer cells, which may lead to tumor regression. However, clinical trials have led to limited success. An alternative approach is to use genetically modified T-cells by altering their receptor for better recognition of tumor antigens. In such an approach, T-cells are collected from patient apheresis using density gradient centrifugation. As resting T-cell lymphocytes are non-dividing, refractory to gene therapy with lentiviral vectors, they need to be stimulated using cytokines such as interleukin-2. T-cells are then exposed to lentiviral vectors with the attached gene for 1–2 days of gene transfer. After transduction by the lentivirus, cells are then stimulated further to obtain a therapeutically effective number of cells. Genetically modified T-cell lymphocytes are then re-infused back into the patient [95]. The high-affinity of modified T-cells in detecting very low levels of tumor antigens is an extremely potent approach against tumor cells. However, it may also destroy normal cells. In another study using T-cell receptor gene-modified cells against melanoma differentiated antigens led to higher responses in patients with malignant melanoma [96]. It also destroyed normal melanocytes leading to vitiligo (skin depigmentation), uveitis, and hearing impairment [97].

Chimeric Antigen Receptor Integrated into T-Lymphocytes

Elimination of malignant cells through host immune system depends largely on α β T-cell receptor that specifically recognize a cell target in the context of the major histocompatibility complex (MHC). Unfortunately, for many B-lineage leukemias and lymphomas, the resident immune system of patients remains incapable of controlling tumor growth, since autologous T-cells lack expression of the required receptors and tumor cells have adapted to evade immunological recognition [98]. It has been demonstrated that a chimeric antigen receptor (CAR) integrated into T-cells from patient (or even from healthy individuals), can directly recognize the CD19 molecule expressed on the cell surface of B-cell malignancies independent of major histocompatibility complex [99]. Recently, CD19-specific chimeric antigen receptor redirected T-lymphocytes have been utilized as gene therapy for patients with B-cell malignancies. One approach is to use a microelectroporator to achieve high throughput non-viral gene transfer of naked DNA plasmid, of in vitro transcribed CAR mRNA into human T cells that had been numerically expanded ex vivo using interleukin-2. After electroporation, a procedure that usually takes about 10 minutes, up to 80% of the passaged T-cells expressed the CD19-specific CAR, with redirected effector function of the genetically manipulated T-cells to specifically lyse CD19+ tumor cells. Preserved T-cells can then be re-infused into patient as an effector form of adoptive immunotherapy [98] [Figure 4]. Similar approaches have been used against other B-lineage restricted antigens such as CD20 in lymphoma, the light chain of human immunoglobulins, or CD30 expressed by Reed-Sternberg Cells in Hodgkin lymphoma [100]. Adding costimulatory endodomain within the chimeric receptors such as CD28, 4-1BB, or their combination, usually leads to enhancement of T-cell functions through the release of interleukin-2, interleukin-7 or interleukin-15 cytokines [100]. Excellent results in patients with B-cell malignancies have been reported [101–103]. CAR-modified allogeneic T-cells, such as those obtained from healthy individuals, have the potential to act as universal effector cells, which can be administered to any patient regardless of MHC type. Such universal effector cells could be used as an 'off-the-shelf' cell-mediated treatment for cancer [104, 105].

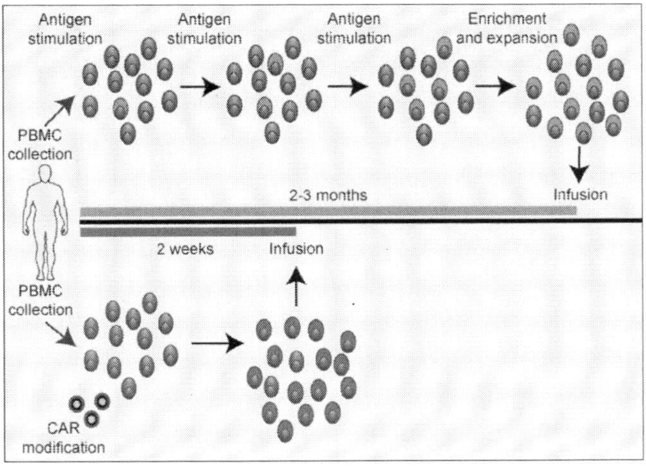

Figure 4. Chimeric antigen receptor modified T-lymphocyte therapy for B-cell malignancies. Generation of tumor-specific T cells by repeated antigen stimulation or genetic modification to express a tumor-targeting receptor. PBMC collected from a patient or healthy individual can be stimulated in vitro with tumor antigen at regular intervals to induce gradual enrichment of antigen-specific T cells (*blue*). Multiple stimulations followed by additional enrichment or expansion strategies are required to ensure sufficient antigen-specific T cells are generated. The entire process may take 2–3 months. In contrast, approaches that utilize genetic modification to redirect T cell specificity to a tumor antigen are much more rapid. PBMC can be collected from a patient or healthy donor and retrovirally or lentivirally transduced to express a tumor-reactive CAR (or TCR). The enriched CAR-modified tumor-reactive T cells (*red*) can be infused into the patient in as little as 1–2 weeks. Abbreviations: PBMC: Peripheral blood mononuclear cells, CAR: Chimeric antigen receptor modified T-lymphocytes. (Courtesy of The International Journal of Hematology, and Springer-Tokyo, Publisher) [102].

Genetically Modified Dendritic Cells

Dendritic cells are the most powerful antigen-presenting cells (APCs) for antigen identification, T-cell costimulation and cytokine production by T lymphocytes [106]. Dendritic cells can be generated from monocytes or a CD34+ cell precursor [107], and can be harvested from peripheral blood using density gradient centrifugation. When cultured in the presence of granulocyte-macrophage colony-stimulating factor (GM-CSF) plus interleukin-4, monocytes develop into immature dendritic cells in 3–5 days. Cells are then exposed to a variety of different stimuli over another 2 days in culture media to become mature dendritic cells, which are more effective in stimulating T-cells [108]. To target a specific tumor, mature dendritic cells are incubated with certain peptides, proteins, or irradiated tumor cells. An alternative approach is genetic modification of dendritic cells through viral and non-viral gene transfer vectors, resulting in dendritic cells that are more potent in antitumor immunity. Dendritic cell vaccines have a favorable safety profile, with toxicities limited to a local inflammatory reaction, flu-like symptoms, and vitiligo-like skin changes [109]. Despite numerous clinical trials, clinical outcomes have been modest. Emphasis has been shifted in using therapy on patients with a lower tumor burden, such as those after surgery, or following successful chemotherapy or radiation therapy. Indeed, significant synergy has been observed for chemotherapy and dendritic cell vaccines [107].

Genetically Modified Tumor Cell Vaccine

Previous attempts to use tumor cells or their products as a vaccine have not been successful. Subsequently, clinical trials have been conducted using methods to increase tumor antigenicity in order to enhance the immune-mediated tumor lysis by T-cells. One approach is to obtain tumor cells, infect them with a viral vector such as recombinant poxvirus that contains multiple costimulatory molecules to enhance the immunogenicity of tumor cells, and subsequently use those modified tumor cells as a vaccine. An alternative approach is to directly administer the poxviral vector into the tumor. Such an approach enhances tumor antigenicity and subsequent antigen-specific T-cell response, leading to an antitumor effect. Other viral vectors include vaccinia virus and recombinant nonreplicating fowlpox virus encoding GM-CSF (TRICOM) [110]. This has led to the development of the Prostvac-VF vaccine (Bavarian Nordic, Inc.) which employs a recombinant vaccinia vector as a primary vaccination, followed by multiple booster vaccinations employing a recombinant fowlpox vector. Both vectors contain the transgenes for prostate-specific antigen (PSA) and multiple T-cell costimulatory molecules (TRICOM). In a randomized clinical study, using the vaccine versus placebo, in 125 patients with metastatic castration resistant prostate cancer, there was an improved survival at 3 years of 30% for the vaccinated group compared to 17% for the placebo patients [18]. A Phase-III clinical trial is presently in progress.

Single-Antigen Plasmid-Based Vaccine

DNA plasmids containing a genetic sequence that encodes a desired antigen with other transcriptional elements have been considered as a mode of cancer vaccine. It readily accesses the nucleus of a transfected cell, transcribed into a peptide or protein, and may lead to cellular and humoral immune response [111]. The technique is considered to be relatively safe compared to viral or bacterial vectors, does not cause infection or autoimmune disorders, and is easy to develop and produce commercially [112]. However, its effectiveness wanes with time. Hence, the need for frequent booster immunizations. Examples of single-antigen plasmid-based vaccines include human prostatic acid phosphatase protein for patients with prostate cancer [113], human epidermal growth factor receptor-2 (HER-2/neu), protooncogene with low-doses of GM-CSF intradermally for patients with metastatic breast cancer [114], and modified carcinoembryonic antigen (CEA) gene fused to a promiscuous tetanus toxoid for colorectal cancer [115]. Although therapy was well tolerated, responses were minimal and transient. Using a multiple-antigens plasmid-based vaccine leads to broadly specific, long lasting, and multifunctional immune stimulation [116]. Improved results were noticed [117, 118].

Genetically Modified Microenvironment

The microenvironment around a tumor plays an important role in tumor progression and metastases. It includes stromal tissue, fibroblasts, and vascular endothelial cells. Interfering with such a microenvironment will lead to tumor regression. The most important target is angiogenesis, which is essential for tumor growth and metastases. It is mediated by tumor-derived pro-angiogenic cytokines, such as the vascular endothelial growth factor and fibroblast growth factor. These factors stimulate the proliferation of microvasculature around a tumor, with subsequent tumor progression and metastases. Compared to the recombinant antivascular endothelial factor antibody "bevacizumab", gene therapy represents an attractive alternative to such drug therapy. Using an anti-angiogenic genes, such as *angiostatin* and *endostatin*, delivered by an adeno-associated virus vector, has led to tumor regression with minimal side effects [24].

OTHER GENE THERAPY APPROACHES IN CANCER MANAGEMENT

As with other modes of cancer therapies, multimodality treatment frequently yields, better results compared to monotherapy. This is similarly true for gene therapy, and is evident when gene therapy is administered after maximum tumor load reduction following radical surgery or successful chemotherapy. Gene therapy has a synergistic effect when combined with chemotherapy, with higher tumor responses and lower therapy-related toxicities.

Gene Directed Enzyme Prodrug Therapy (GDEPT)

This is a new strategy in cancer management that aims to reduce the side effects of chemotherapy. With such an approach, a gene that expresses a nontoxic enzyme into cancer cells is first delivered to the cells, followed by the systemic administration of a pro-drug that can be converted into a toxic compound by the enzyme, leading to selective tumor cell death, with lower adverse effects on normal tissues [119]. Cell-to-cell diffusion of toxic metabolites may damage nearby and adjacent tumor cells (bystander effect) [120]. Release of tumor cell necrotic material in the circulation may activate the immune system in response to the tumor antigen, with subsequent regression of distant tumor cells, such as metastatic nodules (distant bystander effect) [121]. Examples include the use of a retroviral vector, such as suicide gene therapy and herpes simplex virus carrying the thymidine kinase enzyme, to the interior of tumor cells. The enzyme has a 1000-fold greater efficiency to selectively phosphorylate the acyclovir-derived pro-drug ganciclovir [120]. Following the systemic administration of ganciclovir, the drug is metabolized in tumor cells leading to cell death. As the efficacy of such a system is only about 10% of tumor cells, the extent of tumor regression is mainly mediated via bystander effects. The system has been tried in several clinical trials [122]. Replacing ganciclovir with a penciclovir drug, modified to generate radiolabeled analog, will also allow a closer follow-up of therapy results, using high-quality positron emission tomography imaging studies [123].

Cancer Drug-Resistance Gene Transfer

Several studies have used a gene transfer approach that aims to enhance chemotherapy and radiation effects against cancer cells, while protecting normal tissue against therapy mediated toxicities. Such gene transfer may also be used in the protection against HIV virus by making normal cells resistant to viral invasion, or correction of genetic disorders such as sickle cell anemia or metabolic disorders. However, incorporating a new gene into a host stem cell's genome, for the life of an individual, may promote other oncogenes to develop malignant disorders, and may change other adjacent genes, thus creating other medical diseases. Hence, it is a risky approach in gene therapy. Few clinical trials have recently been conducted in this regards. One example is the multidrug-resistant protein-1, which is encoded by the human *ABCBI* gene named as *MDR1* gene. It stimulates the cellular pump to remove cytotoxic drugs from normal cell cytoplasm to the outside, thus protecting normal cells from chemotherapy's side effects, such as with vinca alkaloids, taxanes, epipodophyllotoxins and anthracyclines [124]. The *MDR1* gene is minimally expressed in malignant cells; thus, chemotherapeutic medications entering the cytoplasm will remain at a higher concentration, leading to cell death. Other drug-resistant genes include methyl guanine methyltransferase (*MGMT*) for alkylating chemotherapy [125, 126], and glutathione transferase (*GSTP1*) for cisplatin, doxorubicin, and cyclophosphamide [127, 128, 124].

Theranostic Approach

In a combined diagnostic and therapeutic system (theranostic), gene therapy may also be combined with other diagnostic measures to help diagnose, treat and monitor the response to therapy. For example, a small interfering double-stranded RNA (siRNA) delivery system can be labelled with imaging agents such as dextran-coated superparamagnetic nanoparticles for simultaneous noninvasive imaging of siRNA delivery to tumors, using magnetic resonance imaging (MRI) [59]. The siRNA delivery system can also be labeled with other imaging agents to closely monitor therapy, and may even predict the outcome of therapy long before any anatomical changes [129]. Such molecular diagnostic approaches have been evolving relatively fast in the last few years, and may become an important avenue in cancer diagnosis sometime in the near future [59].

PROBLEMS WITH GENE THERAPY

The most frequent side effects following gene therapy include transient fever and flu-like symptoms [24]. A grade-3 hypersensitivity reaction following intravenous administration is usually transient and managed with the usual supportive measures. Leukocytopenia, and in particular, lymphopenia, may represent cellular redistribution of white blood cells to target tissue such as tumors. Mild transient anemia has also been reported [130]. However, toxicity, mutagenicity and immunogenicity associated with viral vector therapy have raised great concern [12].

Retroviral (such as lentiviruses) mediated gene therapy leads to viral integration into host genome, thus, it may cause mutagenic events with possible second malignancies. This was reported in earlier studies on the murine leukemia retrovirus vector in the treatment of patients with severe combined immunodeficiency and five out of 30 cases developed leukemia [131], though, no second malignancy has been reported so far, in gene therapy for cancer. Such mutagenicity depends on the site of viral insertion. For this reason, the FDA has required all clinical trials involving genomic integrated viral vectors to report and analyzes viral vector insertion sites. Initial methodology was linear amplification mediated polymerase chain reaction [132], but lately, high-throughput DNA sequencing methods have been used [133, 134]. Clinical trials that initially or subsequently show evidence of higher mutagenicity are usually discontinued. Information obtained from such studies is of major significance in designing new and much safer therapeutic approaches [58].

Another major problem with gene therapy for cancer is the resistance to treatment with subsequent tumor recurrences and shorter survival. A potential mechanism is intrinsic, and possibly acquired, tumor cell resistance to therapy-induced cell death (apoptosis) by dysregulation and release of anti-apoptotic inhibitor of apoptosis protein or Bcl-2 proteins [24]. Recently, some pharmaceutical companies have developed several medications such as Novartis-LBH589, cIAP1, and cIAP2 which inhibit the Bcl-2 protein, thus promoting cell death (apoptosis) and tumor regression, prevent or delay tumor resistance, and prolong remission following gene therapy. These medications are presently in clinical trials [24, 135].

REVIEW, CONCLUSIONS

Gene therapy for cancer has evolved relatively fast in the last two decades, and presently, few drugs are commercially available while others are still in clinical trials. Most reports on gene therapy have shown good safety profiles with transient tolerable toxicities. The lack of success in several clinical trials may partly be attributed to patient selection. Similar to initial chemotherapy outcomes thirty years ago, patients with advanced and therapy-resistant malignancies are presently enrolled in gene therapy trials. Perhaps, gene therapy maybe much more successful in patients with earlier stages of malignancies, or in those who have a lower tumor burden. Alternatively, gene therapy may better be used after successful cancer therapy with maximum tumor load reduction, such as after radical surgery, following radiation therapy, or after successful chemotherapy.

In the future, the wide use of patient and tumor genomic analysis as well as the assessment of host humoral and cellular immunity, will facilitate a better selection of the most appropriate gene therapy per patient. Recent progress in developing safe and effective vectors for gene transfer, such as with synthetic viruses and non-viral methods, as well as the success in using autologous and allogenic chimeric antigen receptor integrated T-lymphocytes, even from healthy individuals, as universal effector cells in mediating adoptive immunotherapy, will increase the effectiveness and safety profile of gene therapy. Furthermore, with the advancement in biological research, much cheaper gene vectors will become commercially available, which will make gene therapy readily available to the majority of cancer patients, worldwide. This will transform the future of cancer therapy, from generalized cancer treatment strategies, based on tumor size, nature and location, to a more tailored, individualized cancer therapy, based on the patient's specific genomic constituents, host immune status, and genetic profile of the underlying malignancy. Treatment is expected to be fast, effective, relatively less toxic and inexpensive, with higher cure rates, and may even, cancer prevention.

AUTHORS' INFORMATION

The author is an internist and medical oncologist at a regional cancer center in the United States, with higher qualification from United Kingdom, Canada, and United States, previous academic title at Ohio State University, over thirty five years practice as a medical oncologist, and authored or co-authored over twenty five peer-reviewed papers on cancer management and research.

ABBREVIATIONS

APCs:	Antigen-presenting cells
CEA:	Carcinoembryonic antigen
DNA:	Deoxyribonucleic acid
DTA-H19:	Corynebacterium diphtheriae toxin-A chain

EGFR: Epidermal growth factor receptor

Fox3: Forkhead box 03 transcription factor protein

FDA: The Food and Drug Administration in the United States

GDEPT: Gene directed enzyme prodrug therapy

GM-CSF: Granulocyte macrophage colony stimulating factor

HER-2/neu: Human epidermal growth factor receptor-2

HIV: Human immunodeficiency virus infection

hTERT: Human telomerase reverse transcriptase

IDO: Indoleamine-2,3-dioxygenase

MHC: Major histocompatibility complex

MDR1: Multidrug-resistant protein-1

mRNA: Messenger-RNA

NK: Natural killer cell lymphocytes

PSA: Prostatic acid phosphatase antigen

RISC: Ribonucleic acid induced silencing complex

RNA: Ribonucleic acid

RNAi: Ribonucleic acid interference

siRNA: Small interfering double-stranded RNA

TAAs: Tumor-associated antigens

TCR: T cell antigen receptor

TRICOM: Multiple T-cell costimulatory molecules

VEGF-A: Vascular endothelial growth factor A receptor.

REFERENCES

1. Lander ES, Linton LM, Birren B, Nusbaum C, Zody MC, Baldwin J, Devon K, Dewar K, Doyle M, FitzHugh W, Funke R, Gage D, Harris K, Heaford A, Howland J, Kann L, Lehoczky J, LeVine R, McEwan P, McKernan K, Meldrim J, Mesirov JP, Miranda C, Morris W, Naylor J, Raymond C, Rosetti M, Santos R, Sheridan A, Sougnez C: Initial sequencing and analysis of the human genome. Nature. 2001, 409 (6822): 860-921. 10.1038/35057062.
2. Venter JC, Adams MD, Myers EW, Li PW, Mural RJ, Sutton GG, Smith HO, Yandell M, Evans CA, Holt RA, Gocayne JD, Amanatides P, Ballew RM, Huson DH, Wortman JR, Zhang Q, Kodira CD, Zheng XH, Chen L, Skupski M, Subramanian G, Thomas PD, Zhang J, Gabor Miklos GL, Nelson C, Broder S, Clark AG, Nadeau J, McKusick VA, Zinder N: The sequence of the human genome. Science (New York, NY). 2001, 291 (5507): 1304-1351. 10.1126/science.1058040.
3. Dulbecco R: A turning point in cancer research: sequencing the human genome. Science (New York, NY). 1986, 231 (4742): 1055-1056. 10.1126/science.3945817.
4. Strachan T, Read A: Gene therapy and other molecular genetic-based therapeutic approaches. Human Molecular Genetics. Edited by: Strachan T, Read AP. 1999, New York: Wiley-Liss, 2
5. Miller AD: Retroviral vectors. Curr Top Microbiol Immunol. 1992, 158: 1-24.
6. Weichselbaum RR, Kufe D:Gene therapy of cancer. Lancet. 1997, 349 (Suppl 2): SII10-2.
7. DeVita VT, Rosenberg SA: Two hundred years of cancer research. N Engl J Med. 2012, 366 (23): 2207-2214. 10.1056/NEJMra1204479.
8. Curie P, Curie M, Bémont G: On a new, strongly radioactive substance contained in pitchblende. CR (East Lansing, Mich). 1898, 127: 1215-1217.
9. Lage H: Bacterial delivery of RNAi effectors. Gene Therapy of Cancer. Edited by: Lattime EC, Gerson SL. 2013, San Diego (CA): Elsevier, 67-75. 3
10. Goodman LS, Wintrobe MM, Dameshek W, Goodman MJ, Gilman A, McLennan MT: Landmark article Sept. 21, 1946: nitrogen mustard therapy: use of methyl-bis(beta-chloroethyl)amine hydrochloride and tris(beta-chloroethyl)amine hydrochloride for Hodgkin's disease,

lymphosarcoma, leukemia and certain allied and miscellaneous disorders by Louis S. Goodman, Maxwell M. Wintrobe, William Dameshek, Morton J. Goodman, Alfred Gilman and Margaret T. McLennan. JAMA. 1984, 251 (17): 2255-2261. 10.1001/jama.1984.03340410063036.

11. Farber S, Diamond LK: Temporary remissions in acute leukemia in children produced by folic acid antagonist, 4-aminopteroyl-glutamic acid. N Engl J Med. 1948, 238 (23): 787-793. 10.1056/NEJM194806032382301.
12. Hemminki O, Hemminki A: Oncolytic adenoviruses in the treatment of cancer in humans. Gene Therapy of Cancer. Edited by: Lattime EC, Gerson SL. 2013, San Diego (CA): Elsevier, 153-170. 3
13. Kelly E, Russell SJ: History of oncolytic viruses: genesis to genetic engineering. Mol Ther. 2007, 15 (4): 651-659.
14. Maloney DG, Grillo-Lopez AJ, White CA, Bodkin D, Schilder RJ, Neidhart JA, Janakiraman N, Foon KA, Liles TM, Dallaire BK, Wey K, Royston I, Davis T, Levy R: IDEC-C2B8 (Rituximab) anti-CD20 monoclonal antibody therapy in patients with relapsed low-grade non-Hodgkin's lymphoma. Blood. 1997, 90 (6): 2188-2195.
15. Sheridan C: Gene therapy finds its niche. Nat Biotechnol. 2011, 29 (2): 121-128. 10.1038/nbt.1769.
16. Chiocca EA, Abbed KM, Tatter S, Louis DN, Hochberg FH, Barker F, Kracher J, Grossman SA, Fisher JD, Carson K, Rosenblum M, Mikkelsen T, Olson J, Markert J, Rosenfeld S, Nabors LB, Brem S, Phuphanich S, Freeman S, Kaplan R, Zwiebel J: A phase I open-label, dose-escalation, multi-institutional trial of injection with an E1B-Attenuated adenovirus, ONYX-015, into the peritumoral region of recurrent malignant gliomas, in the adjuvant setting. Mol Ther. 2004, 10 (5): 958-966. 10.1016/j.ymthe.2004.07.021.
17. Block SL, Nolan T, Sattler C, Barr E, Giacoletti KE, Marchant CD, Castellsague X, Rusche SA, Lukac S, Bryan JT, Cavanaugh PF, Reisinger KS, Protocol 016 Study Group: Comparison of the immunogenicity and reactogenicity of a prophylactic quadrivalent human papillomavirus (types 6, 11, 16, and 18) L1 virus-like particle vaccine in male and female adolescents and young adult women. Pediatrics. 2006, 118 (5): 2135-2145. 10.1542/peds.2006-0461.
18. Kantoff PW, Schuetz TJ, Blumenstein BA, Glode LM, Bilhartz DL, Wyand M, Manson K, Panicali DL, Laus R, Schlom J, Dahut WL,

Arlen PM, Gulley JL, Godfrey WR: Overall survival analysis of a phase II randomized controlled trial of a Poxviral-based PSA-targeted immunotherapy in metastatic castration-resistant prostate cancer. J Clin Oncol. 2010, 28 (7): 1099-1105. 10.1200/JCO.2009.25.0597.

19. Robinson DR, Wu YM, Vats P, Su F, Lonigro RJ, Cao X, Kalyana-Sundaram S, Wang R, Ning Y, Hodges L, Gursky A, Siddiqui J, Tomlins SA, Roychowdhury S, Pienta KJ, Kim SY, Roberts JS, Rae JM, Van Poznak CH, Hayes DF, Chugh R, Kunju LP, Talpaz M, Schott AF, Chinnaiyan AM: Activating ESR1 mutations in hormone-resistant metastatic breast cancer. Nat Genet. 2013, 45 (12): 1446-1451. 10.1038/ng.2823.
20. Toy W, Shen Y, Won H, Green B, Sakr RA, Will M, Li Z, Gala K, Fanning S, King TA, Hudis C, Chen D, Taran T, Hortobagyi G, Greene G, Berger M, Baselga J, Chandarlapaty S: ESR1 ligand-binding domain mutations in hormone-resistant breast cancer. Nat Genet. 2013, 45 (12): 1439-1445. 10.1038/ng.2822.
21. Koboldt DC, Fulton RS, McLellan MD, Schmidt H, Kalicki-Veizer J, McMichael JF, Fulton LL, Dooling DJ, Ding L, Mardis ER, Wilson RK, Ally A, Balasundaram M, Butterfield YS, Carlsen R, Carter C, Chu A, Chuah E, Chun HJ, Coope RJ, Dhalla N, Guin R, Hirst C, Hirst M, Holt RA, Lee D, Li HI, Mayo M, Moore RA, Mungall AJ: Comprehensive molecular portraits of human breast tumours. Nature. 2012, 490 (7418): 61-70. 10.1038/nature11412.
22. Shirley S, Heller R, Heller L: Electroporation gene therapy. Gene Therapy of Cancer. Edited by: Lattime EC, Gerson SL. 2013, San Diego (CA): Elsevier, 93-106. 3
23. Baranyi L, Slepushkin V, Dropulic B: Ex vivo gene therapy: utilization of genetic vectors for the generation of genetically modified cell products for therapy. Gene Therapy of Cancer. Edited by: Lattime EC, Gerson SL. 2013, San Diego (CA): Elsevier, 3-18. 3
24. Yuan Z, Pastoriza J, Quinn T, Libutti S: Targeting tumor vasculature using adeno-associated virus page vector coding tumor necrosis factor-a. Gene Therapy of Cancer. Edited by: Lattime EC, Gerson SL. 2013, San Diego (CA): Elsevier, 19-33. 3
25. Zabner J, Fasbender AJ, Moninger T, Poellinger KA, Welsh MJ: Cellular and molecular barriers to gene transfer by a cationic lipid. J Biol Chem. 1995, 270 (32): 18997-19007. 10.1074/jbc.270.32.18997.
26. Tagami T, Suzuki T, Matsunaga M, Nakamura K, Moriyoshi N, Ishida

T, Kiwada H: Anti-angiogenic therapy via cationic liposome-mediated systemic siRNA delivery. Int J Pharm. 2012, 422 (1–2): 280-289.

27. Yang W, Sun T, Cao J, Liu F: Survivin downregulation by siRNA/ cationic liposome complex radiosensitises human hepatoma cells in vitro and in vivo. Int J Radiat Biol. 2010, 86 (6): 445-457. 10.3109/09553001003668006.

28. Wagner E, Plank C, Zatloukal K, Cotten M, Birnstiel ML: Influenza virus hemagglutinin HA-2 N-terminal fusogenic peptides augment gene transfer by transferrin-polylysine-DNA complexes: toward a synthetic virus-like gene-transfer vehicle. Proc Natl Acad Sci U S A. 1992, 89 (17): 7934-7938. 10.1073/pnas.89.17.7934.

29. Soltani F, Sankian M, Hatefi A, Ramezani M: Development of a novel histone H1-based recombinant fusion peptide for targeted non-viral gene delivery. Int J Pharm. 2013, 441 (1–2): 307-315.

30. Di Martino MT, Leone E, Amodio N, Foresta U, Lionetti M, Pitari MR, Cantafio ME, Gulla A, Conforti F, Morelli E, Tomaino V, Rossi M, Negrini M, Ferrarini M, Caraglia M, Shammas MA, Munshi NC, Anderson KC, Neri A, Tagliaferri P, Tassone P: Synthetic miR-34a mimics as a novel therapeutic agent for multiple myeloma: in vitro and in vivo evidence. Clin Cancer Res. 2012, 18 (22): 6260-6270. 10.1158/1078-0432.CCR-12-1708.

31. Soliman M, Nasanit R, Abulateefeh SR, Allen S, Davies MC, Briggs SS, Seymour LW, Preece JA, Grabowska AM, Watson SA, Alexander C: Multicomponent synthetic polymers with viral-mimetic chemistry for nucleic acid delivery. Mol Pharm. 2012, 9 (1): 1-13. 10.1021/ mp200108q.

32. Nie Y, Schaffert D, Rodl W, Ogris M, Wagner E, Gunther M: Dual-targeted polyplexes: one step towards a synthetic virus for cancer gene therapy. J Control Release. 2011, 152 (1): 127-134. 10.1016/j. jconrel.2011.02.028.

33. Hackett PB, Largaespada DA, Switzer KC, Cooper LJ: Evaluating risks of insertional mutagenesis by DNA transposons in gene therapy. Transl Res. 2013, 161 (4): 265-283. 10.1016/j.trsl.2012.12.005.

34. Kwon S, Min J: Genetically engineered Salmonella typhimurium for targeted cancer therapy. Gene Therapy of Cancer. Edited by: Lattime EC, Gerson SL. 2013, San Diego (CA): Elsevier, 443-452. 3

35. Benoit MR, Mayer D, Barak Y, Chen IY, Hu W, Cheng Z, Wang SX,

Spielman DM, Gambhir SS, Matin A: Visualizing implanted tumors in mice with magnetic resonance imaging using magnetotactic bacteria. Clin Cancer Res. 2009, 15 (16): 5170-5177. 10.1158/1078-0432.CCR-08-3206.

36. Baban CK, Cronin M, O'Hanlon D, O'Sullivan GC, Tangney M: Bacteria as vectors for gene therapy of cancer. Bioeng Bugs. 2010, 1 (6): 385-394. 10.4161/bbug.1.6.13146.
37. Thomas CE, Ehrhardt A, Kay MA: Progress and problems with the use of viral vectors for gene therapy. Nat Rev Genet. 2003, 4 (5): 346-358. 10.1038/nrg1066.
38. Kuroda S, Kagawa S, Fujiwara T: Selectively replicating oncolytic adenoviruses combined with chemotherapy, radiotherapy, or molecular targeted therapy for treatment of human cancers. Gene Therapy of Cancer. Edited by: Lattime EC, Gerson SL. 2013, San Diego (CA): Elsevier, 171-183. 3
39. Wu E, Nemerow GR: Virus yoga: the role of flexibility in virus host cell recognition. Trends Microbiol. 2004, 12 (4): 162-169. 10.1016/j.tim.2004.02.005.
40. Mathis JM, Stoff-Khalili MA, Curiel DT: Oncolytic adenoviruses - selective retargeting to tumor cells. Oncogene. 2005, 24 (52): 7775-7791. 10.1038/sj.onc.1209044.
41. Zemp FJ, Corredor JC, Lun X, Muruve DA, Forsyth PA: Oncolytic viruses as experimental treatments for malignant gliomas: using a scourge to treat a devil. Cytokine Growth Factor Rev. 2010, 21 (2–3): 103-117.
42. Balvers R, Gomez-Manzano C, Jiang H, Piya S, Klein S, Lamfers M, Driven C, Fueyo J: Advances on oncolytic virotherapy for brain tumors. Gene Therapy of Cancer. Edited by: Lattime EC, Gerson SL. 2013, San Diego (CA): Elsevier, 137-151. 3
43. Kaliberova LN, Krendelchtchikova V, Harmon DK, Stockard CR, Petersen AS, Markert JM, Gillespie GY, Grizzle WE, Buchsbaum DJ, Kaliberov SA: CRAdRGDflt-IL24 virotherapy in combination with chemotherapy of experimental glioma. Cancer Gene Ther. 2009, 16 (10): 794-805. 10.1038/cgt.2009.23.
44. Nuesch JP, Lacroix J, Marchini A, Rommelaere J: Molecular pathways: rodent parvoviruses–mechanisms of oncolysis and prospects for clinical cancer treatment. Clin Cancer Res. 2012, 18 (13): 3516-3523.

10.1158/1078-0432.CCR-11-2325.

45. White E, Gill S: Selectively replicating Herpes simplex viral vectors. Gene Therapy of Cancer. Edited by: Lattime EC, Gerson SL. 2013, San Diego (CA): Elsevier, 199-211. 3

46. Sharp PM: Origins of human virus diversity. Cell. 2002, 108 (3): 305-312. 10.1016/S0092-8674(02)00639-6.

47. Hu JC, Coffin RS, Davis CJ, Graham NJ, Groves N, Guest PJ, Harrington KJ, James ND, Love CA, McNeish I, Medley LC, Michael A, Nutting CM, Pandha HS, Shorrock CA, Simpson J, Steiner J, Steven NM, Wright D, Coombes RC: A phase I study of OncoVEXGM-CSF, a second-generation oncolytic herpes simplex virus expressing granulocyte macrophage colony-stimulating factor. Clin Cancer Res. 2006, 12 (22): 6737-6747. 10.1158/1078-0432.CCR-06-0759.

48. Liu TC, Zhang T, Fukuhara H, Kuroda T, Todo T, Canron X, Bikfalvi A, Martuza RL, Kurtz A, Rabkin SD: Dominant-negative fibroblast growth factor receptor expression enhances antitumoral potency of oncolytic herpes simplex virus in neural tumors. Clin Cancer Res. 2006, 12 (22): 6791-6799. 10.1158/1078-0432.CCR-06-0263.

49. Bryson P, Wang P: Lentivector vaccines. Gene Therapy of Cancer. Edited by: Lattime EC, Gerson SL. 2013, San Diego (CA): Elsevier, 345-361. 3

50. Breckpot K, Aerts JL, Thielemans K: Lentiviral vectors for cancer immunotherapy: transforming infectious particles into therapeutics. Gene Ther. 2007, 14 (11): 847-862. 10.1038/sj.gt.3302947.

51. Hu B, Yang H, Dai B, Tai A, Wang P: Nonintegrating lentiviral vectors can effectively deliver ovalbumin antigen for induction of antitumor immunity. Hum Gene Ther. 2009, 20 (12): 1652-1664. 10.1089/hum.2009.012.

52. Michelini Z, Negri DR, Baroncelli S, Spada M, Leone P, Bona R, Klotman ME, Cara A: Development and use of SIV-based Integrase defective lentiviral vector for immunization. Vaccine. 2009, 27 (34): 4622-4629. 10.1016/j.vaccine.2009.05.070.

53. Comins C, Simpson G, Relph K, Harrington K, Melcher A, Pandha H: Reoviral therapy for cancer: strategies for improving antitumor efficacy using radio- and chemotherapy. Gene Therapy of Cancer. Edited by: Lattime EC, Gerson SL. 2013, San Diego (CA): Elsevier, 185-198. 3

54. Pachnis V, Brannan CI, Tilghman SM: The structure and expression of

a novel gene activated in early mouse embryogenesis. EMBO J. 1988, 7 (3): 673-681.

55. Matouk I, Raveh E, Ohana P, Lail RA, Gershtain E, Gilon M, De Groot N, Czerniak A, Hochberg A: The increasing complexity of the oncofetal h19 gene locus: functional dissection and therapeutic intervention. Int J Mol Sci. 2013, 14 (2): 4298-4316. 10.3390/ijms14024298.
56. Ohana P, Matouk I, Amit D, Gilon M, Hochberg A: Toxin-based cancer gene therapy under the control of oncofetal H19 regulatory sequences. Gene Therapy of Cancer. Edited by: Lattime EC, Gerson SL. 2013, San Diego (CA): Elsevier, 107-122. 3
57. Li Y, McCadden J, Ferrer F, Kruszewski M, Carducci M, Simons J, Rodriguez R: Prostate-specific expression of the diphtheria toxin A chain (DT-A): studies of inducibility and specificity of expression of prostate-specific antigen promoter-driven DT-A adenoviral-mediated gene transfer. Cancer Res. 2002, 62 (9): 2576-2582.
58. Fung H, Gersson S: Viral insertion site detection and analysis in cancer gene therapy. Gene Therapy of Cancer. Edited by: Lattime EC, Gerson SL. 2013, San Diego (CA): Elsevier, 35-46. 3
59. Gujrati M, Lu Z: Targeted systemic delivery of therapeutic siRNA. Gene Therapy of Cancer. Edited by: Lattime EC, Gerson SL. 2013, San Diego (CA): Elsevier, 47-65. 3
60. Sato Y, Murase K, Kato J, Kobune M, Sato T, Kawano Y, Takimoto R, Takada K, Miyanishi K, Matsunaga T, Takayama T, Niitsu Y: Resolution of liver cirrhosis using vitamin A-coupled liposomes to deliver siRNA against a collagen-specific chaperone. Nat Biotechnol. 2008, 26 (4): 431-442. 10.1038/nbt1396.
61. Davis ME: The first targeted delivery of siRNA in humans via a self-assembling, cyclodextrin polymer-based nanoparticle: from concept to clinic. Mol Pharm. 2009, 6 (3): 659-668. 10.1021/mp900015y.
62. Tabernero J, Shapiro GI, LoRusso PM, Cervantes A, Schwartz GK, Weiss GJ, Paz-Ares L, Cho DC, Infante JR, Alsina M, Gounder MM, Falzone R, Harrop J, White AC, Toudjarska I, Bumcrot D, Meyers RE, Hinkle G, Svrzikapa N, Hutabarat RM, Clausen VA, Cehelsky J, Nochur SV, Gamba-Vitalo C, Vaishnaw AK, Sah DW, Gollob JA, Burris HA: First-in-humans trial of an RNA interference therapeutic targeting VEGF and KSP in cancer patients with liver involvement. Cancer Discov. 2013, 3 (4): 406-417. 10.1158/2159-8290.CD-12-0429.

63. Geng J, Xiao S, Zhang S, Liu C, Li Y, Ji J, Fang Z, Sun Y, Su X, Cai Y, Xu G, Li D, Zhu G, Xu B: Clinical effectiveness of recombinant adenovirus-p53 combined with radiotherapy in advanced soft tissue sarcoma: a report of 37 cases. J Clin Oncol. 2014, 32 (suppl; abstr e21514):
64. Pan JJ, Zhang SW, Chen CB, Xiao SW, Sun Y, Liu CQ, Su X, Li DM, Xu G, Xu B, Lu YY: Effect of recombinant adenovirus-p53 combined with radiotherapy on long-term prognosis of advanced nasopharyngeal carcinoma. J Clin Oncol. 2009, 27 (5): 799-804. 10.1200/JCO.2008.18.9670.
65. Lai SY, Koppikar P, Thomas SM, Childs EE, Egloff AM, Seethala RR, Branstetter BF, Gooding WE, Muthukrishnan A, Mountz JM, Lui VW, Shin DM, Agarwala SS, Johnson R, Couture LA, Myers EN, Johnson JT, Mills G, Argiris A, Grandis JR: Intratumoral epidermal growth factor receptor antisense DNA therapy in head and neck cancer: first human application and potential antitumor mechanisms. J Clin Oncol. 2009, 27 (8): 1235-1242. 10.1200/JCO.2008.17.8251.
66. Greenberger JS: Radioprotection. In Vivo. 2009, 23 (2): 323-336.
67. Cai Y, Bak RO, Mikkelsen JG: Targeted genome editing by lentiviral protein transduction of zinc-finger and TAL-effector nucleases. eLife. 2014, 3: e01911-
68. Uhde-Stone C, Sarkar N, Antes T, Otoc N, Kim Y, Jiang YJ, Lu B: A TALEN-based strategy for efficient bi-allelic miRNA ablation in human cells. RNA (New York, NY). 2014, 20 (6): 948-955. 10.1261/rna.042010.113.
69. Sakuma T, Woltjen K: Nuclease-mediated genome editing: at the front-line of functional genomics technology. Develop Growth Differ. 2014, 56 (1): 2-13. 10.1111/dgd.12111.
70. Cobleigh MA, Vogel CL, Tripathy D, Robert NJ, Scholl S, Fehrenbacher L, Wolter JM, Paton V, Shak S, Lieberman G, Slamon DJ: Multinational study of the efficacy and safety of humanized anti-HER2 monoclonal antibody in women who have HER2-overexpressing metastatic breast cancer that has progressed after chemotherapy for metastatic disease. J Clin Oncol. 1999, 17 (9): 2639-2648.
71. McLaughlin P, Grillo-Lopez AJ, Link BK, Levy R, Czuczman MS, Williams ME, Heyman MR, Bence-Bruckler I, White CA, Cabanillas F, Jain V, Ho AD, Lister J, Wey K, Shen D, Dallaire BK: Rituximab chimeric anti-CD20 monoclonal antibody therapy for relapsed indolent

lymphoma: half of patients respond to a four-dose treatment program. J Clin Oncol. 1998, 16 (8): 2825-2833.

72. Foon KA, Yang XD, Weiner LM, Belldegrun AS, Figlin RA, Crawford J, Rowinsky EK, Dutcher JP, Vogelzang NJ, Gollub J, Thompson JA, Schwartz G, Bukowski RM, Roskos LK, Schwab GM: Preclinical and clinical evaluations of ABX-EGF, a fully human anti-epidermal growth factor receptor antibody. Int J Radiat Oncol Biol Phys. 2004, 58 (3): 984-990. 10.1016/j.ijrobp.2003.09.098.
73. Hurwitz H, Fehrenbacher L, Novotny W, Cartwright T, Hainsworth J, Heim W, Berlin J, Baron A, Griffing S, Holmgren E, Ferrara N, Fyfe G, Rogers B, Ross R, Kabbinavar F: Bevacizumab plus irinotecan, fluorouracil, and leucovorin for metastatic colorectal cancer. N Engl J Med. 2004, 350 (23): 2335-2342. 10.1056/NEJMoa032691.
74. Glassman PM, Balthasar JP: Mechanistic considerations for the use of monoclonal antibodies for cancer therapy. Cancer Biol Med. 2014, 11 (1): 20-33.
75. Boon T, Van der Bruggen P: Human tumor antigens recognized by T lymphocytes. J Exp Med. 1996, 183 (3): 725-729. 10.1084/jem.183.3.725.
76. Steinman RM, Banchereau J: Taking dendritic cells into medicine. Nature. 2007, 449 (7161): 419-426. 10.1038/nature06175.
77. De Souza AP, Bonorino C: Tumor immunosuppressive environment: effects on tumor-specific and nontumor antigen immune responses. Expert Rev Anticancer Ther. 2009, 9 (9): 1317-1332. 10.1586/era.09.88.
78. Gajewski TF, Schreiber H, Fu YX: Innate and adaptive immune cells in the tumor microenvironment. Nat Immunol. 2013, 14 (10): 1014-1022. 10.1038/ni.2703.
79. Mlecnik B, Tosolini M, Kirilovsky A, Berger A, Bindea G, Meatchi T, Bruneval P, Trajanoski Z, Fridman WH, Pages F, Galon J: Histopathologic-based prognostic factors of colorectal cancers are associated with the state of the local immune reaction. J Clin Oncol. 2011, 29 (6): 610-618. 10.1200/JCO.2010.30.5425.
80. Mahmoud SM, Paish EC, Powe DG, Macmillan RD, Grainge MJ, Lee AH, Ellis IO, Green AR: Tumor-infiltrating CD8+ lymphocytes predict clinical outcome in breast cancer. J Clin Oncol. 2011, 29 (15): 1949-1955. 10.1200/JCO.2010.30.5037.

81. Rusakiewicz S, Semeraro M, Sarabi M, Desbois M, Locher C, Mendez R, Vimond N, Concha A, Garrido F, Isambert N, Chaigneau L, Le Brun-Ly V, Dubreuil P, Cremer I, Caignard A, Poirier-Colame V, Chaba K, Flament C, Halama N, Jäger D, Eggermont A, Bonvalot S, Commo F, Terrier P, Opolon P, Emile JF, Coindre JM, Kroemer G, Chaput N, Le Cesne A: Immune infiltrates are prognostic factors in localized gastrointestinal stromal tumors. Cancer Res. 2013, 73 (12): 3499-3510. 10.1158/0008-5472.CAN-13-0371.

82. Rosenberg SA, Dudley ME: Adoptive cell therapy for the treatment of patients with metastatic melanoma. Curr Opin Immunol. 2009, 21 (2): 233-240. 10.1016/j.coi.2009.03.002.

83. Brown IE, Blank C, Kline J, Kacha AK, Gajewski TF: Homeostatic proliferation as an isolated variable reverses CD8+ T cell anergy and promotes tumor rejection. J Immunol. 2006, 177 (7): 4521-4529. 10.4049/jimmunol.177.7.4521.

84. Watkins SK, Zhu Z, Riboldi E, Shafer-Weaver KA, Stagliano KE, Sklavos MM, Ambs S, Yagita H, Hurwitz AA: FOXO3 programs tumor-associated DCs to become tolerogenic in human and murine prostate cancer. J Clin Invest. 2011, 121 (4): 1361-1372. 10.1172/JCI44325.

85. Hodi FS, O'Day SJ, McDermott DF, Weber RW, Sosman JA, Haanen JB, Gonzalez R, Robert C, Schadendorf D, Hassel JC, Akerley W, van den Eertwegh AJ, Lutzky J, Lorigan P, Vaubel JM, Linette GP, Hogg D, Ottensmeier CH, Lebbé C, Peschel C, Quirt I, Clark JI, Wolchok JD, Weber JS, Tian J, Yellin MJ, Nichol GM, Hoos A, Urba WJ: Improved survival with ipilimumab in patients with metastatic melanoma. N Engl J Med. 2010, 363 (8): 711-723. 10.1056/NEJMoa1003466.

86. Wolchok JD, Hodi FS, Weber JS, Allison JP, Urba WJ, Robert C, O'Day SJ, Hoos A, Humphrey R, Berman DM, Lonberg N, Korman AJ: Development of ipilimumab: a novel immunotherapeutic approach for the treatment of advanced melanoma. Annals of the New York Academy of Sciences. 2013, 1291: 1-13. 10.1111/nyas.12180.

87. Burnette BC, Liang H, Lee Y, Chlewicki L, Khodarev NN, Weichselbaum RR, Fu YX, Auh SL: The efficacy of radiotherapy relies upon induction of type i interferon-dependent innate and adaptive immunity. Cancer Res. 2011, 71 (7): 2488-2496. 10.1158/0008-5472.CAN-10-2820.

88. Balachandran VP, Cavnar MJ, Zeng S, Bamboat ZM, Ocuin LM,

Obaid H, Sorenson EC, Popow R, Ariyan C, Rossi F, Besmer P, Guo T, Antonescu CR, Taguchi T, Yuan J, Wolchok JD, Allison JP, DeMatteo RP: Imatinib potentiates antitumor T cell responses in gastrointestinal stromal tumor through the inhibition of Ido. Nat Med. 2011, 17 (9): 1094-1100. 10.1038/nm.2438.

89. Frederick DT, Piris A, Cogdill AP, Cooper ZA, Lezcano C, Ferrone CR, Mitra D, Boni A, Newton LP, Liu C, Peng W, Sullivan RJ, Lawrence DP, Hodi FS, Overwijk WW, Lizée G, Murphy GF, Hwu P, Flaherty KT, Fisher DE, Wargo JA: BRAF inhibition is associated with enhanced melanoma antigen expression and a more favorable tumor microenvironment in patients with metastatic melanoma. Clin Cancer Res. 2013, 19 (5): 1225-1231. 10.1158/1078-0432.CCR-12-1630.

90. Liang H, Deng L, Chmura S, Burnette B, Liadis N, Darga T, Beckett MA, Lingen MW, Witt M, Weichselbaum RR, Fu Y: Radiation-induced equilibrium is a balance between tumor cell proliferation and T cell-mediated killing. J Immunol. 2013, 190 (11): 5874-5881. 10.4049/jimmunol.1202612.

91. Zeng J, See AP, Phallen J, Jackson CM, Belcaid Z, Ruzevick J, Durham N, Meyer C, Harris TJ, Albesiano E, Pradilla G, Ford E, Wong J, Hammers HJ, Mathios D, Tyler B, Brem H, Tran PT, Pardoll D, Drake CG, Lim M: Anti-PD-1 blockade and stereotactic radiation produce long-term survival in mice with intracranial gliomas. Int J Radiat Oncol Biol Phys. 2013, 86 (2): 343-349. 10.1016/j.ijrobp.2012.12.025.

92. Reardon DA, Wucherpfennig KW, Freeman G, Wu CJ, Chiocca EA, Wen PY, Curry WT, Mitchell DA, Fecci PE, Sampson JH, Dranoff G: An update on vaccine therapy and other immunotherapeutic approaches for glioblastoma. Expert Rev Vaccines. 2013, 12 (6): 597-615. 10.1586/erv.13.41.

93. Hinrichs CS, Stevanovic S, Draper L, Somerville R, Wunderlich J, Restifo NP, Sherry R, Giao PQ, Kammula US, Yang JC, Rosenberg SA: HPV-targeted tumor-infiltrating lymphocytes for cervical cancer. J Clin Oncol. 2014, 32 (suppl; abstr LBA3008):

94. Kantoff PW, Higano CS, Shore ND, Berger ER, Small EJ, Penson DF, Redfern CH, Ferrari AC, Dreicer R, Sims RB, Xu Y, Frohlich MW, Schellhammer PF: Sipuleucel-T immunotherapy for castration-resistant prostate cancer. N Engl J Med. 2010, 363 (5): 411-422. 10.1056/NEJMoa1001294.

95. Varela-Rohena A, Carpenito C, Perez EE, Richardson M, Parry RV,

Milone M, Scholler J, Hao X, Mexas A, Carroll RG, June CH, Riley JL: Genetic engineering of T cells for adoptive immunotherapy. Immunol Res. 2008, 42 (1–3): 166-181.

96. Johnson LA, Heemskerk B, Powell DJ, Cohen CJ, Morgan RA, Dudley ME, Robbins PF, Rosenberg SA: Gene transfer of tumor-reactive TCR confers both high avidity and tumor reactivity to nonreactive peripheral blood mononuclear cells and tumor-infiltrating lymphocytes. JImmunol. 2006, 177 (9): 6548-6559. 10.4049/jimmunol.177.9.6548.
97. Wang X, Nishimura M: Genetically engineered (T cell receptor) T cells for adoptive therapy. Gene Therapy of Cancer. Edited by: Lattime EC, Gerson SL. 2013, San Diego (CA): Elsevier, 259-271. 3
98. Choi Y, Yuen C, Maiti SN, Olivares S, Gibbons H, Huls H, Raphael R, Killian TC, Stark DJ, Lee DA, Torikai H, Monticello D, Kelly SS, Kebriaei P, Champlin RE, Biswal SL, Cooper LJ: A high throughput microelectroporation device to introduce a chimeric antigen receptor to redirect the specificity of human T cells. Biomed Microdevices. 2010, 12 (5): 855-863. 10.1007/s10544-010-9440-3.
99. Singh H, Manuri PR, Olivares S, Dara N, Dawson MJ, Huls H, Hackett PB, Kohn DB, Shpall EJ, Champlin RE, Cooper LJ: Redirecting specificity of T-cell populations for CD19 using the sleeping beauty system. Cancer Res. 2008, 68 (8): 2961-2971. 10.1158/0008-5472. CAN-07-5600.
100. Hoyos V, Savoldo B, Quintarelli C, Mahendravada A, Zhang M, Vera J, Heslop HE, Rooney CM, Brenner MK, Dotti G: Engineering CD19-specific T lymphocytes with interleukin-15 and a suicide gene to enhance their anti-lymphoma/leukemia effects and safety. Leukemia. 2010, 24 (6): 1160-1170. 10.1038/leu.2010.75.
101. Xu Y, Zhang M, Ramos CA, Durett A, Liu E, Dakhova O, Liu H, Creighton CJ, Gee AP, Heslop HE, Rooney CM, Savoldo B, Dotti G: Closely related T-memory stem cells correlate with in vivo expansion of CAR.CD19-T cells and are preserved by IL-7 and IL-15. Blood. 2014, 123 (24): 3750-3759. 10.1182/blood-2014-01-552174.
102. Turtle CJ: Chimeric antigen receptor modified T cell therapy for B cell malignancies. Int J Hematol. 2014, 99 (2): 132-140. 10.1007/s12185-013-1490-x.
103. Kochenderfer JN, Dudley ME, Feldman SA, Wilson WH, Spaner DE, Maric I, Stetler-Stevenson M, Phan GQ, Hughes MS, Sherry RM, Yang JC, Kammula US, Devillier L, Carpenter R, Nathan DA, Morgan

RA, Laurencot C, Rosenberg SA: B-cell depletion and remissions of malignancy along with cytokine-associated toxicity in a clinical trial of anti-CD19 chimeric-antigen-receptor-transduced T cells. Blood. 2012, 119 (12): 2709-2720. 10.1182/blood-2011-10-384388.

104. Marcus A, Eshhar Z: Allogeneic chimeric antigen receptor-modified cells for adoptive cell therapy of cancer. Expert Opin Biol Ther. 2014, 14 (7): 947-954. 10.1517/14712598.2014.900540.

105. Torikai H, Reik A, Liu PQ, Zhou Y, Zhang L, Maiti S, Huls H, Miller JC, Kebriaei P, Rabinovitch B, Lee DA, Champlin RE, Bonini C, Naldini L, Rebar EJ, Gregory PD, Holmes MC, Cooper LJ: A foundation for universal T-cell based immunotherapy: T cells engineered to express a CD19-specific chimeric-antigen-receptor and eliminate expression of endogenous TCR. Blood. 2012, 119 (24): 5697-5705. 10.1182/blood-2012-01-405365.

106. Boudreau JE, Bonehill A, Thielemans K, Wan Y: Engineering dendritic cells to enhance cancer immunotherapy. Mol Ther. 2011, 19 (5): 841-853. 10.1038/mt.2011.57.

107. Pinder-Schenck M, Antonia S: Genetically modified dendritic cell vaccines for solid tumors. Gene Therapy of Cancer. Edited by: Lattime EC, Gerson SL. 2013, San Diego (CA): Elsevier, 273-282. 3

108. Jonuleit H, Giesecke-Tuettenberg A, Tuting T, Thurner-Schuler B, Stuge TB, Paragnik L, Kandemir A, Lee PP, Schuler G, Knop J, Enk AH: A comparison of two types of dendritic cell as adjuvants for the induction of melanoma-specific T-cell responses in humans following intranodal injection. Int J Cancer. 2001, 93 (2): 243-251. 10.1002/ijc.1323.

109. Tsao H, Millman P, Linette GP, Hodi FS, Sober AJ, Goldberg MA, Haluska FG: Hypopigmentation associated with an adenovirus-mediated gp100/MART-1-transduced dendritic cell vaccine for metastatic melanoma. Arch Dermatol. 2002, 138 (6): 799-802.

110. Marshall JL, Gulley JL, Arlen PM, Beetham PK, Tsang KY, Slack R, Hodge JW, Doren S, Grosenbach DW, Hwang J, Fox E, Odogwu L, Park S, Panicali D, Schlom J: Phase I study of sequential vaccinations with fowlpox-CEA(6D)-TRICOM alone and sequentially with vaccinia-CEA(6D)-TRICOM, with and without granulocyte-macrophage colony-stimulating factor, in patients with carcinoembryonic antigen-expressing carcinomas. J Clin Oncol. 2005, 23 (4): 720-731. 10.1200/JCO.2005.10.206.

111. Wolff JA, Budker V: The mechanism of naked DNA uptake and expression. Adv Genet. 2005, 54: 3-20.

112. Kutzler MA, Weiner DB: DNA vaccines: ready for prime time?. Nat Rev Genet. 2008, 9 (10): 776-788. 10.1038/nrg2432.

113. McNeel DG, Dunphy EJ, Davies JG, Frye TP, Johnson LE, Staab MJ, Horvath DL, Straus J, Alberti D, Marnocha R, Liu G, Eickhoff JC, Wilding G: Safety and immunological efficacy of a DNA vaccine encoding prostatic acid phosphatase in patients with stage D0 prostate cancer. J Clin Oncol. 2009, 27 (25): 4047-4054. 10.1200/JCO.2008.19.9968.

114. Norell H, Poschke I, Charo J, Wei WZ, Erskine C, Piechocki MP, Knutson KL, Bergh J, Lidbrink E, Kiessling R: Vaccination with a plasmid DNA encoding HER-2/neu together with low doses of GM-CSF and IL-2 in patients with metastatic breast carcinoma: a pilot clinical trial. J Transl Med. 2010, 8: 53-10.1186/1479-5876-8-53.

115. Staff C, Mozaffari F, Haller BK, Wahren B, Liljefors M: A Phase I safety study of plasmid DNA immunization targeting carcinoembryonic antigen in colorectal cancer patients. Vaccine. 2011, 29 (39): 6817-6822. 10.1016/j.vaccine.2010.12.063.

116. Rosa DS, Ribeiro SP, Almeida RR, Mairena EC, Postol E, Kalil J, Cunha-Neto E: A DNA vaccine encoding multiple HIV CD4 epitopes elicits vigorous polyfunctional, long-lived CD4+ and CD8+ T cell responses. PLoS One. 2011, 6 (2): e16921-10.1371/journal.pone.0016921.

117. Garcia F, Petry KU, Muderspach L, Gold MA, Braly P, Crum CP, Magill M, Silverman M, Urban RG, Hedley ML, Beach KJ: ZYC101a for treatment of high-grade cervical intraepithelial neoplasia: a randomized controlled trial. Obstet Gynecol. 2004, 103 (2): 317-326. 10.1097/01.AOG.0000110246.93627.17.

118. Auci D, Cecil D, Herendeen D, Broussard E, Liao J, Holt G, Disis M: Clinical application of plasmid-based cancer vaccines. Gene Therapy of Cancer. Edited by: Lattime EC, Gerson SL. 2013, San Diego (CA): Elsevier, 335-343. 3

119. Karjoo Z, Ganapathy V, Hatefi A: Gene-directed enzyme prodrug cancer therapy. Gene Therapy of Cancer. Edited by: Lattime EC, Gerson SL. 2013, San Diego (CA): Elsevier, 77-91. 3

120. Nicholas TW, Read SB, Burrows FJ, Kruse CA: Suicide gene therapy with Herpes simplex virus thymidine kinase and ganciclovir is enhanced

with connexins to improve gap junctions and bystander effects. Histol Histopathol. 2003, 18 (2): 495-507.

121. Agard C, Ligeza C, Dupas B, Izembart A, El Kouri C, Moullier P, Ferry N: Immune-dependent distant bystander effect after adenovirus-mediated suicide gene transfer in a rat model of liver colorectal metastasis. Cancer Gene Ther. 2001, 8 (2): 128-136. 10.1038/sj.cgt.7700281.
122. Edelstein ML, Abedi MR, Wixon J, Edelstein RM: Gene therapy clinical trials worldwide 1989-2004-an overview. J Gene Med. 2004, 6 (6): 597-602. 10.1002/jgm.619.
123. Alauddin MM, Shahinian A, Gordon EM, Bading JR, Conti PS: Preclinical evaluation of the penciclovir analog 9-(4-[(18)F]fluoro-3-hydroxymethylbutyl)guanine for in vivo measurement of suicide gene expression with PET. J Nucl Med. 2001, 42 (11): 1682-1690.
124. Lin Y, Gerson S: Clinical trials using LB-P140K-MGMT for gliomas. Gene Therapy of Cancer. Edited by: Lattime EC, Gerson SL. 2013, San Diego (CA): Elsevier, 379-391. 3
125. Maier P, Herskind C, Fleckenstein K, Spier I, Laufs S, Zeller WJ, Fruehauf S, Wenz F: MDR1 gene transfer using a lentiviral SIN vector confers radioprotection to human CD34+ hematopoietic progenitor cells. Radiat Res. 2008, 169 (3): 301-310. 10.1667/RR1067.1.
126. Maier P, Spier I, Laufs S, Veldwijk MR, Fruehauf S, Wenz F, Zeller WJ: Chemoprotection of human hematopoietic stem cells by simultaneous lentiviral overexpression of multidrug resistance 1 and O(6)-methylguanine-DNA methyltransferase(P140K). Gene Ther. 2010, 17 (3): 389-399. 10.1038/gt.2009.133.
127. Arai T, Miyoshi Y, Kim SJ, Akazawa K, Maruyama N, Taguchi T, Tamaki Y, Noguchi S: Association of GSTP1 expression with resistance to docetaxel and paclitaxel in human breast cancers. Eur J Surg Oncol. 2008, 34 (7): 734-738. 10.1016/j.ejso.2007.07.008.
128. Oguri T, Fujiwara Y, Katoh O, Daga H, Ishikawa N, Fujitaka K, Yamasaki M, Yokozaki M, Isobe T, Ishioka S, Yamakido M: Glutathione S-transferase-pi gene expression and platinum drug exposure in human lung cancer. Cancer Lett. 2000, 156 (1): 93-99. 10.1016/S0304-3835(00)00447-X.
129. Matranga C, Tomari Y, Shin C, Bartel DP, Zamore PD: Passenger-strand cleavage facilitates assembly of siRNA into Ago2-containing

RNAi enzyme complexes. Cell. 2005, 123 (4): 607-620. 10.1016/j.cell.2005.08.044.

130. Seiradake E, Henaff D, Wodrich H, Billet O, Perreau M, Hippert C, Mennechet F, Schoehn G, Lortat-Jacob H, Dreja H, Ibanes S, Kalatzis V, Wang JP, Finberg RW, Cusack S, Kremer EJ: The cell adhesion molecule "CAR" and sialic acid on human erythrocytes influence adenovirus in vivo biodistribution. PLoS Pathog. 2009, 5 (1): e1000277-10.1371/journal.ppat.1000277.
131. Howe SJ, Mansour MR, Schwarzwaelder K, Bartholomae C, Hubank M, Kempski H, Brugman MH, Pike-Overzet K, Chatters SJ, de Ridder D, Gilmour KC, Adams S, Thornhill SI, Parsley KL, Staal FJ, Gale RE, Linch DC, Bayford J, Brown L, Quaye M, Kinnon C, Ancliff P, Webb DK, Schmidt M, von Kalle C, Gaspar HB, Thrasher AJ: Insertional mutagenesis combined with acquired somatic mutations causes leukemogenesis following gene therapy of SCID-X1 patients. J Clin Invest. 2008, 118 (9): 3143-3150. 10.1172/JCI35798.
132. Schmidt M, Zickler P, Hoffmann G, Haas S, Wissler M, Muessig A, Tisdale JF, Kuramoto K, Andrews RG, Wu T, Kiem HP, Dunbar CE, von Kalle C: Polyclonal long-term repopulating stem cell clones in a primate model. Blood. 2002, 100 (8): 2737-2743. 10.1182/blood-2002-02-0407.
133. Bartholomae CC, Glimm H, Von Kalle C, Schmidt M: Insertion site pattern: global approach by linear amplification-mediated PCR and mass sequencing. Methods Mol Biol. 2012, 859: 255-265. 10.1007/978-1-61779-603-6_15.
134. Gillet NA, Malani N, Melamed A, Gormley N, Carter R, Bentley D, Berry C, Bushman FD, Taylor GP, Bangham CR: The host genomic environment of the provirus determines the abundance of HTLV-1-infected T-cell clones. Blood. 2011, 117 (11): 3113-3122. 10.1182/blood-2010-10-312926.
135. Novartis: Investigational New Drug Development Program. 2014, [http://www.novartisoncology.com/ct/pipelineHome]

10

CRISPR to the Rescue: Advances in Gene Editing for the FMR1 Gene

Carolyn M. Yrigollen [1] **and Beverly L. Davidson** [1,2]

[1]The Raymond G. Perelman Center of Cellular and Molecular Therapeutics, Children's Hospital of Philadelphia, Philadelphia, PA 19104, USA
[2]Department of Pathology and Laboratory Medicine, University of Pennsylvania, Philadelphia, PA 19104, USA

ABSTRACT

Gene-editing using Clustered Regularly Interspaced Short Palindromic Repeats (CRISPR) is promising as a potential therapeutic strategy for many genetic disorders. CRISPR-based therapies are already being assessed in clinical trials, and evaluation of this technology in Fragile X syndrome has been performed by a number of groups. The findings from these studies and the advancement of CRISPR-based technologies are insightful as the field continues towards treatments and cures of Fragile X-Associated Disorders

Citation: Yrigollen, C.M.; Davidson, B.L. CRISPR to the Rescue: Advances in Gene Editing for the FMR1 Gene. Brain Sci. 2019, 9, 17. https://doi.org/10.3390/brainsci9010017.

(FXADs). In this review, we summarize reports using CRISPR-editing strategies to target Fragile X syndrome (FXS) molecular dysregulation, and highlight how differences in FXS and Fragile X-associated Tremor/Ataxia Syndrome (FXTAS) might alter treatment strategies for each syndrome. We discuss the various modifications and evolutions of the CRISPR toolkit that expand its therapeutic potential, and other considerations for moving these strategies from bench to bedside. The rapidly growing field of CRISPR therapeutics is providing a myriad of approaches to target a gene, pathway, or transcript for modification. As cures for FXADs have remained elusive, CRISPR opens new avenues to pursue.

Keywords: Fragile X syndrome 1; Fragile X-associated Tremor/Ataxia Syndrome 2; CRISPR 3; Trinucleotide Repeat 4; Gene editing

INTRODUCTION

The expansion of the CGG trinucleotide repeat within the 5' untranslated region (UTR) of the Fragile X Mental Retardation 1 (*FMR1*) gene is the predominant cause of Fragile X syndrome (FXS), and the only known cause of Fragile X-associated Tremor/Ataxia Syndrome (FXTAS) and Fragile X-associated Primary Ovarian Insufficiency (FXPOI) [1]. The trinucleotide repeat is normally between 5 and 44 CGG repeats in length and interspersed with up to 4 AGG interruptions.

Premutation carriers have 55–200 CGG repeats which lead to misregulation of the *FMR1* gene in several ways. Carriers of the premutation allele have elevated *FMR1* mRNA levels but also have a reduction in the translational efficiency of the *FMR1* encoded protein FMRP. *FMR1* transcripts harboring premutation length CGG repeats can give rise to Repeat-Associated Non-ATG (RAN) homopolypeptides (FMRpolyA, FMRpolyG, and FMRpolyR), some of these RAN translation products with homopolymeric amino acid tracts have been shown to be toxic [2,3,4]. Additionally, RNA toxicity can result from the long CGG tracts with *FMR1* transcripts forming stable hairpin structures and binding with proteins. This aberrant protein-RNA interaction results in protein sequestration and mislocalization, impairing normal cellular processes and leading to the formation of intranuclear inclusions [5,6,7]. Premutation carriers are at risk of developing the neurodegenerative disorder FXTAS. The age of onset for FXTAS is typically greater than 55 years of age with core features including intention tremor, gait ataxia, executive dysfunction, and neuropathy. Post mortem evaluation of FXTAS patients has identified the presence of intranuclear inclusions throughout the brain [8]. Women have a lower risk of developing FXTAS but are at risk for developing the reproductive disorder FXPOI.

Individuals with more than 200 CGG repeats, categorized as a full mutation allele, are diagnosed with FXS when this mutation becomes hypermethylated leading to a loss of FMRP. Women with FXS are often less severely affected than men; women have a protective second copy of the *FMR1* gene that is expressed in cells when the full mutation resides on the inactive X chromosome. FXS is a neurodevelopmental disorder when the full mutation allele is aberrantly methylated and transcriptionally silenced. *FMR1* epigenetic changes occur early during embryonic development and cause intellectual disability, facial dysmorphia, macroorchidism, and hyperextensible joints, which are diagnosed early in childhood [9].

Both FXS and FXTAS are neurological disorders with no disease modifying therapies. Clustered Regularly Interspaced Short Palindromic Repeats (CRISPR) is a relatively new gene editing technology which has quickly proven effective in correcting a number of pathological mutations in model systems and is advancing to clinical trials [10]. In this review, the current state of CRISPR as a tool to treat Fragile X-Associated Disorders (FXADs) will be presented.

CRISPR

The utility of CRISPR as a genome editor in eukaryotic cells was first reported in 2013 [11,12,13]; since these publications, the technology has quickly evolved and now offers a multitude of modified CRISPR associated (Cas) enzymes capable of an array of genetic modifications [14,15]. CRISPR, which was discovered as an adaptive immune system for bacteria [16], uses Cas enzymes complexed to RNA to identify invading virus and phage DNA. The "memory" mechanism, which occurred during a previous invasion with a similar species, resulted in short DNA sequences from the invaders being stored between short palindromic repeats. When these sequences are expressed into RNA, they are resolved into a structured RNA fragment that is loaded onto the Cas nuclease. The approximately 20 bp RNA sequence that corresponds to the DNA target is accessible to genomic DNA as the nuclease scans the genetic code. When the guideRNA complements with a DNA strand the nuclease changes conformation to a catalytically opened state that cleaves the foreign DNA [17]. To use CRISPR as a genome editor, the endogenous system has been adapted by using synthesized guideRNAs that direct the Cas nucleases to the genomic region of interest. After DNA cleavage, the double stranded breaks at the target site are repaired by one of the host cells' DNA damage response mechanisms, non-homologous end joining (NHEJ) or homology directed repair (HDR) [12].

Cas9 from *Streptococcus pyogenes* (SpCas9) was the first CRISPR enzyme shown to edit eukaryotic DNA using synthetic guideRNAs (Figure 1a). It was successfully adapted for use in eukaryotic cells because cleavage could be achieved using only Cas9 and two short RNA molecules. Since these first reports, novel CRISPR based tools have been developed with a range of functions and advantages [14]. Modified Cas9 enzymes can now be completely deactivated from cleaving DNA while retaining their binding activity (dCas9; Figure 1d) [18,19]. When coupled to repressive proteins, the guideRNAs guide the Cas9 to the appropriate site to repress transcription (dCas9-HDAC; dCas9KRAB; Figure 1e) [20]. Variants have also been made that generate single stranded breaks (Cas9 nickases; Figure 1b) [21], increase transcription (dCas9-Suntag, dCas9-p300, dCas9-VP64, dCas9-SAM; Figure 1g–i) [20,22,23,24], label DNA (dCas9-GFP; Figure 1f), or directly convert cytosine nucleotides to thymine, or guanine to adenine (Base Editors; Figure 1k) [25,26]. Modifications to SpCas9 and other newly characterized Cas enzymes have also improved the specificity of the nuclease to a target sequence (High Fidelity Cas9) [27,28,29], altered the required protospacer adjacent motif (PAM) motifs (Figure 1j) [30,31], and demonstrated RNA targeting capability (C2c2; Figure 1l) [32].

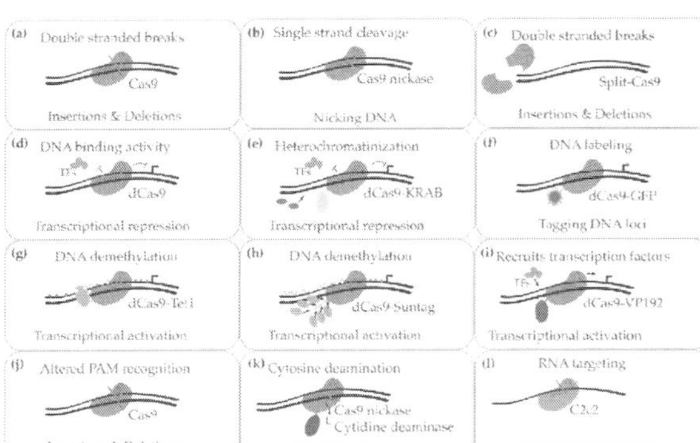

Figure 1. Overview of Clustered Regularly Interspersed Short Palindromic Repeats (CRISPR) technologies with therapeutic utility. (a) Wildtype SpCas9 is directed to the target site for editing through guideRNAs possessing sequences complementary to the DNA region of interest. When bound to the target site,

the active Cas9 induces a double stranded DNA break. When repaired, this can induce the formation of indels. (b) Mutations that inactivate either the RuvC or HNH domain in Cas9 (Cas9 nickase) impacts nuclease such that cleavage in only one strand can occur; the other DNA strand remains intact. (c) Expressing split Cas9 that complexes into a functional nuclease when complementing the target DNA sequence. (d) Mutations in both RuvC and HNH domains deactivate Cas9; the 'dead' Cas9 retains the ability to bind target sequences but is incapable of generating single or double stranded breaks. When Cas9 is directed to genomic regions near start transcription start sites, expression can be inhibited because normal transcription factor binding sites are blocked. (e) Fusing a Krueppel-associated box (KRAB) domain onto dCas9 recruits chromatin remodeling factors that elicit heterochromatinization of the target locus, further reducing transcription (f) GFP fused to Cas9 is used as a molecular beacon to monitor specific regions of the genome in vitro and in vivo. (g–i) Fusing demethylases or transcriptional activators onto dCas9 to upregulate target genes. Suntag and VP192 use multiple copies of the activating domains to induce higher upregulation. (j) SpCas9 nuclease mutations to alter PAM recognition motifs. Wildtype Cas9 recognizes NGG, and mutation variants recognize NGCG, NGAG, NGAN, and NGNG. (k) Cytosine deaminase fused onto Cas9 nickase for conversion of cytosine to thymine without inducing double stranded breaks. (l) Cas enzymes that target RNA include C2c2, renamed Cas13a.

RECENTLY REPORTED CRISPR-BASED THERAPIES

The first study published using CRISPR to edit *FMR1* targeted the full mutation allele in human induced pluripotent stem cells (iPSCs; Table 1). These cells, harboring an epigenetically silenced full mutation, were electroporated with plasmid DNA encoding SpCas9 and a guideRNA designed to target 47 base pairs upstream of the start of the trinucleotide repeat [33]. Complete deletion of the CGG repeats were observed in iPSCs. These cells, clonally expanded post-editing had a reported 2–3% editing efficiency. The edited iPSCs were shown to reactivate *FMR1* expression to levels similar to the control cells with normal CGG repeat alleles. The authors further reported a loss of methylation at the promoter following CRISPR mediated deletion of the full mutation CGG repeats, and sustained *FMR1* expression was present after reprogramming the iPSCs into neuronal cells [33]. The edited cells that were differentiated into mature neurons stained positive for FMRP compared to a lack of FMRP positive cells in the unedited parental lines. Differences in genes expressed in edited and unedited neurons showed a reduction in three glutamate receptor genes,

GRIA1, *GRIN2B*, and *GRIN3A* following deletion of the CGG repeat, consistent with restoration of FMRP expression.

Table 1. Summary of Fragile X syndrome (FXS) CRISPR gene editing studies.

Study	**Cells or Tissue**	**Host**	**Delivery**	**Target Sequence**	**Nuclease Used**	**Outcome**
Park et al., *Cell Reports* 2015	iPSC	Human	Electro-poration	47 bp upstream of CGG repeat	SpCas9	Deletion of CGG repeats in 2–3% clonally expanded cells; reactivation of the *FMR1* gene; sustained *FMR1* reactivation upon differentiation of iPSCs into neurons.
Xie et al., *PLoS ONE* 2016	iPSC	Human	Nucleo-fection	40 bp upstream and 35 bp down-stream of CGG repeat	SpCas9	Deletion of CGG repeats in 20% of cells; variability in reactivaton among edited clones. Reactivation of *FMR1* persisted 50 days in culture.
Liu et al., *Cell* 2018	iPSC	Human	Lenti virus	CGG repeats	dCas9-Tet1	Reactivation of *FMR1* and demethylation of the CGG repeat locus in iPSCs and derived neurons. Lower reactivation efficiency in neurons. Edited iPSCs were reprogrammed into neurons and engrafted into mouse brain; half of neurons actively expressed FMRP 3 months post transplantation.
Haenfler et al., *Frontiers of Molecular Neuroscience* 2018	ESC and neuronal progenitor cells	Human	Lipid mediated transfection	Promoter and CGG repeats	dCas9-VP192	Reactivation of *FMR1* but low FMRP expression with CGG targeting. Higher reactivation efficiency with CGG targeting guideRNAs than promoter targeting.
Lee et al., *Nature Biomedical Engineering* 2018	Striatum	Mouse	Gold nanoparticles	*Grm5*	Cas9 and Cpf1	Reduction of *Grm5* expression; phenotypic rescue of marble burying and jumping behaviors.

Xie and colleagues [34] used a similar strategy to delete the CGG repeats from HEK 293 cells and iPSCs using nucleofection of CRISPR plasmids with guideRNAs targeting 40 bps upstream and 35 bps downstream of the trinucleotide repeat (Table 1). With dual guides, the editing efficiency increased to 20% in iPSCs. Importantly, this study replicated that a silenced full mutation could be reactivated by deletion of the CGG repeats in iPSCs, and this transcriptional activation was stable for a prolonged time in culture (50 days post reactivation). However, it was reported by Xie et al. that not all of the clonal lines reactivated following CRISPR editing. Evaluation of epigenetic status showed that reactivated clones had decreased methylation levels at the CpG sites at the promoter region and adjacent to the CGG repeat locus, while clones that lacked reactivation showed similar methylation levels as unedited iPSC clones. This variability in epigenetic modifications and *FMR1* expression in CGG deleted cells was hypothesized to be the result of incomplete DNA remethylation following DNA replication. As such, more actively replicating cell types would be expected to undergo more efficient *FMR1* reactivation following CRISPR editing of the CGG repeat, while nondividing cell types will be less prone to this demethylation. While promising, neither study showed direct editing in differentiated cells, and further investigation into the reactivation capabilities of nondividing neurons is warranted.

CRISPR has also been used to epigenetically modify the *FMR1* full mutation outright [35]. In the first study of this kind, catalytically deactivated SpCas9 (dCas9) was fused to Tet1, an enzyme that induces demethylation of cytosines to create a methylation eraser (dCas9-Tet1). When dCas9-Tet1 and guideRNAs designed to target the CGG repeats were introduced into iPSCs with lentivirus, epigenetic changes occurred at the *FMR1* locus. As expected, gene expression occurred without sequence modifications of the trinucleotide repeat. Edited FXS iPSCs showed *FMR1* expression levels that were 90% of that measured in a control human embryonic stem cell. There was also DNA hypomethylation of the CpG island adjacent to the repeats, and histone modifications including H3 lysine 27 acetylation, H3 lysine 4 trimethylation, and a decrease of H3 lysine 9 trimethylation at the promoter of *FMR1*. The histone modifications reactivated transcription, and this was maintained for over 35 days in culture. When the edited iPSCs were derived into neurons *FMR1* remained transcriptionally active and the electrophysiological hyperactive firing rate phenotypes were rescued. Gene expression changes were described for 41 identified off target genes to be less than 4-fold upregulated in the edited neurons, while one gene *RGPD1* had a 9-fold increase in gene expression. In contrast, *FMR1* was reported to have a 481-fold increase in gene expression following epigenetic editing. When engrafted into mouse brains, the edited cells expressed FMRP for 3 months post-transplantation. This is the first in vivo analysis of ex vivo edited, transplanted neuron. These findings demonstrate the utility of targeting epigenetic-editors to the CGG repeat locus. A critical question in using gene editing technologies for FXS is whether a constitutively active dCas9-Tet1 is necessary for long term reactivation at the *FMR1* locus. Epigenetic editing of iPSC derived neurons was shown to be less efficient than in iPSCs, *FMR1* mRNA levels were restored to 45% that of control neurons and a 20% decrease in CpG methylation at the *FMR1* promoter was achieved post editing. These differences could have been the technical limitations of isolating the edited neurons or differences in mechanisms to demethylation between cell types. The authors also reported a rescue of hyperactivity in the edited FXS neurons compared to their unedited counterparts.

In another approach, Haenfler et al. [36] fused dCas9 to multiple VP16 transcriptional activator domains (dCas9-VP192) to drive expression of *FMR1* without altering the genetic sequence (Table 1). Human embryonic stem cells harboring a silenced full mutation allele were transfected in vitro with dCas9-VP192 and guideRNAs targeting either the *FMR1* promoter region or the ~800 CGG trinucleotide repeats. Cells transfected with the transcriptional activator targeting the CGG repeats showed robust transcriptional reactivation but only modest gains in FMRP. The CGG targeting guideRNAs were reported to induce higher transcriptional activity than guideRNAs that targeted the promoter region. The ability of the CGG targeting guideRNAs to target within the trinucleotide repeats multiple times might explain the stronger reactivation of *FMR1*, as more VP192 activating domains are recruited to the gene. They also transfected neuronal progenitor cells derived from the fully methylated hESC with the CGG targeting dCas9-VP192 and reactivated *FMR1* and increased transcriptional expression. Again, there was a limited increase in FMRP. This work highlights a key limitation of reactivating a full mutation; long CGG repeats have been shown to decrease the translational efficiency of *FMR1* [37]. The molecular implications of expressing long CGG repeat tracts is also an area that needs further investigation, as 55–200 repeats can give rise to FXTAS and individuals with an unmethylated full mutation have been reported to develop FXTAS symptoms [38,39]. Thus it will be important to understand if expressing more than 200 repeats will increase the risk of developing FXTAS or increase the severity of clinical pathology.

The first in vivo gene-editing in animal models of FXS was recently reported by Lee et al. This strategy assessed both Cas9 and Cpf1 for their efficiency to knockout gene expression in vivo in mice. The nucleases were delivered to specific regions of the brain using gold-nanoparticles with editing in neurons, astrocytes, and microglia shown to decrease reporter gene expression. In an alternative gene editing strategy to targeting *Fmr1*, the metabotropic glutamate receptor (mGluR5) encoding gene *Grm5* was targeted for knockdown with Cas9 in the striatum of *Fmr1* knockout mice (Table 1) [40]. *Grm5* was targeted because of the over-activation of mGluR5-dependent signaling present in FXS and other autism spectrum disorders. The efficiency of editing *Grm5* was measured to be 14.6%, resulting in a 40–50% reduction in the encoded protein. Repression in mGluR5-dependent signaling resulted in a rescue of a marble burying phenotype in *Fmr1*$^{-/y}$ mice as well as jumping behaviors, both considered measures of hyperactivity and stereotypy. While this study demonstrated in vivo gene-editing in an animal model of FXS, it did not target the *FMR1* locus directly. However, the work did establish a potential strategy for treating autism. Although the authors used CRISPR editing to treat FXS-like phenotypes in mice, they did so by inducing a loss of function mutation in a second gene rather than correct the initial null mutation in *Fmr1*; the authors suggest that while clinical trials that used drugs to reduce mGluR5 activity were disappointing, using CRISPR to target the overactivation may be more effective. Of note, the delivery strategy resulted in a focused treatment area, which is necessary when knocking out an important regulator of plasticity. However, it has limitations if multiple regions or a large area of the brain must be edited for therapeutic effect, an important consideration in translating from mice to humans. Notably, gold nanoparticles have been shown to have cellular toxicity at high or repeated doses [41,42].

Other trinucleotide repeat disorders are also being targeted with CRISPR-based therapies [43]. One example is the Huntingtin gene (*HTT*), which has been corrected in vitro using patient derived fibroblast cells by targeting only the expanded allele for deletion of the CAG repeat and knocked down expression. This strategy was also used in vivo in a mouse model of the expanded CAG repeat, with similar decreased expression of the gene and encoded protein as was seen in cell lines [44]. An in vivo mouse model of Friedreich Ataxia showed partial increased expression of the frataxin (*FXN*) gene when GAA trinucleotide repeats were deleted from intron 1 with Cas9 [45].

The modifications, fusions, and evolutions of CRISPR for therapeutics opens the field to numerous strategies for using gene-editing to treat and cure FXADs. For example, RNA targeting nucleases can be used to target the CGG repeat in FXTAS and off-target activity can be decreased using split Cas9 or Cas9 nickases that require two guideRNAs to complement the genomic sequence near each other [46, 47]. The SpCas9 nuclease has also been genetically modified to decrease off-target activity while maintaining high editing activity at the target loci [28, 29]. These designer nucleases can improve the safety of these therapies by minimizing unwanted DNA damage. Cas9 enzymes with alternative PAM motifs are also available, and these make more of the genome accessible for CRISPR modification [31].

FURTHER CONSIDERATIONS FOR GENE-EDITING FOR FXTAS AND FXS

Delivery

Delivering CRISPR-based therapeutics into the brain and specifically neuronal cells is essential for FXS and FXTAS therapy. Several strategies to deliver gene therapies to the brain, some with success in model systems, are described below. Also see a more comprehensive review by Cwetsch, Pinto [48].

Transducing neurons with Adeno Associated Virus (AAV) is the most common method to deliver CRISPR editors in vivo. AAVs have high transduction efficiency and low immunogenicity profiles compared to many other viral vectors, contributing to their broad use in gene therapy applications [49]. AAVs can package approximately 4.8 kb of foreign DNA sequence, which is only 600 bp larger than the coding sequence of SpCas9. This leaves minimal room for cassettes to drive guideRNAs and Cas9 expression. Strategies to bypass this limitation include dual vector systems where the nuclease and guideRNAs are packaged into separate AAV vectors [50] or splitting Cas9 into two fragments that dimerize in the presence of a small molecule (Figure 1c) [51]. Alternative enzymes that are encoded by sequences smaller than the 4.2 kB sequence for SpCas9 include the 3.2 kB *Staphylococcus aureus* Cas9, which is small enough to allow for cis expression of one or two guideRNAs [30,52].

AAV capsids vary in their tropism, defined as their ability to bind to and be internalized by specific cell types. AAV serotype 9 has been used to deliver an *FMR1* transgene into *Fmr1* knockout neonate mice via intracerebroventricular injection [53]. These experiments showed neuronal expression of FMRP in several brain regions including the cortex, hippocampus, and striatum. The cortex and hippocampus had the highest transduction efficiency as measured by the number of FMRP-positive cells for a given area. However, FMRP was not detected beyond the midbrain and cerebellum. Together with AAV9, AAV1, AAV2, AAV5 and AAV8 have been reported to express transgenes throughout the brain [54,55]. Developing genetically engineered AAV serotype variants that increase overall transduction efficiency or specificity is an active area of research [56,57].

Lentiviruses have also been engineered to deliver CRISPR components into cells. Unlike AAVs, that remain mostly episomal; lentiviruses integrate into the host genome, a mechanism that can increase the risk of insertional mutagenesis. An advantage of lentiviruses for gene editing is that they have a packaging capacity of approximately 9 kb. Lentiviruses have been used to deliver CGG targeting CRISPR components into iPSCs in vitro for gene editing as discussed above [35].

Lipid nanoparticles (LNPs) are a non-viral strategy to deliver CRISPR components in vivo. In the liver, high editing efficiency was shown after a single intravenous injection of CRISPR-mRNA containing LNPs [42], or LNPs containing plasmids encoding Cas9 and guideRNAs [58,59]. Currently, efficient editing in the brain using LNP-based systems have not been shown. However, *Fmr1* knockout mice have been focally transfected with gold nanoparticles, as described above [40].

The main advantage of non-viral delivery is that the Cas nuclease is present and functional for a limited time. In recent studies comparing plasmid expression systems or RNPs of Cas9/guide RNAs, Behlke and colleagues showed that transient Cas9 exposure minimizes off target activity [60]. Additionally, transiently expressing Cas and guideRNAs provides the availability of the full CRISPR toolkit for editing new targets without interference from a previous treatment.

The route of CRISPR delivery into the brain has to be considered. Some modalities will allow for widespread delivery but require larger dosages to be administered (i.e., intraventricular injection) while others can provide for targeting a precise region (i.e., intracerebellar or cerebral injection). Localized injections will minimize the amount of therapeutic administered while having a high coverage of cells in the targeted area. Such a delivery strategy will also narrow the functional range of the therapy. This is unlikely to be useful for FXS and FXTAS therapy where pathology occurs in many brain regions [61,62]. FXS is associated with increased white matter and gray matter volume in the thalamus, frontal and temporal lobe, cerebellum and caudate by magnetic resonance imaging (MRI). Histopathological analysis of brains from individuals with FXS identified abnormalities in the hippocampus and cerebellar vermis [63]. Patients with FXTAS often have increased T2 intensity by MRI in the middle cerebellar peduncles (MCPs), this is referred to as the MCP sign and occurs in approximately 60% of diagnosed cases of FXTAS. Other hallmarks of the disorder include cerebellar and cerebral atrophy and a thinning of the corpus callosum. Hippocampal and amygdala structural differences are less obvious and not consistently seen across studies. Balancing precision and spread of the therapeutic components will be important for successful clinical results, and will be heavily dependent on the delivery vehicle [64].

Another important consideration is when to deliver the gene-editors. The timeframe for treatment will be different depending on whether FXS or FXTAS is being treated. While FXS symptoms occur early during development and early interventions show the greatest outcomes [65], FXTAS is a late onset and incompletely penetrant disorder. Premutation carriers may never develop FXTAS or require medical intervention, but we are yet unclear whether features of FXTAS can be improved post onset and what treatment windows will be beneficial.

Off-Target Editing

The main concerns of researchers developing CRISPR therapies are the off-target and undesired editing effects permanently incorporated into patients' somatic cells. Off target editing has been studied in vitro and in vivo using a variety of genome-wide protocols (i.e., Genome-wide, Unbiased Identification of DSBs Enabled by Sequencing [GUIDE-seq] [66]; High-Throughput Genome-Wide Translocation Sequencing [HTGTS] [67]; Integrative-Deficient Lentiviral Vectors [IDLV] [68]; Digenome-sequencing [Digenome-seq] [69,70]; Circularization for In vitro Reporting of Cleavage Effects by sequencing [CIRCLE-seq] [51]; Selective enrichment and Identification of Tagged genomic DNA Ends by sequencing [SITE-seq] [71]; Breaks Labeling, Enrichment on Streptavidin, and Sequencing/Breaks Labeling In Situ and Sequencing [BLESS/BLISS] [30,72,73], and others [74]). These off-target detection screens have demonstrated that prediction of sites prone to erroneous cleavage by CRISPR is challenging and impacted by the guideRNA sequence, tissue or cell type, the specific nuclease used, and the method of delivery. Off-target activity is balanced with on-target editing efficiency. A high on-target, low off-target ratio is of course the holy grail of therapeutic editing. To reduce off-target binding of guide RNAs, most current guideRNA design tools use algorithms that allow minimizing the number of mismatches of off-target sequences in the host genome to the on-target sequence. Some algorithms also consider chromatin accessibility and guideRNA stability in vivo, although this information changes with cell type [75,76,77].

Characterizing and reducing off-target effects are critical to move any gene-editing therapy into the clinic. None of the recently published *FMR1* editing studies have surveyed off-target editing using one of the genome wide unbiased methods [51,66,68,69,72,73], with predicted off-targets shown to have little overlap with off targets identified using these approaches. Off-target loci have been shown to occur with as many as 6 mismatches to the 20 nucleotide guideRNA sequence [78], and are significantly reduced with transient Cas9 expression, either by delivery of mRNA encoding the nuclease and guideRNAs, or delivery of the complexed RNPs [79]. Self-destroying CRISPR (KamiCas9) systems has also been shown to limit the time Cas9 is present in treated cells, reducing off-target accumulation [80].

Editing systems must be empirically tested for each gene, and possibly for each cell type, of interest. In general, multiple guideRNAs are tested and their on- and off-target editing efficiencies evaluated. A consideration of off-target effects is whether a biological consequence (e.g., truncated protein) would result from misdirected editing and how well such an outcome is tolerated by the cell type.

CRISPR-editing of the *FMR1* gene may erroneously repair the on-target sequence resulting in large deletions, incorporation of plasmid or viral DNA sequences, or fusion of *FMR1* with other genes at other loci. A non-biased genetic analysis of repaired sequences will quantify which mistakes are prone to occur and at what frequency. Large deletions extending beyond the start codon, or upstream of the transcriptional start site, would disrupt *FMR1* expression or function. Incorporation of foreign DNA sequences may dysregulate *FMR1* reducing therapeutic rescue post editing [81,82]. Fusions of genes after editing could create novel, immunogenic or toxic proteins or protein chimeras that could alter the function of the cell or even be tumorigenic [83,84]. While these possibilities seem dire, careful preclinical testing in relevant model systems would identify problematic editing components such that alternative approaches could be developed and tested.

One of the unique considerations of gene-editing for FXADs therapy is the presence of functional and dysfunctional CGG repeats in females. Females are genetically protected by the presence of a normal CGG repeat allele in a portion of their cells [85, 86]. It is important to avoid editing the normal allele-containing cells, while targeting premutation or full mutation expressing cells. Liu et al suggest that preferential editing at the expanded allele occurs because the CGG repeats provide more targetable sequence compared to the few CGG repeat in the normal allele or at other genomic loci Liu, Wu [35]. While possible, this requires more thorough investigation using single cell RNA seq analysis of cells after CRISPR/Cas9 editing of a mixture of cells harboring normal and expanded repeat sequences.

CRISPR Limitations

There are several additional limitations when implementing the CRISPR technology as a therapeutic strategy. The stringency of CRISPR nucleases to a PAM sequence can prevent researchers from targeting a specific genomic locus because there are no available PAM sites within that region, this results in changing the gene-editing strategy to conform to the currently accessible sequences. However, CRISPR tools are being developed to broaden the sites that Cas9 and other nucleases recognize [31]. The incomplete editing of CRISPR in somatic cells can result in mosaicism of desired and off target modifications in vivo. These mosaic modifications can result in unpredictable outcomes [87]. As the field continues to progress, we will have an improved perspective on how often these sporadic events occur and the severity they have in vivo.

Ethical Considerations

With the rapid progress of CRISPR in the biomedical field, keeping pace with the ethical issues that arise is important. Like clinical trials that are currently underway for new pharmaceuticals and gene therapies, gene-editing has the potential for unforeseen outcomes and side effects. Because the therapy is a permanent modification of the genome, these unexpected effects can carry a greater burden to the participant. Treatment of an embryo, child, or patient incapable of understanding the risks associated with the therapy mean informed consent cannot be obtained. In a disorder as variable as FXS, balancing the potential benefit of treating a patient during early development with the ability to assess the severity of the individual's symptoms later in life comes into question. Mosaic off-target mutations can result in unpredictable side effects to the therapy in a single participant. Additionally, as access to gene-editing therapies becomes available, equal access to these technologies across society will be essential to not disenfranchise sectors of our society [88]. The ethical issues that arise as genome editing in humans continues to advance are important and more each come with serious consequences can result from how these issues are resolved. As we move forward with developing genome editing therapies that have the potential to greatly improve the quality of life of individuals within the Fragile X community, it is important to also move forward in our understanding of these issues and decide how to address these ethical considerations.

FINAL REMARKS

The recent successes that have been achieved using CRISPR for genetic and epigenetic editing and the advances being made to deliver, control, and modify Cas nucleases are driving rapid development of editing-based gene therapies for disease treatment. The continued developments in the field, even for other disorders, will allow us to optimize CRISPR technology for FXADs with regards to efficacy and safety. Finally, CRISPR/Cas9 technologies may also provide tools to develop an in vivo model of the *FMR1* epigenetically silenced full mutation, which to date does not exist. Thus, the technological advancement of CRISPR/Cas9 has infiltrated many sectors of biomedical research and provides one of the more promising approaches for FXS and FXTAS therapy.

AUTHOR CONTRIBUTIONS

C.M.Y. and B.L.D. wrote the paper.

FUNDING

This research received no external funding.

ACKNOWLEDGMENTS

This work was supported by: the Burroughs Wellcome Fund Postdoctoral Enrichment Program and the University of Pennsylvania Academic Diversity Fellowship and the Children's Hospital of Philadelphia Research Institute.

Conflicts of Interest

B.L.D. is a founder of Spark Therapeutics and Talee Bio, and serves on the Scientific Advisory Board of Sarepta Therapeutics, Prevail Therapeutics, Intellia Therapeutics and Homology Medicines.

REFERENCES

1. Oostra, B.A.; Willemsen, R. Fmr1: A gene with three faces. *Biochim. Biophys. Acta* 2009, *1790*, 467–477.
2. Hukema, R.K.; Buijsen, R.A.; Schonewille, M.; Raske, C.; Severijnen, L.A.; Nieuwenhuizen-Bakker, I.; Verhagen, R.F.; van Dessel, L.; Maas, A.; Charlet-Berguerand, N.; et al. Reversibility of neuropathology and motor deficits in an inducible mouse model for fxtas. *Hum. Mol. Genet.* 2015, *24*, 4948–4957.
3. Sellier, C.; Buijsen, R.A.M.; He, F.; Natla, S.; Jung, L.; Tropel, P.; Gaucherot, A.; Jacobs, H.; Meziane, H.; Vincent, A.; et al. Translation of expanded cgg repeats into fmrpolyg is pathogenic and may contribute to fragile x tremor ataxia syndrome. *Neuron* 2017, *93*, 331–347.
4. Todd, P.K.; Oh, S.Y.; Krans, A.; He, F.; Sellier, C.; Frazer, M.; Renoux, A.J.; Chen, K.C.; Scaglione, K.M.; Basrur, V.; et al. Cgg repeat-associated translation mediates neurodegeneration in fragile x tremor ataxia syndrome. *Neuron* 2013, *78*, 440–455.
5. Muslimov, I.A.; Patel, M.V.; Rose, A.; Tiedge, H. Spatial code recognition in neuronal rna targeting: Role of rna-hnrnp a2 interactions. *J. Cell Biol.* 2011, *194*, 441–457.
6. Morriss, G.R.; Cooper, T.A. Protein sequestration as a normal function of long noncoding rnas and a pathogenic mechanism of rnas containing nucleotide repeat expansions. *Hum. Genet.* 2017, *136*, 1247–1263.
7. Sellier, C.; Rau, F.; Liu, Y.; Tassone, F.; Hukema, R.K.; Gattoni, R.; Schneider, A.; Richard, S.; Willemsen, R.; Elliott, D.J.; et al. Sam68 sequestration and partial loss of function are associated with splicing alterations in fxtas patients. *EMBO J.* 2010, *29*, 1248–1261.
8. Hagerman, P.J.; Hagerman, R.J. Fragile x-associated tremor/ataxia syndrome. *Ann. N. Y. Acad. Sci.* 2015, *1338*, 58–70.
9. Hagerman, R.J.; Berry-Kravis, E.; Hazlett, H.C.; Bailey, D.B., Jr.; Moine, H.; Kooy, R.F.; Tassone, F.; Gantois, I.; Sonenberg, N.; Mandel, J.L.; et al. Fragile x syndrome. *Nat. Rev. Dis. Primers* 2017, *3*, 17065.
10. Bak, R.O.; Gomez-Ospina, N.; Porteus, M.H. Gene editing on center stage. *Trends Genet.* 2018, *34*, 600–611.
11. Cong, L.; Ran, F.A.; Cox, D.; Lin, S.; Barretto, R.; Habib, N.; Hsu, P.D.; Wu, X.; Jiang, W.; Marraffini, L.A.; et al. Multiplex genome engineering using crispr/cas systems. *Science* 2013, *339*, 819–823.

12. Jinek, M.; East, A.; Cheng, A.; Lin, S.; Ma, E.; Doudna, J. Rna-programmed genome editing in human cells. *Elife* 2013, *2*, e00471.
13. Mali, P.; Yang, L.; Esvelt, K.M.; Aach, J.; Guell, M.; DiCarlo, J.E.; Norville, J.E.; Church, G.M. Rna-guided human genome engineering via cas9. *Science* 2013, *339*, 823–826.
14. Adli, M. The crispr tool kit for genome editing and beyond. *Nat. Commun.* 2018, *9*, 1911.
15. Hsu, P.D.; Lander, E.S.; Zhang, F. Development and applications of crispr-cas9 for genome engineering. *Cell* 2014, *157*, 1262–1278.
16. Barrangou, R.; Fremaux, C.; Deveau, H.; Richards, M.; Boyaval, P.; Moineau, S.; Romero, D.A.; Horvath, P. Crispr provides acquired resistance against viruses in prokaryotes. *Science* 2007, *315*, 1709–1712.
17. Jiang, F.; Doudna, J.A. Crispr-cas9 structures and mechanisms. *Annu. Rev. Biophys.* 2017, *46*, 505–529.
18. Larson, M.H.; Gilbert, L.A.; Wang, X.; Lim, W.A.; Weissman, J.S.; Qi, L.S. Crispr interference (crispri) for sequence-specific control of gene expression. *Nat. Protoc.* 2013, *8*, 2180–2196.
19. Qi, L.S.; Larson, M.H.; Gilbert, L.A.; Doudna, J.A.; Weissman, J.S.; Arkin, A.P.; Lim, W.A. Repurposing crispr as an rna-guided platform for sequence-specific control of gene expression. *Cell* 2013, *152*, 1173–1183.
20. Gilbert, L.A.; Larson, M.H.; Morsut, L.; Liu, Z.; Brar, G.A.; Torres, S.E.; Stern-Ginossar, N.; Brandman, O.; Whitehead, E.H.; Doudna, J.A.; et al. Crispr-mediated modular rna-guided regulation of transcription in eukaryotes. *Cell* 2013, *154*, 442–451.
21. Shen, B.; Zhang, W.; Zhang, J.; Zhou, J.; Wang, J.; Chen, L.; Wang, L.; Hodgkins, A.; Iyer, V.; Huang, X.; et al. Efficient genome modification by crispr-cas9 nickase with minimal off-target effects. *Nat. Methods* 2014, *11*, 399–402.
22. Tanenbaum, M.E.; Gilbert, L.A.; Qi, L.S.; Weissman, J.S.; Vale, R.D. A protein-tagging system for signal amplification in gene expression and fluorescence imaging. *Cell* 2014, *159*, 635–646.
23. Konermann, S.; Brigham, M.D.; Trevino, A.E.; Joung, J.; Abudayyeh, O.O.; Barcena, C.; Hsu, P.D.; Habib, N.; Gootenberg, J.S.; Nishimasu, H.; et al. Genome-scale transcriptional activation by an engineered crispr-cas9 complex. *Nature* 2015, *517*, 583–588.

24. Hilton, I.B.; D'Ippolito, A.M.; Vockley, C.M.; Thakore, P.I.; Crawford, G.E.; Reddy, T.E.; Gersbach, C.A. Epigenome editing by a crispr-cas9-based acetyltransferase activates genes from promoters and enhancers. *Nat. Biotechnol.* 2015, *33*, 510–517.
25. Kim, Y.B.; Komor, A.C.; Levy, J.M.; Packer, M.S.; Zhao, K.T.; Liu, D.R. Increasing the genome-targeting scope and precision of base editing with engineered cas9-cytidine deaminase fusions. *Nat. Biotechnol.* 2017, *35*, 371–376.
26. Komor, A.C.; Kim, Y.B.; Packer, M.S.; Zuris, J.A.; Liu, D.R. Programmable editing of a target base in genomic DNA without double-stranded DNA cleavage. *Nature* 2016, *533*, 420–424.
27. Vakulskas, C.A.; Dever, D.P.; Rettig, G.R.; Turk, R.; Jacobi, A.M.; Collingwood, M.A.; Bode, N.M.; McNeill, M.S.; Yan, S.; Camarena, J.; et al. A high-fidelity cas9 mutant delivered as a ribonucleoprotein complex enables efficient gene editing in human hematopoietic stem and progenitor cells. *Nat. Med.* 2018, *24*, 1216–1224.
28. Slaymaker, I.M.; Gao, L.; Zetsche, B.; Scott, D.A.; Yan, W.X.; Zhang, F. Rationally engineered cas9 nucleases with improved specificity. *Science* 2016, *351*, 84–88.
29. Kleinstiver, B.P.; Pattanayak, V.; Prew, M.S.; Tsai, S.Q.; Nguyen, N.T.; Zheng, Z.; Joung, J.K. High-fidelity crispr-cas9 nucleases with no detectable genome-wide off-target effects. *Nature* 2016, *529*, 490–495.
30. Ran, F.A.; Cong, L.; Yan, W.X.; Scott, D.A.; Gootenberg, J.S.; Kriz, A.J.; Zetsche, B.; Shalem, O.; Wu, X.; Makarova, K.S.; et al. In vivo genome editing using staphylococcus aureus cas9. *Nature* 2015, *520*, 186–191.
31. Kleinstiver, B.P.; Prew, M.S.; Tsai, S.Q.; Topkar, V.V.; Nguyen, N.T.; Zheng, Z.; Gonzales, A.P.; Li, Z.; Peterson, R.T.; Yeh, J.R.; et al. Engineered crispr-cas9 nucleases with altered pam specificities. *Nature* 2015, *523*, 481–485.
32. Cox, D.B.T.; Gootenberg, J.S.; Abudayyeh, O.O.; Franklin, B.; Kellner, M.J.; Joung, J.; Zhang, F. Rna editing with crispr-cas13. *Science* 2017, *358*, 1019–1027.
33. Park, C.Y.; Halevy, T.; Lee, D.R.; Sung, J.J.; Lee, J.S.; Yanuka, O.; Benvenisty, N.; Kim, D.W. Reversion of fmr1 methylation and silencing by editing the triplet repeats in fragile x ipsc-derived neurons. *Cell Rep.* 2015, *13*, 234–241.

34. Xie, N.; Gong, H.; Suhl, J.A.; Chopra, P.; Wang, T.; Warren, S.T. Reactivation of fmr1 by crispr/cas9-mediated deletion of the expanded cgg-repeat of the fragile x chromosome. *PLoS ONE* 2016, *11*, e0165499.
35. Liu, X.S.; Wu, H.; Krzisch, M.; Wu, X.; Graef, J.; Muffat, J.; Hnisz, D.; Li, C.H.; Yuan, B.; Xu, C.; et al. Rescue of fragile x syndrome neurons by DNA methylation editing of the fmr1 gene. *Cell* 2018, *172*, 979–992 e976.
36. Haenfler, J.M.; Skariah, G.; Rodriguez, C.M.; Monteiro da Rocha, A.; Parent, J.M.; Smith, G.D.; Todd, P.K. Targeted reactivation of fmr1 transcription in fragile x syndrome embryonic stem cells. *Front. Mol. Neurosci.* 2018, *11*, 282.
37. Peprah, E.; He, W.; Allen, E.; Oliver, T.; Boyne, A.; Sherman, S.L. Examination of fmr1 transcript and protein levels among 74 premutation carriers. *J. Hum. Genet.* 2010, *55*, 66–68.
38. Loesch, D.Z.; Sherwell, S.; Kinsella, G.; Tassone, F.; Taylor, A.; Amor, D.; Sung, S.; Evans, A. Fragile x-associated tremor/ataxia phenotype in a male carrier of unmethylated full mutation in the fmr1 gene. *Clin. Genet.* 2012, *82*, 88–92.
39. Santa Maria, L.; Pugin, A.; Alliende, M.A.; Aliaga, S.; Curotto, B.; Aravena, T.; Tang, H.T.; Mendoza-Morales, G.; Hagerman, R.; Tassone, F. Fxtas in an unmethylated mosaic male with fragile x syndrome from chile. *Clin. Genet.* 2014, *86*, 378–382.
40. Lee, B.; Lee, K.; Panda, S.; Gonzales-Rojas, R.; Chong, A.; Bugay, V.; Park, H.M.; Brenner, R.; Murthy, N.; Lee, H.Y. Nanoparticle delivery of crispr into the brain rescues a mouse model of fragile x syndrome from exaggerated repetitive behaviours. *Nat. Biomed. Eng.* 2018, *2*, 497–507.
41. Ferreira, G.K.; Cardoso, E.; Vuolo, F.S.; Galant, L.S.; Michels, M.; Goncalves, C.L.; Rezin, G.T.; Dal-Pizzol, F.; Benavides, R.; Alonso-Nunez, G.; et al. Effect of acute and long-term administration of gold nanoparticles on biochemical parameters in rat brain. *Mater. Sci. Eng. C Mater. Biol. Appl.* 2017, *79*, 748–755.
42. Finn, J.D.; Smith, A.R.; Patel, M.C.; Shaw, L.; Youniss, M.R.; van Heteren, J.; Dirstine, T.; Ciullo, C.; Lescarbeau, R.; Seitzer, J.; et al. A single administration of crispr/cas9 lipid nanoparticles achieves robust and persistent in vivo genome editing. *Cell Rep.* 2018, *22*, 2227–2235.
43. Simpson, B.P.; Davidson, B.L. Crispr-cas gene editing for neurological

disease. In *Nervous System Drug Delivery: Principles and Practice*; Lonser, R., Sarntinoranont, M., Bankiewicz, K., Eds.; Academic Press/ Elsevier: San Diego, CA, USA, 2019; in press. [Google Scholar]

44. Monteys, A.M.; Ebanks, S.A.; Keiser, M.S.; Davidson, B.L. Crispr/ cas9 editing of the mutant huntingtin allele in vitro and in vivo. *Mol. Ther* 2017, *25*, 12–23.
45. Ouellet, D.L.; Cherif, K.; Rousseau, J.; Tremblay, J.P. Deletion of the gaa repeats from the human frataxin gene using the crispr-cas9 system in yg8r-derived cells and mouse models of friedreich ataxia. *Gene Ther.* 2017, *24*, 265–274.
46. Tsai, S.Q.; Wyvekens, N.; Khayter, C.; Foden, J.A.; Thapar, V.; Reyon, D.; Goodwin, M.J.; Aryee, M.J.; Joung, J.K. Dimeric crispr rna-guided foki nucleases for highly specific genome editing. *Nat. Biotechnol.* 2014, *32*, 569–576.
47. Dai, X.; Chen, X.; Fang, Q.; Li, J.; Bai, Z. Inducible crispr genome-editing tool: Classifications and future trends. *Crit. Rev. Biotechnol.* 2018, *38*, 573–586.
48. Cwetsch, A.W.; Pinto, B.; Savardi, A.; Cancedda, L. In vivo methods for acute modulation of gene expression in the central nervous system. *Prog. Neurobiol.* 2018, *168*, 69–85.
49. Naso, M.F.; Tomkowicz, B.; Perry, W.L., 3rd; Strohl, W.R. Adeno-associated virus (aav) as a vector for gene therapy. *BioDrugs* 2017, *31*, 317–334.
50. Nelson, C.E.; Hakim, C.H.; Ousterout, D.G.; Thakore, P.I.; Moreb, E.A.; Castellanos Rivera, R.M.; Madhavan, S.; Pan, X.; Ran, F.A.; Yan, W.X.; et al. In vivo genome editing improves muscle function in a mouse model of duchenne muscular dystrophy. *Science* 2016, *351*, 403–407.
51. Tsai, S.Q.; Nguyen, N.T.; Malagon-Lopez, J.; Topkar, V.V.; Aryee, M.J.; Joung, J.K. Circle-seq: A highly sensitive in vitro screen for genome-wide crispr-cas9 nuclease off-targets. *Nat. Methods* 2017, *14*, 607–614.
52. Friedland, A.E.; Baral, R.; Singhal, P.; Loveluck, K.; Shen, S.; Sanchez, M.; Marco, E.; Gotta, G.M.; Maeder, M.L.; Kennedy, E.M.; et al. Characterization of staphylococcus aureus cas9: A smaller cas9 for all-in-one adeno-associated virus delivery and paired nickase applications. *Genome Biol.* 2015, *16*, 257.

53. Gholizadeh, S.; Arsenault, J.; Xuan, I.C.; Pacey, L.K.; Hampson, D.R. Reducedphenotypicseverityfollowingadeno-associatedvirus-mediated fmr1 gene delivery in fragile x mice. *Neuropsychopharmacology* 2014, *39*, 3100–3111.

54. Aschauer, D.F.; Kreuz, S.; Rumpel, S. Analysis of transduction efficiency, tropism and axonal transport of aav serotypes 1, 2, 5, 6, 8 and 9 in the mouse brain. *PLoS ONE* 2013, *8*, e76310.

55. Markakis, E.A.; Vives, K.P.; Bober, J.; Leichtle, S.; Leranth, C.; Beecham, J.; Elsworth, J.D.; Roth, R.H.; Samulski, R.J.; Redmond, D.E., Jr. Comparative transduction efficiency of aav vector serotypes 1-6 in the substantia nigra and striatum of the primate brain. *Mol. Ther.* 2010, *18*, 588–593.

56. Bartel, M.A.; Weinstein, J.R.; Schaffer, D.V. Directed evolution of novel adeno-associated viruses for therapeutic gene delivery. *Gene Ther.* 2012, *19*, 694–700.

57. Rincon, M.Y.; de Vin, F.; Duque, S.I.; Fripont, S.; Castaldo, S.A.; Bouhuijzen-Wenger, J.; Holt, M.G. Widespread transduction of astrocytes and neurons in the mouse central nervous system after systemic delivery of a self-complementary aav-php.B vector. *Gene Ther.* 2018, *25*, 83–92.

58. Jiang, C.; Mei, M.; Li, B.; Zhu, X.; Zu, W.; Tian, Y.; Wang, Q.; Guo, Y.; Dong, Y.; Tan, X. A non-viral crispr/cas9 delivery system for therapeutically targeting hbv DNA and pcsk9 in vivo. *Cell Res.* 2017, *27*, 440–443.

59. Miller, J.B.; Zhang, S.; Kos, P.; Xiong, H.; Zhou, K.; Perelman, S.S.; Zhu, H.; Siegwart, D.J. Non-viral crispr/cas gene editing in vitro and in vivo enabled by synthetic nanoparticle co-delivery of cas9 mrna and sgrna. *Angew. Chem. Int. Ed. Engl.* 2017, *56*, 1059–1063.

60. Kim, S.; Kim, D.; Cho, S.W.; Kim, J.; Kim, J.S. Highly efficient rna-guided genome editing in human cells via delivery of purified cas9 ribonucleoproteins. *Genome Res.* 2014, *24*, 1012–1019.

61. Brown, S.S.; Stanfield, A.C. Fragile x premutation carriers: A systematic review of neuroimaging findings. *J. Neurol. Sci.* 2015, *352*, 19–28.

62. Romano, D.; Nicolau, M.; Quintin, E.M.; Mazaika, P.K.; Lightbody, A.A.; Cody Hazlett, H.; Piven, J.; Carlsson, G.; Reiss, A.L. Topological methods reveal high and low functioning neuro-phenotypes within fragile x syndrome. *Hum. Brain Mapp.* 2014, *35*, 4904–4915.

63. Greco, C.M.; Navarro, C.S.; Hunsaker, M.R.; Maezawa, I.; Shuler, J.F.; Tassone, F.; Delany, M.; Au, J.W.; Berman, R.F.; Jin, L.W.; et al. Neuropathologic features in the hippocampus and cerebellum of three older men with fragile x syndrome. *Mol. Autism* 2011, *2*, 2.
64. Ramanathan, S.; Archunan, G.; Sivakumar, M.; Tamil Selvan, S.; Fred, A.L.; Kumar, S.; Gulyas, B.; Padmanabhan, P. Theranostic applications of nanoparticles in neurodegenerative disorders. *Int. J. Nanomed.* 2018, *13*, 5561–5576.
65. Winarni, T.I.; Schneider, A.; Borodyanskara, M.; Hagerman, R.J. Early intervention combined with targeted treatment promotes cognitive and behavioral improvements in young children with fragile x syndrome. *Case Rep. Genet.* 2012, *2012*, 280813.
66. Tsai, S.Q.; Zheng, Z.; Nguyen, N.T.; Liebers, M.; Topkar, V.V.; Thapar, V.; Wyvekens, N.; Khayter, C.; Iafrate, A.J.; Le, L.P.; et al. Guide-seq enables genome-wide profiling of off-target cleavage by crispr-cas nucleases. *Nat. Biotechnol.* 2015, *33*, 187–197.
67. Frock, R.L.; Hu, J.; Meyers, R.M.; Ho, Y.J.; Kii, E.; Alt, F.W. Genome-wide detection of DNA double-stranded breaks induced by engineered nucleases. *Nat. Biotechnol.* 2015, *33*, 179–186.
68. Wang, X.; Wang, Y.; Wu, X.; Wang, J.; Wang, Y.; Qiu, Z.; Chang, T.; Huang, H.; Lin, R.J.; Yee, J.K. Unbiased detection of off-target cleavage by crispr-cas9 and talens using integrase-defective lentiviral vectors. *Nat. Biotechnol* 2015, *33*, 175–178.
69. Kim, D.; Bae, S.; Park, J.; Kim, E.; Kim, S.; Yu, H.R.; Hwang, J.; Kim, J.I.; Kim, J.S. Digenome-seq: Genome-wide profiling of crispr-cas9 off-target effects in human cells. *Nat. Methods* 2015, *12*, 237–243.
70. Kim, D.; Kim, S.; Kim, S.; Park, J.; Kim, J.S. Genome-wide target specificities of crispr-cas9 nucleases revealed by multiplex digenome-seq. *Genome Res.* 2016, *26*, 406–415.
71. Cameron, P.; Fuller, C.K.; Donohoue, P.D.; Jones, B.N.; Thompson, M.S.; Carter, M.M.; Gradia, S.; Vidal, B.; Garner, E.; Slorach, E.M.; et al. Mapping the genomic landscape of crispr-cas9 cleavage. *Nat. Methods* 2017, *14*, 600–606.
72. Crosetto, N.; Mitra, A.; Silva, M.J.; Bienko, M.; Dojer, N.; Wang, Q.; Karaca, E.; Chiarle, R.; Skrzypczak, M.; Ginalski, K.; et al. Nucleotide-resolution DNA double-strand break mapping by next-generation sequencing. *Nat. Methods* 2013, *10*, 361–365.

73. Yan, W.X.; Mirzazadeh, R.; Garnerone, S.; Scott, D.; Schneider, M.W.; Kallas, T.; Custodio, J.; Wernersson, E.; Li, Y.; Gao, L.; et al. Bliss is a versatile and quantitative method for genome-wide profiling of DNA double-strand breaks. *Nat. Commun.* 2017, *8*, 15058.

74. Zischewski, J.; Fischer, R.; Bortesi, L. Detection of on-target and off-target mutations generated by crispr/cas9 and other sequence-specific nucleases. *Biotechnol. Adv.* 2017, *35*, 95–104.

75. Singh, R.; Kuscu, C.; Quinlan, A.; Qi, Y.; Adli, M. Cas9-chromatin binding information enables more accurate crispr off-target prediction. *Nucleic Acids Res.* 2015, *43*, e118.

76. Moreno-Mateos, M.A.; Vejnar, C.E.; Beaudoin, J.D.; Fernandez, J.P.; Mis, E.K.; Khokha, M.K.; Giraldez, A.J. Crisprscan: Designing highly efficient sgrnas for crispr-cas9 targeting in vivo. *Nat. Methods* 2015, *12*, 982–988.

77. Cui, Y.; Xu, J.; Cheng, M.; Liao, X.; Peng, S. Review of crispr/cas9 sgrna design tools. *Interdiscip Sci.* 2018, *10*, 455–465.

78. Haeussler, M.; Schonig, K.; Eckert, H.; Eschstruth, A.; Mianne, J.; Renaud, J.B.; Schneider-Maunoury, S.; Shkumatava, A.; Teboul, L.; Kent, J.; et al. Evaluation of off-target and on-target scoring algorithms and integration into the guide rna selection tool crispor. *Genome Biol.* 2016, *17*, 148.

79. Hendel, A.; Bak, R.O.; Clark, J.T.; Kennedy, A.B.; Ryan, D.E.; Roy, S.; Steinfeld, I.; Lunstad, B.D.; Kaiser, R.J.; Wilkens, A.B.; et al. Chemically modified guide rnas enhance crispr-cas genome editing in human primary cells. *Nat. Biotechnol.* 2015, *33*, 985–989.

80. Merienne, N.; Vachey, G.; de Longprez, L.; Meunier, C.; Zimmer, V.; Perriard, G.; Canales, M.; Mathias, A.; Herrgott, L.; Beltraminelli, T.; et al. The self-inactivating kamicas9 system for the editing of cns disease genes. *Cell Rep.* 2017, *20*, 2980–2991.

81. Zelensky, A.N.; Schimmel, J.; Kool, H.; Kanaar, R.; Tijsterman, M. Inactivation of pol theta and c-nhej eliminates off-target integration of exogenous DNA. *Nat. Commun.* 2017, *8*, 66.

82. Hendel, A.; Kildebeck, E.J.; Fine, E.J.; Clark, J.; Punjya, N.; Sebastiano, V.; Bao, G.; Porteus, M.H. Quantifying genome-editing outcomes at endogenous loci with smrt sequencing. *Cell Rep.* 2014, *7*, 293–305.

83. Torres, R.; Martin, M.C.; Garcia, A.; Cigudosa, J.C.; Ramirez, J.C.; Rodriguez-Perales, S. Engineering human tumour-associated

chromosomal translocations with the rna-guided crispr-cas9 system. *Nat. Commun.* 2014, *5*, 3964.

84. Hendel, A.; Fine, E.J.; Bao, G.; Porteus, M.H. Quantifying on- and off-target genome editing. *Trends Biotechnol.* 2015, *33*, 132–140.
85. Pretto, D.; Yrigollen, C.M.; Tang, H.T.; Williamson, J.; Espinal, G.; Iwahashi, C.K.; Durbin-Johnson, B.; Hagerman, R.J.; Hagerman, P.J.; Tassone, F. Clinical and molecular implications of mosaicism in fmr1 full mutations. *Front. Genet.* 2014, *5*, 318.
86. Pretto, D.I.; Mendoza-Morales, G.; Lo, J.; Cao, R.; Hadd, A.; Latham, G.J.; Durbin-Johnson, B.; Hagerman, R.; Tassone, F. Cgg allele size somatic mosaicism and methylation in fmr1 premutation alleles. *J. Med. Genet.* 2014, *51*, 309–318.
87. Lee, H.; Kim, J.S. Unexpected crispr on-target effects. *Nat. Biotechnol.* 2018, *36*, 703–704.
88. Brokowski, C.; Adli, M. Crispr ethics: Moral considerations for applications of a powerful tool. *J. Mol. Biol.* 2019, *431*, 88–101.

11

Correction of Monogenic and Common Retinal Disorders with Gene Therapy

Jesse D. Sengillo [1,2,3,], Sally Justus [1,2,], Thiago Cabral [1,2,4,5] and Stephen H. Tsang [1,2,6]

[1]Jonas Children's Vision Care, and Bernard & Shirlee Brown Glaucoma Laboratory, Department of Ophthalmology, Columbia University Medical Center, New York, NY, 10032 USA
[2]Edward S. Harkness Eye Institute, New York-Presbyterian Hospital, New York, NY, 10032 USA
[3]State University of New York Downstate Medical Center, Brooklyn, NY, 11203 USA
[4]Department of Ophthalmology, Federal University of Espírito Santo, Vitória, 29075-910 Brazil
[5]Department of Ophthalmology, Federal University of São Paulo, São Paulo, 04021-001 Brazil
[6]Department of Pathology & Cell Biology, Institute of Human Nutrition, College of Physicians and Surgeons, Columbia University, New York, NY, 10032 USA

Citation: Sengillo, J.D.; Justus, S.; Cabral, T.; Tsang, S.H. Correction of Monogenic and Common Retinal Disorders with Gene Therapy. Genes 2017, 8, 53. https://doi.org/10.3390/genes8020053.

ABSTRACT

The past decade has seen major advances in gene-based therapies, many of which show promise for translation to human disease. At the forefront of research in this field is ocular disease, as the eye lends itself to gene-based interventions due to its accessibility, relatively immune-privileged status, and ability to be non-invasively monitored. A landmark study in 2001 demonstrating successful gene therapy in a large-animal model for Leber congenital amaurosis set the stage for translation of these strategies from the bench to the bedside. Multiple clinical trials have since initiated for various retinal diseases, and further improvements in gene therapy techniques have engendered optimism for alleviating inherited blinding disorders. This article provides an overview of gene-based strategies for retinal disease, current clinical trials that engage these strategies, and the latest techniques in genome engineering, which could serve as the next frontline of therapeutic interventions.

KEYWORDS: gene therapy, genome engineering, CRISPR, retinal degeneration, inherited retinal disease

INTRODUCTION

Gene therapy holds the potential to revolutionize physicians' ability to apply precision medicine. Depending on the inheritance pattern of a given genetic defect, gene therapy may involve the delivery of genetic material that either silences a pathogenic mutation or supplements patients with a wild type copy of a mutated gene. Enthusiasm for such strategies grew quickly, until the death of a clinical trial patient following a gene therapy-based intervention for ornithine transcarbamylase (OTC) deficiency [1], in addition to other adverse outcomes following gene therapy [2], prompted careful reevaluation of the ethical and safety considerations involved with such strategies. Vector choice and delivery strategies came under close scrutiny, and additional testing in animal models was emphasized as an essential pre-clinical safety measure.

In 1996, a team that included Drs. Bennett, Maguire and Gouras presented the first study to use gene transfer to rescue retinal degeneration in a mouse model of retinitis pigmentosa (RP) [3]. Subretinal injection of a recombinant adenovirus carrying wild type cDNA of the phosphodiesterase β subunit (*βPDE*) effectively delayed photoreceptor degeneration in the *rd* mouse model, which harbors biallelic mutations in the β subunit of rod cGMP-phosphodiesterase. Later, in 2001, injection of an adeno-associated virus (AAV) delivering subretinal *RPE65* to an *RPE65*$^{-/-}$ canine model was reported to delay disease progression and even restore visual function [4]. These studies provided an initial preclinical proof-of-principle for such strategies to be translated to clinical trials. Multiple groups subsequently launched clinical trials to deliver subretinal *RPE65* via an AAV vector to patients with Leber congenital amaurosis [5,6,7], reinvigorating research in gene therapy.

The eye has been at the forefront of developments in gene-based technology because it possesses several unique anatomical and physiological properties. The eye and retina are easily accessible to physicians, and in clinical trials, the untreated eye can serve as a control if the condition progresses symmetrically [8,9]. The optical transparency of the eye allows in vivo, longitudinal, and non-invasive monitoring of disease progression and treatment response [10]. Moreover, the retina can be viewed with impressively high resolution, on the scale of microns, with spectral domain optical coherence tomography (SD-OCT), enabling visualization and delineation of the individual layers of the retina [10,11]. Retinal imaging continues to improve, and OCT imaging can now be acquired *en face*, with images oriented parallel to the plane of the retina as opposed to perpendicular, as in the traditional SD-OCT system [11,12,13,14,15,16]. These improvements can lead to an even deeper understanding of previously well-characterized diseases and open new doors for less-studied conditions. OCT, as well as other imaging modalities such as fundus autofluorescence, have allowed investigators to characterize retinal disease and identify potential outcome measures for disease progression and treatment efficacy, supplementing traditional tests of visual function [10,17,18,19,20,21,22,23, 24,25,26,27]. Another factor that makes the eye more approachable to gene therapy is its relatively immune-privileged status. This is maintained by a lack of lymphatics, immunomodulatory factors in the vitreous humor, and antigen-presenting cells, hence leading to a significantly weaker immune response to novel antigens [28].

The advantages the eye confers for studying gene-based therapies has accelerated research initiatives in this area, and today, there are multiple clinical trials utilizing such strategies in the treatment of retinal disease (Table 1). This article will review some of the parameters that must be considered for gene delivery in the eye, provide examples of applications of these techniques in current clinical trials, and present an overview of some of the latest advances in genome surgery and their future applications to treating inherited retinal dystrophies.

Table 1. Gene therapy trials for retinal diseases, with relevant articles cited.

Disease	NCT ID & Sponsor	Phase	Intervention	Citations
Leber Congenital Amaurosis	NCT00999609 NCT01208389 NCT00516477 Spark Therapeutics (Philadelphia, PA, USA)	III I, II I	Subretinal AAV2-hRPE65v2	[7,29, 30, 31, 32,33]
	NCT00643747 University College, London (London, UK)	I, II	Subretinal rAAV2/2.hR-PE65p.hRPE65 (tgAAG76)	[5,34, 35, 36]
	NCT00749957 Applied Genetic Technologies Corp. (Alachua, FL, USA)	I, II	Subretinal rAAV2-CB-hRPE65	[37]
	NCT01496040 Nantes University Hospital (Nantes, France)	I, II	Subretinal rAAV2/4.hRPE65	
	NCT02781480 University College, London (London, UK)	I	Subretinal AAV2/5 OPTIRPE65	[38]
	NCT00481546 University of Pennsylvania (Philadelphia, PA, USA)	I	Subretinal rAAV2-CBSB-hRPE65	[39, 40, 41, 42]
Choroideremia	NCT02077361 Ian M. MacDonald	I, II	Subretinal rAAV2.REP1	
	NCT02553135 Byron Lam	II	Subretinal AAV2-REP1	
	NCT01461213 NightstaRx (London, UK)	I, II	Subretinal AAV2.REP1	[43, 44, 45]
	NCT02671539 STZ Eyetrial (Tübingen, Germany)	II	Subretinal rAAV2.REP1	
	NCT02341807 Spark Therapeutics (Philadelphia, PA, USA)	I, II	Subretinal AAV2-hCHM	

Achromatopsia	NCT02599922 Applied Genetic Technologies Corp. (Alachua, FL, USA)	I, II	Subretinal rAAV2tYF-PR1.7-hCNGB3	[46,47,48]
	NCT02610582 STZ eyetrial (Tübingen, Germany)	I, II	Subretinal rAAV.hCNGA3	
X-linked Retinoschisis	NCT02317887 National Eye Institute (Bethesda, MD, USA)	I, II	Intravitreal AAV8-scRS/IRB-PhRS	[49,50,51,52]
	NCT02416622 Applied Genetic Technologies Corp. (Alachua, FL, USA)	I, II	Intravitreal rAAV2tYF-CB-hRS1	[53,54]
Leber Hereditary Optic Neuropathy	NCT02652767 NCT02652780 GenSight Biologics (Paris, France)	III III	Intravitreal GS010; Sham intravitreal injection	[55]
	NCT02161380 John Guy	I	Intravitreal scAAV2-P1ND4v2	[56,57,58]
	NCT01267422 Bin Li	--	Intravitreal rAAV2-ND4	[59]
Retinitis Pigmentosa	NCT02556736 Retrosense Therapeutics (Ann Arbor, MI, USA)	I, II	Intravitreal RST-001	[60,61]
	NCT01482195 Fowzan Alkuraya	I	Subretinal rAAV2-VMD2-hMERTK	[62,63]
Usher Syndrome	NCT01505062 NCT02065011 Sanofi (Paris, France)	I, II I, II	Subretinal EIAV-CMV-MYO7A (UshStat)	[64]
Stargardt Disease	NCT01736592 NCT01367444 Sanofi (Paris, France)	I, II I, II	Subretinal SAR422459	[65,66,67]

Neovascular Age-Related Macular Degeneration	NCT01494805 Lions Eye Institute (Perth, WA, Australia)	I, II	Subretinal rAAV. sFlt-1; Control (ranibizumab alone)	[68,69,70]
	NCT01024998 Genzyme, a Sanofi company (Cambridge, MA, USA)	I	Intravitreal AAV2-sFLT01	[71,72,73]
	NCT01301443 NCT01678872 Oxford BioMedica (Oxford, UK)	I I	Subretinal RetinoStat	[73,74,75]

VECTOR CHOICE

While there exist a variety of techniques for delivering genes into target cells, the most popular options for use in patients include viral-mediated strategies typically involving one of two viruses: AAVs or lentiviruses. AAV, a single-stranded DNA vector belonging to the family *Parvoviridae*, remains the vector of choice for most ocular gene therapy applications due to several advantageous properties: AAVs elicit less immunogenic reactions than that observed from administration of adenoviral vectors; they exhibit minimal self-replication; and they can last for up to six years after a single administration [76,77]. Furthermore, a recombinant AAV (rAAV) capsid can be selected or altered to suit different experimental needs and to optimize transfection [78,79,80]. While each of these characteristics are advantageous for gene-based therapy, some challenges remain.

A known limitation of AAVs is their carrying capacity, which is approximately 4.5 kb, making gene-based delivery difficult for longer genes [81]. Researchers studying delivery of genes larger than 5 kb have circumvented this issue by utilizing dual AAV strategies. Yan et al. were the first to employ a dual AAV strategy, effectively delivering the erythropoietin genomic locus to mouse muscle tissue [82]. A similar study reported that this dual AAV technique could be applied to the retina [83], and subsequent studies successfully delivered the *MYO7A* gene in vitro and in vivo. Encouragingly, co-infected cells showed recombination of the two halves, leading to a full-length gene product [84,85]. Another strategy for delivering large genes is with lentiviral vectors. Lentiviruses are retroviruses that can carry up to 8 kb of transgene material [86]. Large genes such as *MYO7A* (Usher syndrome) and *ABCA4* (Stargardt disease) have been approached successfully with lentiviruses in pre-clinical studies [64,67] and are currently being tested in clinical trials (Table 1).

VECTOR DELIVERY

Subretinal Injection

Most gene therapy trials for retinal degenerations involve subretinal injection of the vector-containing therapeutic (Table 1). To accomplish this, a transient retinal detachment is made intraoperatively to create a subretinal space to place the injection. This detachment resolves spontaneously [9]. The viral vector subsequently occupies the subretinal space and infects target cells of the neurosensory retina or retinal pigment epithelium (RPE). Subretinal injections are particularly localized, which has been postulated to affect transfection and may explain the variable patterns of functional vision improvement that were observed in the clinical trials of Leber congenital amaurosis (LCA) [41,87]. For example, portions of the detachment that reattach first (such as the edges) may have lower transfection [41,87,88]. Although there is no proven superior alternative to subretinal injection for most retinal degenerations at this time, patients may benefit from more homogeneously distributed and less invasive strategies [87].

Intravitreal Injection

The intravitreal AAV2 virus that is currently used predominantly infects the ganglion cell layer that lies close to the vitreous cavity [89,90]. This is in comparison to subretinal injections, in which cells in the area of the induced bleb, such as photoreceptors and RPE, are transduced [91]. For therapies targeting the inner retina, such as in Leber hereditary optic neuropathy (LHON), the intravitreal route of administration may prove successful in clinical trials (Table 1). The intravitreal approach is also less invasive, thus potentially decreasing the chance for complications [92]. This may be advantageous for retinas that are particularly at risk for retinal detachment, such as in X-linked retinoschisis (XLRS), which is caused by mutations in retinoschisin 1 (*RS1*) [93]. Furthermore, RS1 protein is normally secreted and may impart its effects on areas that are more distant to that which is transduced by a subretinal injection [93,94]. Although the eye is considered to be immune-privileged, previous experiments in mice suggest that the vitreous compartment may respond differently to AAV2 compared to the subretinal space [95,96]. Li et al. found that intravitreal injection of AAV2-

CBA-PEDF resulted in a humoral immune response which prevented therapeutic effects upon readministration in a mouse model for choroidal neovascularization [96]. The same effect was not observed via subretinal administration. In LHON and XLRS pre-clinical studies, however, favorable safety and biodistribution were reported for single injections of AAV-containing gene therapy administered to the vitreous, corroborating findings in phase I trials for LHON that report no inflammatory response of the vitreous or anterior chamber (Table 1) [54,56,57,58]. Biodistribution of gene-based therapeutics delivered to the vitreous remains complex, but pre-clinical and phase 1 clinical trial results are encouraging.

Non-Viral Strategies

To date, non-viral gene delivery strategies have not been used in human retina, but they deserve mention due to their success and continued improvements in animal models. Electroporation is a strategy developed by Matsuda and Cepko [97] in which DNA is directly injected in the subretinal space without being packaged into a vector. A non-invasive, high voltage pulse is subsequently administered, leading to changes in membrane potential and permeability, and thus cellular uptake of the nucleic acid [98]. This method is especially useful for quickly testing promoter activity and the effect of protein overexpression in models of retinal degeneration [97]. Previously thought to have little therapeutic potential in humans, electroporation has advanced significantly, particularly through suprachoroidal injection of DNA followed by ab externo electroporation, which is less invasive and results in a large area of transfection [99].

Nanoparticles (NPs) have also been studied for the purpose of gene therapy delivery, mainly through polymers, liposomes, and peptide-compacted DNA, which has been extensively reviewed elsewhere [100,101]. Compared to AAV technology, NPs may be more cost-effective but have lower transfection efficiency [102,103]. Additional advantages of NPs over AAVs include a carrying capacity up to 20 kbp and a lower likelihood of complications, as there is less potential for insertional mutagenesis. This technology, like electroporation, has not been translated to clinical trials but has shown progress as an approach to ocular gene therapy [104,105,106,107,108,109,110,111,112].

UPDATE ON GENE THERAPY TRIALS FOR RETINAL DISEASE

Leber Congenital Amaurosis

LCA causes a severe early-onset retinal dystrophy with an estimated prevalence of 2–3 per 100,000 births [113]. As previously stated, gene therapy trials for *RPE65*-associated LCA are arguably among the major accomplishments in modern medicine (Table 1). RPE65 is a protein required for recycling visual pigment in the retinal pigment epithelium [114]. Mutations in *RPE65* are one cause of LCA, which typically manifests as severe visual dysfunction at birth with a pigmented retina, wandering nystagmus, and amaurotic pupils [115]. After efficacy was shown in a naturally occurring canine model, multiple clinical trials confirmed that an AAV vector delivering *RPE65* was safe and improved visual function during the follow-up of LCA patients [4,5,6,7,29,31,32,33,40,116,117,118,119]. Spark Therapeutics (Philadelphia, PA, USA) is sponsoring a phase III trial for bilateral subretinal delivery of AAV containing *RPE65*. In an October 2015 press release, Spark stated the trial met its primary endpoint of improved bilateral mobility test change score and two secondary endpoints, which included favorable full-field light sensitivity and mobility test change scores for the first treated eye compared to baseline. Visual acuity was a secondary outcome measure, but it was not met.

Choroideremia

Approximately 1 in 50,000 individuals are affected by choroideremia (CHM), an X-linked retinal dystrophy [120]. Affected males typically present with night blindness and a progressively constricting visual field, similar to retinitis pigmentosa [121]. A mutation in *CHM*, which expresses REP-1, leads to chorioretinal degeneration starting in the mid-periphery. A pale fundus is observed as the sclera illuminates through the degenerated layers of RPE and choriocapillaries. Female carriers may exhibit changes on electroretinography despite no visual complaints [120,122]. Multiple phase I/II studies to test the efficacy of rAAV-REP1 delivery are ongoing (Table 1). Initial findings reported by MacLaren et al. [43] suggested improved visual function in two of six patients, even in the context of iatrogenic retinal detachment that is required intraoperatively (NCT01461213). Increased sensitivity measured via microperimetry in the treated eyes correlated with vector dose per area of live retina. A follow-up study showed sustained improvement in visual acuity of the same two patients after 3.5 years, while the control eyes continued to progress [123]. Choroideremia patients possibly have less surgical risk when treated earlier, as one treated patient had intraoperative complications from extensive retinal detachment. Hence, treatment may be more appropriate when there is less cell loss and a more intact retinal structure [124].

Achromatopsia

Patients with achromatopsia, also known as rod monochromatism, typically experience decreased visual acuity, nystagmus, photophobia, decreased color vision, and a central visual field defect. The condition is estimated to affect 1 in 30,000 individuals [125,126]. On OCT, an optical gap can be seen, representing foveal hypoplasia [126]. This condition is inherited recessively, and multiple genes of the phototransduction pathway have been implicated. Most commonly, the cyclic nucleotide-gated ion channel B3 gene (*CNGB3*) is mutated in patients of European descent [127]. *CNGB3* encodes a subunit of the CNG channel, which closes in the presence of a light stimulus, hyperpolarizing the photoreceptor [128]. Applied Genetic Technologies Corporation (AGTC) (Alachua, FL, USA) is sponsoring a phase I/II trial delivering rAAV2tYF-PR1.7-hCNGB3 to achromatopsia patients (Table 1). Additionally, another phase I/II gene therapy clinical trial has launched in Tübingen, Germany for patients with achromatopsia caused by defects in a similar gene, *CNGA3* (Table 1).

X-Linked Retinoschisis

XLRS is a common cause of macular degeneration in young males and is estimated to affect between 1 in 5000 and 1 in 25,000 individuals [129]. Characteristic symptoms include central vision loss and mildly decreased visual acuity due to foveomacular schisis. A spoke-wheel appearance encompassing the fovea is seen on fundoscopy, and OCT reveals cysts throughout the macula, including the fovea [130]. Two phase I/II clinical trials sponsored by the National Eye Institute (NEI) and AGTC are ongoing (Table 1). Each trial is testing therapeutic delivery of the causative gene, *RS1*, utilizing a different vector and promoter. Both trials are delivering the vector via an intravitreal injection, which is likely more ideal, as the retina in XLRS patients is fragile, and the foveomacular schisis does not provide ideal anatomy for subretinal injection [94]. Retina specialists should carefully assess XLRS patients for enrollment in clinical trials, as patients' visual function may not be sufficiently reduced to meet inclusion criteria.

Leber Hereditary Optic Neuropathy

LHON is an inherited mitochondrial disease that causes degeneration of the retinal ganglion cells and is thus characterized by optic atrophy. The prevalence is estimated to be approximately 1 in 31,000 to 50,000 among Northern European populations, and the three most common genetic causes are mutations in *ND1*, *ND4*, or *ND6*, which code for a key enzyme in cellular respiration in mitochondria [131]. Current vectors of gene therapy cannot faithfully deliver therapeutic genes directly into the mitochondria. Researchers overcame this challenge by demonstrating that allotopically expressed *ND4* delivered into the vitreous via an AAV vector was effective in a mouse model of LHON [132,133,134]. This strategy has subsequently been translated to a clinical trial sponsored by GenSight Biologics (Paris, France). Initial results from the open-label phase I study reported no outstanding adverse events following administration of GS010, an AAV vector carrying normal *ND4*, in patients with *ND4*-associated LHON. At the 90-day follow-up appointment, two patients had an increase in visual acuity, and three patients were unchanged from baseline [58]. In the phase III study, which is currently recruiting subjects, the effects of intravitreal GS010 injection will be compared to a sham-control.

Retinitis Pigmentosa

RP is a genetically heterogenous disease with nearly 80 genes associated with the condition, which can be inherited in multiple fashions, namely dominantly, recessively, or in X-linked or mitochondrial forms [135]. Affecting approximately 1 in 4000 individuals, RP presents at various ages with night blindness as a first symptom. Degeneration typically starts at the mid-periphery and progresses towards the macula, corresponding to a constricting visual field [136]. Homozygous or compound heterozygous mutations in the mer receptor tyrosine kinase (*MERTK*) gene are a rare cause of autosomal recessive RP, which is being approached with gene therapy at the King Khaled Eye Specialist Hospital and King Faisal Specialist Hospital & Research Center (Riyadh, Saudi Arabia). Results of a phase I trial administering rAAV2-VMD2-hMERTK to one eye in each of six patients with *MERTK*-associated RP were recently reported [62]. Safety of the vector was confirmed, and three patients experienced a post-surgical increase in visual acuity; however, this was transient in two of the patients, whose improvements returned to baseline by the two-year follow-up appointment. Another trial for clinically-diagnosed RP patients (mutation- and gene-independent) is assessing the efficacy of RST-001, an intravitreally administered gene therapy that expresses a photoswitch, channelrhodopsin-2, which allows transfected cells of the inner retina to respond to light even after photoreceptors have degenerated (Table 1). Proof-of-principle studies in mice and marmosets supported the initiation of a phase I/II clinical trial sponsored by Retrosense Therapeutics (Ann Arbor, MI, USA) [60,61].

Usher Syndrome

Syndromic retinitis pigmentosa associated with neurosensory hearing loss, Usher syndrome (USH), occurs with a prevalence of approximately 3–6 per 100,000 individuals, although this may be an underestimate [137,138]. In the absence of genetic testing, Usher syndrome can be diagnosed clinically if irreversible hearing loss occurs in the setting of retinitis pigmentosa and a family history that suggests an autosomal recessive inheritance pattern. Usher syndrome type 1 (USH1) is characterized by profound hearing loss at birth, early-onset retinitis pigmentosa, and possibly decreased vestibular function. Mutations in myosin 7a (*MYO7A*) can cause USH1 and are currently being studied in gene therapy clinical trials. Normally, MYO7A is involved in protein and organelle transport across photoreceptor cilia [139]. Because *MYO7A* is large and difficult to package in AAV, the drug UshStat® was developed by Sanofi (Paris, France) to deliver the therapeutic gene in a lentivirus vector [64]. Phase II of this trial is projected to finish in 2018.

Stargardt Disease

Affecting approximately 1 in 10,000 individuals, Stargardt disease is an inherited macular degeneration associated with dysfunction in the ATP-binding cassette, subfamily A, member 4 (*ABCA4*) gene [140]. Bi-allelic pathogenic mutation of *ABCA4* causes accumulation of toxic byproducts in the RPE, which subsequently leads to cell degeneration [140]. Sanofi is studying the efficacy of SAR422459, a lentiviral vector providing addition of the normally functioning gene (Table 1). Completion of phase II is anticipated in 2017. In the meantime, Parker et al. have defined criteria for statistically significant changes in functional and structural measurements in Stargardt disease progression [65].

Neovascular Age-Related Macular Degeneration

Gene therapy has also been applied to acquired retinal disease (Table 1). Age-related macular degeneration (AMD) is the leading cause of visual impairment in the United States, and the neovascular form, or wet AMD (wAMD), is characterized by abnormal choroidal vessel proliferation behind the macula [68,141]. Over-secretion of vascular endothelial growth factor (VEGF) plays a key role in the pathogenesis of neovascular AMD, and the current standard of care includes the administration of intravitreal anti-VEGF inhibitors [142]. Gene therapy has a theoretical advantage, as long-lasting expression would reduce or eliminate the need for repeated injections of anti-VEGF inhibitors in patients. Lions Eye Institute (Perth, WA, Australia) is sponsoring a trial for subretinal delivery of rAAV.sFLT-1, and no serious ocular or systemic adverse events due to the vector were reported in phase I/IIa trials [68,70]. Two patients experienced mild eye and anterior chamber inflammation suspected to have been caused by rAAV.sFLT-1 administration, although the inflammation did resolve. While the cohort receiving gene therapy required fewer additional ranibizumab injections, post-hoc analysis revealed no significant differences in outcome measures between groups in the phase 2a study. This may be in part due to the low number of patients enrolled ($n = 32$), and future trials would be necessary to determine if rAAV.sFLT-1 is an efficacious adjunct to ranibizumab [70]. Similarly, Genzyme, a Sanofi company (Cambridge, MA, USA) is sponsoring a phase I safety and tolerability trial assessing intravitreal delivery of the same anti-VEGF gene therapy vector. Oxford BioMedica (Oxford, UK) has also initiated clinical trials for subretinal RetinoStat® (GEM study), a lentivirus containing angiostatic genes, angiostatin and endostatin, to prevent vessel growth and subsequent vision decline (Table 1). Campochiaro et al. reported results of the GEM study, which showed vector-containing gene therapy delivered by subretinal injection was well tolerated, and expression levels were maintained at last measurement, which was more than 2.5 years in eight subjects and more than 4 years in two subjects [75]. It remains to be seen if gene therapy will provide a viable treatment option and/or adjunct for wAMD. If efficacy is demonstrated, this may represent a paradigm shift in ophthalmic medicine, as gene therapy has traditionally been applied to monogenic diseases [70].

IMPORTANCE OF NATURAL DISEASE CHARACTERIZATION

The advent of clinical trials for inherited retinal disease has necessitated natural disease history studies for various degenerations. Although gene therapy trials use the contralateral eye as a control, it is important to study the rate of disease progression and the degree of symmetry between each eye to validate using the untreated eye as a control. Assessing disease progression rates from large cohort studies is additionally useful for identifying optimal functional and structural outcome measures [10]. Moreover, results from these studies aid investigators in deciding which patients are the best candidates for gene therapy. Although it can be difficult to complete a well-powered study given the low sample size of patients with rare degenerative diseases, several observational cohort studies are being attempted for a variety of retinal degenerations (Table 2).

Table 2. Clinical trials for natural disease characterization of retinal degenerations.

	NCT ID & Sponsor	Design (Gene of Interest)
Leber Congenital Amaurosis	NCT02575430 QLT Inc. (Vancouver, BC, Canada)	Retrospective (*LRAT* & *RPE65)*
	NCT02714816 MeiraGTx UK II Ltd. (New York, NY, USA)	Prospective (*RPE65*)
	NCT00422721 Nantes University Hospital (Nantes, France)	Prospective (*RPE65*)
Choroideremia	NCT02994368 4D Molecular Therapeutics (Emeryville, CA, USA)	Prospective (REP-1)
Achromatopsia	NCT01846052 Applied Genetic Technologies Corp. (Alachua, FL, USA)	Prospective (*CNGB3*)
Leber Hereditary Optic Neuropathy	NCT02796274 Santhera Pharmaceuticals (Liestal, Switzerland)	Retrospective (G11778A, G3460A, and T14484C mtDNA mutations)
X-linked Retinoschisis	NCT00055029 National Eye Institute (Bethesda, MD, USA)	Prospective (*RS1*)

Retinitis Pigmentosa	NCT02759952 STZ Eyetrial (Tübingen, Germany)	Prospective (*PDE6A*)
	NCT01021982 Sheba Medical Center (Ramat Gan, Israel)	Prospective
	NCT01949623 Johns Hopkins University (Baltimore, MD, USA)	Prospective biomarker study
	NCT02926092 Shire (Dublin, Ireland)	Prospective (*RHO*)
	NCT00784901 National Taiwan University Hospital (Taipei City, Taiwan)	Retrospective
Usher Syndrome	NCT00106743 National Eye Institute (Bethesda, MD, USA)	Prospective
Stargardt Disease	NCT01977846 Foundation Fighting Blindness Clinical Research Institute (Columbia, MD, USA)	Retrospective & Prospective (*ABCA4*)
	NCT02410122 John Hopkins University (Baltimore, MD, USA)	Prospective (*PROM1*)
	NCT01736293 National Eye Institute (Bethesda, MD, USA)	Prospective (*ABCA4*)

CRISPR/CAS: OPHTHALMIC GENOME SURGERY OF THE FUTURE?

Clustered regularly interspaced short palindromic repeats (CRISPR) are present in bacterial and archaeal genomes and function to provide immunity against invading viruses [143,144,145,146]. The CRISPR locus contains integrated pieces of foreign DNA and encodes for a long transcript that is ultimately processed intracellularly to become CRISPR-derived RNA (crRNA). Consequently, crRNAs contain complementary sequences to the previously invading foreign DNA, and they can mediate a highly specific, nucleic-acid based adaptive immune system by complexing with CRISPR-associated (Cas) nucleases that induce double strand breaks (DSBs) [147]. Also within the CRISPR locus of a strain of *Streptococcus pyogenes*, there exists a necessary noncoding *trans*-activating crRNA (tracrRNA), which investigators combined with crRNA to form a chimeric crRNA-tracrRNA hybrid, termed guide RNA (gRNA) [148]. Cas9 was shown to effectively be directed to specific target sequences by custom-made gRNA in vitro, introducing a new genome-modifying toolkit which has revolutionized the field of genetic engineering [148].

Once a double-strand break occurs in the eukaryotic genome, either non-homologous end joining (NHEJ) or homology-directed repair (HDR) subsequently occurs to prevent cell death. NHEJ occurs with less fidelity, as the two broken ends are directly reconnected with the possibility of insertions and deletions, termed 'indels', often yielding a dysfunctional gene product. In the case of HDR, the homologous chromosome serves as a template to ensure the broken strands contain the correct sequence before ligation [149]. DSBs caused by the CRISPRn/Cas system (where 'n' denotes traditional CRISPR, namely to distinguish it from newer techniques, which will be discussed in subsequent text) at a desired location induce these endogenous DNA repair mechanisms. Thus, applications of CRISPRn/Cas have made use of the NHEJ and HDR that occurs after induction of DSBs. NHEJ occurs more frequently compared to HDR, making the gene 'knock-out' strategy more readily feasible thus far [150]. However, there is also 'knock-in' strategies that supplement a donor template in addition to the gRNA and Cas protein, which consequently promotes gene repair via the HDR pathway [151]. Previous genomic engineering also required HDR to occur; however, these methods were dependent on the low incidence of DNA damage, whereas CRISPRn creates DSBs with impressive specificity [152].

Applications to Retinal Disease

Retinal disease has been at the forefront of previous gene-based therapy research and, for the same reasons, may prove to be amenable to CRISPRn/Cas in vivo applications. Of note, Bakondi et al. used CRISPRn/Cas9 to ablate the mutant rhodopsin gene containing an *S334ter* mutation in a rat model of dominant retinitis pigmentosa. Sub-retinal delivery of gRNA and Cas9 with subsequent electroporation was sufficient to ablate the gain-of-function mutation and proved highly specific, as function of the wild type allele was unaffected. Phenotypic rescue was evident with immunohistology, showing robust photoreceptor and retinal synapse preservation [153]. This study marked the first instance where a dominantly inherited mutant allele was amenable to ablation through CRISPRn/Cas in vivo. In addition to this rescue experiment, CRISPRn/Cas has already been used to validate disease-causing pathogenic gene mutations in the retina [154,155] and identify genes important for normal retinal function and development [156,157,158].

Repurposing the prokaryotic CRISPRn/Cas adaptive immune system to perform gene-editing in human cells has also been accomplished in vitro [159,160,161,162], representing a major milestone in molecular biology [150]. For example, induced pluripotent stem cells (iPSCs) obtained from skin fibroblasts of a patient with X-linked retinitis pigmentosa caused by an *RPGR* point mutation were subjected to CRISPRn/Cas9 gene editing, producing corrected graft cells [163]. In theory, these corrected iPSCs can be used for autologous cell transplantation, which will circumvent the need for immunosuppression. For mature retinal cells, in vivo delivery experiments have been done with AAV2, which provided Cas9 and sgRNA that targeted the yellow fluorescent protein (*YFP*) gene expressed in a transgenic mouse [164]. This strategy effectively decreased the number of cells expressing YFP in the retina without affecting retinal function, suggesting that applying CRISPRn/Cas to the mature retina in vivo is feasible.

Despite rapid developments in CRISPR/Cas genome engineering, various challenges persist. Off-target effects remain a concern for future therapeutic applications, as mutagenesis of non-desired sites can occur at a high frequency [165]. To circumvent this issue, a catalytic domain of Cas9 was reengineered in such a way that the nuclease could function as a nickase, which induces a single strand break (SSB) that, in the event of an off-targeting event, will likely be less deleterious (Cas9n) [166,167,168]. To obtain a DSB at the desired location using Cas9n, two gRNAs are used that complement opposite strands of the target sequence. This improves specificity by more than an order of magnitude [166,167,168]. Another study optimized the reduction of off-target effects in human cells, characterizing the effects of base pair mismatch location and number in the gRNA, and the advantages of titrating the dosage of Cas9 and gRNA [169]. Delivery poses an additional challenge for the clinical translation of CRISPR/Cas to human disease. While the most frequently used Cas protein has a cDNA sequence that can be packaged into AAV, its small carrying capacity does not allow much additional room for other elements, as in the case of gene 'knock-in' where a donor template is required.

New Developments in CRISPR-Based Genome Engineering

While CRISPR in its traditional form has become a staple of most genetic labs, its utility has significantly diversified as researchers began developing ways to repurpose the technology. Recently, a strategy for reversible CRISPR-based gene editing was developed and termed CRISPR interference (CRISPRi). This technique employs a catalytically dead Cas (dCas) protein, which can bind but not cleave DNA [170]. Thus, the dCas–sgRNA complex binds to the target gene and impedes transcription by preventing RNA polymerase from binding to the promoter region or from elongating a transcript [170]. In microbial organisms like *Escherichia coli*, simply replacing Cas with dCas is typically sufficient to prevent transcription, but in mammalian cells, the efficiency of inhibition is significantly enhanced by fusing the dCas with a repressor domain, the Krüppel-associated box (KRAB) domain being one commonly used repressor [171]. This arrangement creates a highly programmable system that can alter gene expression at the transcription level. The technique holds many similarities to the conventionally used RNA interference (RNAi) system, but it has been shown to be highly sensitive to mismatches between the sgRNA and target sequence, thus reducing off-targeting effects compared to RNAi [172,173].

While KRAB domains can be fused with dCas, other molecules have been added to the endonuclease to alter its activity in the opposite fashion. By fusing transcription activating domains with dCas, gene expression can be enhanced rather than inhibited. This technique is known as CRISPR-on or more commonly, CRISPR activation (CRISPRa) [174], and it typically involves fusion of dCas with transactivator domain VP16, among others [175,176,177]. By localizing the dCas to the promoter region of a gene, VP16 can upregulate the gene's expression. While only minimal activation has been obtained using this strategy, designing multiple sgRNAs for the same promoter region has been shown to significantly augment efficiency [178]. Efforts are underway to further enhance the efficiency of activation by recruiting additional activation domains, for example, and there remains great enthusiasm for the future applications of this technique.

One of the newest developments in CRISPR arose from researchers seeking a means to create single-nucleotide alterations without inducing DSBs [179,180]. They did so by fusing dCas with a rodent cytidine deaminase called APOBEC1, which converts cytosine to uracil. Cells transfected with the new dCas/deaminase fusion protein undergo a single cytosine to uracil conversion, which is subsequently corrected to thymine by DNA correction mechanisms or during replication. Komor et al. created several generations of the fusion protein to optimize its efficiency, and they dubbed the final generation BE3. The optimized BE3 protein replaces dCas with a nickase, which, as mentioned previously, is a catalytically active Cas protein that induces only a SSB. Although dCas results in the fewest number of insertions and deletions (indels) in the genome due to the lack of a cleavage event, nickase-induced SSBs also result in much fewer indels compared to the number accrued following a DSB. The rationale for using a nickase over dCas was that by nicking the non-edited homologous strand, the cell perceives the strand to be newly synthesized, which signals for DNA repair proteins to correct it for errors using the edited target strand as a template. In this manner, both strands will be edited despite only one having been cleaved. BE3 editing efficiency was further enhanced by fusing Cas nickase with an inhibitor of the naturally occurring base excision repair process, which removes uracil nucleotides from DNA. The overall efficiency of the system combined with the reduction in the frequency of indels and off-targeting effects represent a considerable advancement and evolution from the traditional application of CRISPR.

When juxtaposing all four techniques, CRISPRi and CRISPRa distinguish themselves from traditional CRISPRn/Cas or BE3 approaches because they act at the transcriptome level, preserving the original genome. This allows the techniques to be reversible, which confers both benefits and detriments. The transient effects of CRISPRi/a impart a considerable advantage over CRISPRn in obtaining FDA approval for use in clinical trials, since the patient's own genome will remain intact and unaltered. However, for patients with inherited retinal conditions, for example, multiple subretinal injections to re-administer the Cas/sgRNA-containing vectors could be harmful and impractical. A long-term, single intervention would be preferable for these patients, in which case CRISPRn or the BE3 system might be a stronger alternative. Additionally, despite the many advantages of the BE3 technology over traditional CRISPRn/Cas, such as reduced off-targeting effects and higher overall efficiency of editing, the current BE3 strategy is limited in its applicability. Until conversions can be made between adenine and guanine as well, the breadth of diseases that can be corrected using this technique is restricted to a select subset. However, CRISPR technology has continued to advance at a rapid pace, and it is likely that soon these techniques will be employed extensively in genetics research and across multiple fields of medicine.

ACKNOWLEDGMENTS

Jonas Children's Vision Care, and Bernard & Shirlee Brown Glaucoma Laboratory are supported by the National Institutes of Health [5P30EY019007, R01EY018213, R01EY024698, R01EY026682, R21AG050437], National Cancer Institute Core [5P30CA013696], the Research to Prevent Blindness (RPB) Physician-Scientist Award, unrestricted funds from RPB, New York, NY, USA. J.D.S. is supported by the RPB Medical Student Research Fellowship. T.C. is supported by the International Council of Ophthalmology—Retina Research Foundation Helmerich Fellowship, honoring W. H. Helmerich III. S.H.T. is a member of the RD-CURE Consortium and is supported by the Tistou and Charlotte Kerstan Foundation, the Schneeweiss Stem Cell Fund, New York State [C029572], the Foundation Fighting Blindness New York Regional Research Center Grant [C-NY05-0705-0312], the Joel Hoffman Fund, the Professor Gertrude Rothschild Stem Cell Foundation, and the Gebroe Family Foundation.

AUTHOR CONTRIBUTIONS

J.D.S. and S.J. performed literature searches and wrote the manuscript. T.C. assisted in writing the manuscript. S.H.T. oversaw all aspects of manuscript preparation and holds final responsibility for contained information and the decision to submit for publication.

CONFLICTS OF INTEREST

The authors declare no conflicts of interest.

REFERENCES

1. Raper, S.E.; Chirmule, N.; Lee, F.S.; Wivel, N.A.; Bagg, A.; Gao, G.P.; Wilson, J.M.; Batshaw, M.L. Fatal systemic inflammatory response syndrome in a ornithine transcarbamylase deficient patient following adenoviral gene transfer. *Mol. Genet. Metab.* 2003, *80*, 148–158.
2. Kohn, D.B.; Sadelain, M.; Glorioso, J.C. Occurrence of leukaemia following gene therapy of X-linked SCID. *Nat. Rev. Cancer* 2003, *3*, 477–488.
3. Bennett, J.; Tanabe, T.; Sun, D.; Zeng, Y.; Kjeldbye, H.; Gouras, P.; Maguire, A.M. Photoreceptor cell rescue in retinal degeneration (*rd*) mice by in vivo gene therapy. *Nat. Med.* 1996, *2*, 649–654.
4. Acland, G.M.; Aguirre, G.D.; Ray, J.; Zhang, Q.; Aleman, T.S.; Cideciyan, A.V.; Pearce-Kelling, S.E.; Anand, V.; Zeng, Y.; Maguire, A.M.; et al. Gene therapy restores vision in a canine model of childhood blindness. *Nat. Genet.* 2001, *28*, 92–95.
5. Bainbridge, J.W.; Smith, A.J.; Barker, S.S.; Robbie, S.; Henderson, R.; Balaggan, K.; Viswanathan, A.; Holder, G.E.; Stockman, A.; Tyler, N.; et al. Effect of gene therapy on visual function in Leber's congenital amaurosis. *N. Engl. J. Med.* 2008, *358*, 2231–2239.
6. Hauswirth, W.W.; Aleman, T.S.; Kaushal, S.; Cideciyan, A.V.; Schwartz, S.B.; Wang, L.; Conlon, T.J.; Boye, S.L.; Flotte, T.R.; Byrne, B.J.; et al. Treatment of Leber congenital amaurosis due to *RPE65* mutations by ocular subretinal injection of adeno-associated virus gene vector: Short-term results of a phase I trial. *Hum. Gene Ther.* 2008, *19*, 979–990.
7. Maguire, A.M.; Simonelli, F.; Pierce, E.A.; Pugh, E.N., Jr.; Mingozzi, F.; Bennicelli, J.; Banfi, S.; Marshall, K.A.; Testa, F.; Surace, E.M.; et al. Safety and efficacy of gene transfer for Leber's congenital amaurosis. *N. Engl. J. Med.* 2008, *358*, 2240–2248.
8. Boye, S.; Boye, S. Genetic therapies for inheritied retinal disease: Adeno-associated viruses have shown promise. *Retin. Physician* 2015, *12*, 19–24.
9. Lin, M.K.; Tsai, Y.T.; Tsang, S.H. Emerging treatments for retinitis pigmentosa: Genes and stem cells, as well as new electronic and medical therapies, are gaining ground. *Retin. Physician* 2015, *12*, 52–70.
10. Sujirakul, T.; Lin, M.K.; Duong, J.; Wei, Y.; Lopez-Pintado, S.; Tsang,

S.H. Multimodal imaging of central retinal disease progression in a 2-year mean follow-up of retinitis pigmentosa. *Am. J. Ophthalmol.* 2015, *160*, 786–798.

11. Liu, G.; Liu, X.; Li, H.; Du, Q.; Wang, F. Optical coherence tomographic analysis of retina in retinitis pigmentosa patients. *Ophthalmic Res.* 2016, *56*, 111–122.
12. Chalam, K.V.; Sambhav, K. Optical coherence tomography angiography in retinal diseases. *J. Ophthalmic Vis. Res.* 2016, *11*, 84–92.
13. Choi, W.; Mohler, K.J.; Potsaid, B.; Lu, C.D.; Liu, J.J.; Jayaraman, V.; Cable, A.E.; Duker, J.S.; Huber, R.; Fujimoto, J.G. Choriocapillaris and choroidal microvasculature imaging with ultrahigh speed OCT angiography. *PLoS ONE* 2013, *8*, e81499.
14. Kim, D.Y.; Fingler, J.; Zawadzki, R.J.; Park, S.S.; Morse, L.S.; Schwartz, D.M.; Fraser, S.E.; Werner, J.S. Optical imaging of the chorioretinal vasculature in the living human eye. *Proc. Natl. Acad. Sci. USA* 2013, *110*, 14354–14359.
15. Schwartz, D.M.; Fingler, J.; Kim, D.Y.; Zawadzki, R.J.; Morse, L.S.; Park, S.S.; Fraser, S.E.; Werner, J.S. Phase-variance optical coherence tomography: A technique for noninvasive angiography. *Ophthalmology* 2014, *121*, 180–187.
16. Spaide, R.F.; Klancnik, J.M., Jr.; Cooney, M.J. Retinal vascular layers imaged by fluorescein angiography and optical coherence tomography angiography. *JAMA Ophthalmol.* 2015, *133*, 45–50.
17. Fakin, A.; Jarc-Vidmar, M.; Glavac, D.; Bonnet, C.; Petit, C.; Hawlina, M. Fundus autofluorescence and optical coherence tomography in relation to visual function in Usher syndrome type 1 and 2. *Vis. Res.* 2012, *75*, 60–70.
18. Greenstein, V.C.; Duncker, T.; Holopigian, K.; Carr, R.E.; Greenberg, J.P.; Tsang, S.H.; Hood, D.C. Structural and functional changes associated with normal and abnormal fundus autofluorescence in patients with retinitis pigmentosa. *Retina* 2012, *32*, 349–357.
19. Lenassi, E.; Saihan, Z.; Cipriani, V.; Le Quesne Stabej, P.; Moore, A.T.; Luxon, L.M.; Bitner-Glindzicz, M.; Webster, A.R. Natural history and retinal structure in patients with Usher syndrome type 1 owing to *MYO7A* mutation. *Ophthalmology* 2014, *121*, 580–587.
20. Lima, L.H.; Burke, T.; Greenstein, V.C.; Chou, C.L.; Cella, W.; Yannuzzi, L.A.; Tsang, S.H. Progressive constriction of the hyperautofluorescent

ring in retinitis pigmentosa. *Am. J. Ophthalmol.* 2012, *153*, 718–727.

21. Oishi, A.; Ogino, K.; Makiyama, Y.; Nakagawa, S.; Kurimoto, M.; Yoshimura, N. Wide-field fundus autofluorescence imaging of retinitis pigmentosa. *Ophthalmology* 2013, *120*, 1827–1834.
22. Popovic, P.; Jarc-Vidmar, M.; Hawlina, M. Abnormal fundus autofluorescence in relation to retinal function in patients with retinitis pigmentosa. *Graefes Arch. Clin. Exp. Ophthalmol.* 2005, *243*, 1018–1027.
23. Ramachandran, R.; Zhou, L.; Locke, K.G.; Birch, D.G.; Hood, D.C. A comparison of methods for tracking progression in X-linked retinitis pigmentosa using frequency domain OCT. *Transl. Vis. Sci. Technol.* 2013, *2*, 5.
24. Robson, A.G.; Lenassi, E.; Saihan, Z.; Luong, V.A.; Fitzke, F.W.; Holder, G.E.; Webster, A.R. Comparison of fundus autofluorescence with photopic and scotopic fine matrix mapping in patients with retinitis pigmentosa: 4- to 8-year follow-up. *Investig. Ophthalmol. Vis. Sci.* 2012, *53*, 6187–6195.
25. Robson, A.G.; Saihan, Z.; Jenkins, S.A.; Fitzke, F.W.; Bird, A.C.; Webster, A.R.; Holder, G.E. Functional characterisation and serial imaging of abnormal fundus autofluorescence in patients with retinitis pigmentosa and normal visual acuity. *Br. J. Ophthalmol.* 2006, *90*, 472–479.
26. Sujirakul, T.; Davis, R.; Erol, D.; Zhang, L.; Schillizzi, G.; Royo-Dujardin, L.; Shen, S.; Tsang, S. Bilateral concordance of the fundus hyperautofluorescent ring in typical retinitis pigmentosa patients. *Ophthalmic Genet.* 2015, *36*, 113–122.
27. Wakabayashi, T.; Sawa, M.; Gomi, F.; Tsujikawa, M. Correlation of fundus autofluorescence with photoreceptor morphology and functional changes in eyes with retinitis pigmentosa. *Acta Ophthalmol.* 2010, *88*, e177–e183.
28. Sonoda, K.H.; Sakamoto, T.; Qiao, H.; Hisatomi, T.; Oshima, T.; Tsutsumi-Miyahara, C.; Exley, M.; Balk, S.P.; Taniguchi, M.; Ishibashi, T. The analysis of systemic tolerance elicited by antigen inoculation into the vitreous cavity: Vitreous cavity-associated immune deviation. *Immunology* 2005, *116*, 390–399.
29. Bennett, J.; Wellman, J.; Marshall, K.A.; McCague, S.; Ashtari, M.; DiStefano-Pappas, J.; Elci, O.U.; Chung, D.C.; Sun, J.; Wright, J.F.; et

al. Safety and durability of effect of contralateral-eye administration of AAV2 gene therapy in patients with childhood-onset blindness caused by *RPE65* mutations: A follow-on phase 1 trial. *Lancet* 2016, *388*, 661–672.

30. Ashtari, M.; Cyckowski, L.L.; Monroe, J.F.; Marshall, K.A.; Chung, D.C.; Auricchio, A.; Simonelli, F.; Leroy, B.P.; Maguire, A.M.; Shindler, K.S.; et al. The human visual cortex responds to gene therapy-mediated recovery of retinal function. *J. Clin. Investig.* 2011, *121*, 2160–2168.
31. Bennett, J.; Ashtari, M.; Wellman, J.; Marshall, K.A.; Cyckowski, L.L.; Chung, D.C.; McCague, S.; Pierce, E.A.; Chen, Y.; Bennicelli, J.L.; et al. AAV2 gene therapy readministration in three adults with congenital blindness. *Sci. Transl. Med.* 2012, *4*, 120ra115.
32. Maguire, A.M.; High, K.A.; Auricchio, A.; Wright, J.F.; Pierce, E.A.; Testa, F.; Mingozzi, F.; Bennicelli, J.L.; Ying, G.S.; Rossi, S.; et al. Age-dependent effects of *RPE65* gene therapy for Leber's congenital amaurosis: A phase 1 dose-escalation trial. *Lancet* 2009, *374*, 1597–1605.
33. Simonelli, F.; Maguire, A.M.; Testa, F.; Pierce, E.A.; Mingozzi, F.; Bennicelli, J.L.; Rossi, S.; Marshall, K.; Banfi, S.; Surace, E.M.; et al. Gene therapy for Leber's congenital amaurosis is safe and effective through 1.5 years after vector administration. *Mol. Ther.* 2010, *18*, 643–650.
34. Bainbridge, J.W.; Mehat, M.S.; Sundaram, V.; Robbie, S.J.; Barker, S.E.; Ripamonti, C.; Georgiadis, A.; Mowat, F.M.; Beattie, S.G.; Gardner, P.J.; et al. Long-term effect of gene therapy on Leber's congenital amaurosis. *N. Engl. J. Med.* 2015, *372*, 1887–1897.
35. Ripamonti, C.; Henning, G.B.; Robbie, S.J.; Sundaram, V.; van den Born, L.I.; Casteels, I.; de Ravel, T.J.; Moore, A.T.; Smith, A.J.; Bainbridge, J.W.; et al. Spectral sensitivity measurements reveal partial success in restoring missing rod function with gene therapy. *J. Vis.* 2015, *15*, 20.
36. Stieger, K. Tgaag76, an adeno-associated virus delivered gene therapy for the potential treatment of vision loss caused by *RPE65* gene abnormalities. *Curr. Opin. Mol. Ther.* 2010, *12*, 471–477.
37. Weleber, R.G.; Pennesi, M.E.; Wilson, D.J.; Kaushal, S.; Erker, L.R.; Jensen, L.; McBride, M.T.; Flotte, T.R.; Humphries, M.; Calcedo, R.; et al. Results at 2 years after gene therapy for *RPE65*-deficient Leber congenital amaurosis and severe early-childhood-onset retinal

dystrophy. *Ophthalmology* 2016, *123*, 1606–1620.

38. Georgiadis, A.; Duran, Y.; Ribeiro, J.; Abelleira-Hervas, L.; Robbie, S.J.; Sunkel-Laing, B.; Fourali, S.; Gonzalez-Cordero, A.; Cristante, E.; Michaelides, M.; et al. Development of an optimized AAV2/5 gene therapy vector for Leber congenital amaurosis owing to defects in *RPE65*. *Gene Ther.* 2016, *23*, 857–862.
39. Cideciyan, A.V.; Aguirre, G.K.; Jacobson, S.G.; Butt, O.H.; Schwartz, S.B.; Swider, M.; Roman, A.J.; Sadigh, S.; Hauswirth, W.W. Pseudo-fovea formation after gene therapy for *RPE65*-LCA. *Investig. Ophthalmol. Vis. Sci.* 2014, *56*, 526–537.
40. Jacobson, S.G.; Cideciyan, A.V.; Ratnakaram, R.; Heon, E.; Schwartz, S.B.; Roman, A.J.; Peden, M.C.; Aleman, T.S.; Boye, S.L.; Sumaroka, A.; et al. Gene therapy for leber congenital amaurosis caused by *RPE65* mutations: Safety and efficacy in 15 children and adults followed up to 3 years. *Arch. Ophthalmol.* 2012, *130*, 9–24.
41. Jacobson, S.G.; Cideciyan, A.V.; Roman, A.J.; Sumaroka, A.; Schwartz, S.B.; Heon, E.; Hauswirth, W.W. Improvement and decline in vision with gene therapy in childhood blindness. *N. Engl. J. Med.* 2015, *372*, 1920–1926.
42. Jacobson, S.G.; Boye, S.L.; Aleman, T.S.; Conlon, T.J.; Zeiss, C.J.; Roman, A.J.; Cideciyan, A.V.; Schwartz, S.B.; Komaromy, A.M.; Doobrajh, M.; et al. Safety in nonhuman primates of ocular AAV2-*RPE65*, a candidate treatment for blindness in Leber congenital amaurosis. *Hum. Gene Ther.* 2006, *17*, 845–858.
43. MacLaren, R.E.; Groppe, M.; Barnard, A.R.; Cottriall, C.L.; Tolmachova, T.; Seymour, L.; Clark, K.R.; During, M.J.; Cremers, F.P.; Black, G.C.; et al. Retinal gene therapy in patients with choroideremia: Initial findings from a phase 1/2 clinical trial. *Lancet* 2014, *383*, 1129–1137.
44. Seitz, I.P.; Zhour, A.; Kohl, S.; Llavona, P.; Peter, T.; Wilhelm, B.; Zrenner, E.; Ueffing, M.; Bartz-Schmidt, K.U.; Fischer, M.D. Multimodal assessment of choroideremia patients defines pre-treatment characteristics. *Graefes Arch. Clin. Exp. Ophthalmol.* 2015, *253*, 2143–2150.
45. Xue, K.; Oldani, M.; Jolly, J.K.; Edwards, T.L.; Groppe, M.; Downes, S.M.; MacLaren, R.E. Correlation of optical coherence tomography and autofluorescence in the outer retina and choroid of patients with choroideremia. *Investig. Ophthalmol. Vis. Sci.* 2016, *57*, 3674–3684.

46. Komaromy, A.M.; Alexander, J.J.; Rowlan, J.S.; Garcia, M.M.; Chiodo, V.A.; Kaya, A.; Tanaka, J.C.; Acland, G.M.; Hauswirth, W.W.; Aguirre, G.D. Gene therapy rescues cone function in congenital achromatopsia. *Hum. Mol. Genet.* 2010, *19*, 2581–2593.

47. Ye, G.J.; Budzynski, E.; Sonnentag, P.; Nork, T.M.; Miller, P.E.; McPherson, L.; Ver Hoeve, J.N.; Smith, L.M.; Arndt, T.; Mandapati, S.; et al. Safety and biodistribution evaluation in *CNGB3*-deficient mice of rAAV2tYF-PR1.7-hCNGB3, a recombinant AAV vector for treatment of achromatopsia. *Hum. Gene Ther. Clin. Dev.* 2016, *27*, 27–36.

48. Ye, G.J.; Budzynski, E.; Sonnentag, P.; Nork, T.M.; Miller, P.E.; Sharma, A.K.; Ver Hoeve, J.N.; Smith, L.M.; Arndt, T.; Calcedo, R.; et al. Safety and biodistribution evaluation in cynomolgus macaques of rAAV2tYF-PR1.7-hCNGB3, a recombinant AAV vector for treatment of achromatopsia. *Hum. Gene Ther. Clin. Dev.* 2016, *27*, 37–48.

49. Marangoni, D.; Wu, Z.; Wiley, H.E.; Zeiss, C.J.; Vijayasarathy, C.; Zeng, Y.; Hiriyanna, S.; Bush, R.A.; Wei, L.L.; Colosi, P.; et al. Preclinical safety evaluation of a recombinant AAV8 vector for X-linked retinoschisis after intravitreal administration in rabbits. *Hum. Gene Ther. Clin. Dev.* 2014, *25*, 202–211.

50. Marangoni, D.; Bush, R.A.; Zeng, Y.; Wei, L.L.; Ziccardi, L.; Vijayasarathy, C.; Bartoe, J.T.; Palyada, K.; Santos, M.; Hiriyanna, S.; et al. Ocular and systemic safety of a recombinant AAV8 vector for X-linked retinoschisis gene therapy: GLP studies in rabbits and *RS1*-KO mice. *Mol. Ther. Methods Clin. Dev.* 2016, *5*, 16011.

51. Bush, R.A.; Zeng, Y.; Colosi, P.; Kjellstrom, S.; Hiriyanna, S.; Vijayasarathy, C.; Santos, M.; Li, J.; Wu, Z.; Sieving, P.A. Preclinical dose-escalation study of intravitreal AAV-*RS1* gene therapy in a mouse model of X-linked retinoschisis: Dose-dependent expression and improved retinal structure and function. *Hum. Gene Ther.* 2016, *27*, 376–389.

52. Zeng, Y.; Petralia, R.S.; Vijayasarathy, C.; Wu, Z.; Hiriyanna, S.; Song, H.; Wang, Y.X.; Sieving, P.A.; Bush, R.A. Retinal structure and gene therapy outcome in retinoschisin-deficient mice assessed by spectral-domain optical coherence tomography. *Investig. Ophthalmol. Vis. Sci.* 2016, *57*, OCT277–OCT287.

53. Ye, G.J.; Budzynski, E.; Sonnentag, P.; Miller, P.E.; Sharma, A.K.; Ver Hoeve, J.N.; Howard, K.; Knop, D.R.; Neuringer, M.; McGill, T.; et al. Safety and biodistribution evaluation in cynomolgus macaques of

rAAV2tYF-CB-hRS1, a recombinant adeno-associated virus vector expressing retinoschisin. *Hum. Gene Ther. Clin. Dev.* 2015, *26*, 165–176.

54. Ye, G.J.; Conlon, T.; Erger, K.; Sonnentag, P.; Sharma, A.K.; Howard, K.; Knop, D.R.; Chulay, J.D. Safety and biodistribution evaluation of rAAV2tYF-CB-hRS1, a recombinant adeno-associated virus vector expressing retinoschisin, in *RS1*-deficient mice. *Hum. Gene Ther. Clin. Dev.* 2015, *26*, 177–184.
55. Cwerman-Thibault, H.; Augustin, S.; Lechauve, C.; Ayache, J.; Ellouze, S.; Sahel, J.A.; Corral-Debrinski, M. Nuclear expression of mitochondrial *ND4* leads to the protein assembling in complex I and prevents optic atrophy and visual loss. *Mol. Ther. Methods Clin. Dev.* 2015, *2*, 15003.
56. Koilkonda, R.; Yu, H.; Talla, V.; Porciatti, V.; Feuer, W.J.; Hauswirth, W.W.; Chiodo, V.; Erger, K.E.; Boye, S.L.; Lewin, A.S.; et al. LHON gene therapy vector prevents visual loss and optic neuropathy induced by G11778A mutant mitochondrial DNA: Biodistribution and toxicology profile. *Investig. Ophthalmol. Vis. Sci.* 2014, *55*, 7739–7753.
57. Koilkonda, R.D.; Yu, H.; Chou, T.H.; Feuer, W.J.; Ruggeri, M.; Porciatti, V.; Tse, D.; Hauswirth, W.W.; Chiodo, V.; Boye, S.L.; et al. Safety and effects of the vector for the Leber hereditary optic neuropathy gene therapy clinical trial. *JAMA Ophthalmol.* 2014, *132*, 409–420.
58. Feuer, W.J.; Schiffman, J.C.; Davis, J.L.; Porciatti, V.; Gonzalez, P.; Koilkonda, R.D.; Yuan, H.; Lalwani, A.; Lam, B.L.; Guy, J. Gene therapy for Leber hereditary optic neuropathy: Initial results. *Ophthalmology* 2016, *123*, 558–570.
59. Wan, X.; Pei, H.; Zhao, M.J.; Yang, S.; Hu, W.K.; He, H.; Ma, S.Q.; Zhang, G.; Dong, X.Y.; Chen, C.; et al. Efficacy and safety of rAAV2-ND4 treatment for Leber's hereditary optic neuropathy. *Sci. Rep.* 2016, *6*, 21587.
60. Bi, A.; Cui, J.; Ma, Y.P.; Olshevskaya, E.; Pu, M.; Dizhoor, A.M.; Pan, Z.H. Ectopic expression of a microbial-type rhodopsin restores visual responses in mice with photoreceptor degeneration. *Neuron* 2006, *50*, 23–33.
61. Ivanova, E.; Hwang, G.S.; Pan, Z.H.; Troilo, D. Evaluation of AAV-mediated expression of Chop2-GFP in the marmoset retina. *Investig. Ophthalmol. Vis. Sci.* 2010, *51*, 5288–5296.

62. Ghazi, N.G.; Abboud, E.B.; Nowilaty, S.R.; Alkuraya, H.; Alhommadi, A.; Cai, H.; Hou, R.; Deng, W.T.; Boye, S.L.; Almaghamsi, A.; et al. Treatment of retinitis pigmentosa due to *MERTK* mutations by ocular subretinal injection of adeno-associated virus gene vector: Results of a phase I trial. *Hum. Genet.* 2016, *135*, 327–343.
63. Conlon, T.J.; Deng, W.T.; Erger, K.; Cossette, T.; Pang, J.J.; Ryals, R.; Clement, N.; Cleaver, B.; McDoom, I.; Boye, S.E.; et al. Preclinical potency and safety studies of an AAV2-mediated gene therapy vector for the treatment of *MERTK* associated retinitis pigmentosa. *Hum. Gene Ther. Clin. Dev.* 2013, *24*, 23–28.
64. Zallocchi, M.; Binley, K.; Lad, Y.; Ellis, S.; Widdowson, P.; Iqball, S.; Scripps, V.; Kelleher, M.; Loader, J.; Miskin, J.; et al. EIAV-based retinal gene therapy in the *shaker1* mouse model for Usher syndrome type 1B: Development of UshStat. *PLoS ONE* 2014, *9*, e94272.
65. Parker, M.A.; Choi, D.; Erker, L.R.; Pennesi, M.E.; Yang, P.; Chegarnov, E.N.; Steinkamp, P.N.; Schlechter, C.L.; Dhaenens, C.M.; Mohand-Said, S.; et al. Test-retest variability of functional and structural parameters in patients with Stargardt disease participating in the SAR422459 gene therapy trial. *Transl. Vis. Sci. Technol.* 2016, *5*, 10.
66. Binley, K.; Widdowson, P.; Loader, J.; Kelleher, M.; Iqball, S.; Ferrige, G.; de Belin, J.; Carlucci, M.; Angell-Manning, D.; Hurst, F.; et al. Transduction of photoreceptors with equine infectious anemia virus lentiviral vectors: Safety and biodistribution of stargen for Stargardt disease. *Investig. Ophthalmol. Vis. Sci.* 2013, *54*, 4061–4071.
67. Kong, J.; Kim, S.R.; Binley, K.; Pata, I.; Doi, K.; Mannik, J.; Zernant-Rajang, J.; Kan, O.; Iqball, S.; Naylor, S.; et al. Correction of the disease phenotype in the mouse model of Stargardt disease by lentiviral gene therapy. *Gene Ther.* 2008, *15*, 1311–1320.
68. Rakoczy, E.P.; Lai, C.M.; Magno, A.L.; Wikstrom, M.E.; French, M.A.; Pierce, C.M.; Schwartz, S.D.; Blumenkranz, M.S.; Chalberg, T.W.; Degli-Esposti, M.A.; et al. Gene therapy with recombinant adeno-associated vectors for neovascular age-related macular degeneration: 1 year follow-up of a phase 1 randomised clinical trial. *Lancet* 2015, *386*, 2395–2403.
69. Constable, I.J.; Blumenkranz, M.S.; Schwartz, S.D.; Barone, S.; Lai, C.M.; Rakoczy, E.P. Gene therapy for age-related macular degeneration. *Asia Pac. J. Ophthalmol.* 2016, *5*, 300–303.

70. Constable, I.J.; Pierce, C.M.; Lai, C.M.; Magno, A.L.; Degli-Esposti, M.A.; French, M.A.; McAllister, I.L.; Butler, S.; Barone, S.B.; Schwartz, S.D.; et al. Phase 2a randomized clinical trial: Safety and post hoc analysis of subretinal rAAV.sFlt-1 for wet age-related macular degeneration. *EBioMedicine* 2016, *14*, 168–175.
71. Lukason, M.; DuFresne, E.; Rubin, H.; Pechan, P.; Li, Q.; Kim, I.; Kiss, S.; Flaxel, C.; Collins, M.; Miller, J.; et al. Inhibition of choroidal neovascularization in a nonhuman primate model by intravitreal administration of an AAV2 vector expressing a novel anti-VEGF molecule. *Mol. Ther.* 2011, *19*, 260–265.
72. Maclachlan, T.K.; Lukason, M.; Collins, M.; Munger, R.; Isenberger, E.; Rogers, C.; Malatos, S.; Dufresne, E.; Morris, J.; Calcedo, R.; et al. Preclinical safety evaluation of AAV2-sFLT01—A gene therapy for age-related macular degeneration. *Mol. Ther.* 2011, *19*, 326–334.
73. Pecen, P.E.; Kaiser, P.K. Current phase 1/2 research for neovascular age-related macular degeneration. *Curr. Opin. Ophthalmol.* 2015, *26*, 188–193.
74. Binley, K.; Widdowson, P.S.; Kelleher, M.; de Belin, J.; Loader, J.; Ferrige, G.; Carlucci, M.; Esapa, M.; Chipchase, D.; Angell-Manning, D.; et al. Safety and biodistribution of an equine infectious anemia virus-based gene therapy, retinostat(®), for age-related macular degeneration. *Hum. Gene Ther.* 2012, *23*, 980–991.
75. Campochiaro, P.A.; Lauer, A.K.; Sohn, E.H.; Mir, T.A.; Naylor, S.; Anderton, M.C.; Kelleher, M.; Harrop, R.; Ellis, S.; Mitrophanous, K.A. Lentiviral vector gene transfer of endostatin/angiostatin for macular degeneration (GEM) study. *Hum. Gene Ther.* 2016, *28*, 99–111.
76. Daya, S.; Berns, K.I. Gene therapy using adeno-associated virus vectors. *Clin. Microbiol. Rev.* 2008, *21*, 583–593.
77. Solinis, M.A.; del Pozo-Rodriguez, A.; Apaolaza, P.S.; Rodriguez-Gascon, A. Treatment of ocular disorders by gene therapy. *Eur. J. Pharm. Biopharm.* 2015, *95*, 331–342.
78. Liu, M.M.; Tuo, J.; Chan, C.C. Republished review: Gene therapy for ocular diseases. *Postgrad. Med. J.* 2011, *87*, 487–495.
79. Sarra, G.M.; Stephens, C.; Schlichtenbrede, F.C.; Bainbridge, J.W.; Thrasher, A.J.; Luthert, P.J.; Ali, R.R. Kinetics of transgene expression in mouse retina following sub-retinal injection of recombinant adeno-

associated virus. *Vis. Res.* 2002, *42*, 541–549.

80. Garoon, R.B.; Stout, J.T. Update on ocular gene therapy and advances in treatment of inherited retinal diseases and exudative macular degeneration. *Curr. Opin. Ophthalmol.* 2016, *27*, 268–273.
81. Hirsch, M.L.; Wolf, S.J.; Samulski, R.J. Delivering transgenic DNA exceeding the carrying capacity of AAV vectors. *Methods Mol. Biol.* 2016, *1382*, 21–39.
82. Yan, Z.; Zhang, Y.; Duan, D.; Engelhardt, J.F. Trans-splicing vectors expand the utility of adeno-associated virus for gene therapy. *Proc. Natl. Acad. Sci. USA* 2000, *97*, 6716–6721.
83. Reich, S.J.; Auricchio, A.; Hildinger, M.; Glover, E.; Maguire, A.M.; Wilson, J.M.; Bennett, J. Efficient trans-splicing in the retina expands the utility of adeno-associated virus as a vector for gene therapy. *Hum. Gene Ther.* 2003, *14*, 37–44.
84. Dyka, F.M.; Boye, S.L.; Chiodo, V.A.; Hauswirth, W.W.; Boye, S.E. Dual adeno-associated virus vectors result in efficient in vitro and in vivo expression of an oversized gene, *MYO7A*. *Hum. Gene Ther. Methods* 2014, *25*, 166–177.
85. Trapani, I.; Colella, P.; Sommella, A.; Iodice, C.; Cesi, G.; de Simone, S.; Marrocco, E.; Rossi, S.; Giunti, M.; Palfi, A.; et al. Effective delivery of large genes to the retina by dual AAV vectors. *EMBO Mol. Med.* 2014, *6*, 194–211.
86. Semple-Rowland, S.L.; Berry, J. Use of lentiviral vectors to deliver and express bicistronic transgenes in developing chicken embryos. *Methods* 2014, *66*, 466–473.
87. Cepko, C.L.; Vandenberghe, L.H. Retinal gene therapy coming of age. *Hum. Gene Ther.* 2013, *24*, 242–244.
88. Jacobson, S.G.; Cideciyan, A.V.; Peshenko, I.V.; Sumaroka, A.; Olshevskaya, E.V.; Cao, L.; Schwartz, S.B.; Roman, A.J.; Olivares, M.B.; Sadigh, S.; et al. Determining consequences of retinal membrane guanylyl cyclase (RetGC1) deficiency in human Leber congenital amaurosis en route to therapy: Residual cone-photoreceptor vision correlates with biochemical properties of the mutants. *Hum. Mol. Genet.* 2013, *22*, 168–183.
89. Provost, N.; Le Meur, G.; Weber, M.; Mendes-Madeira, A.; Podevin, G.; Cherel, Y.; Colle, M.A.; Deschamps, J.Y.; Moullier, P.; Rolling, F. Biodistribution of rAAV vectors following intraocular administration:

Evidence for the presence and persistence of vector DNA in the optic nerve and in the brain. *Mol. Ther.* 2005, *11*, 275–283.

90. Yin, L.; Greenberg, K.; Hunter, J.J.; Dalkara, D.; Kolstad, K.D.; Masella, B.D.; Wolfe, R.; Visel, M.; Stone, D.; Libby, R.T.; et al. Intravitreal injection of AAV2 transduces macaque inner retina. *Investig. Ophthalmol. Vis. Sci.* 2011, *52*, 2775–2783.
91. Bennett, J.; Maguire, A.M.; Cideciyan, A.V.; Schnell, M.; Glover, E.; Anand, V.; Aleman, T.S.; Chirmule, N.; Gupta, A.R.; Huang, Y.; et al. Stable transgene expression in rod photoreceptors after recombinant adeno-associated virus-mediated gene transfer to monkey retina. *Proc. Natl. Acad. Sci. USA* 1999, *96*, 9920–9925.
92. Yu-Wai-Man, P. Genetic manipulation for inherited neurodegenerative diseases: Myth or reality? *Br. J. Ophthalmol.* 2016, *100*, 1322–1331.
93. Park, T.K.; Wu, Z.; Kjellstrom, S.; Zeng, Y.; Bush, R.A.; Sieving, P.A.; Colosi, P. Intravitreal delivery of AAV8 retinoschisin results in cell type-specific gene expression and retinal rescue in the *RS1*-KO mouse. *Gene Ther.* 2009, *16*, 916–926.
94. Byrne, L.C.; Ozturk, B.E.; Lee, T.; Fortuny, C.; Visel, M.; Dalkara, D.; Schaffer, D.V.; Flannery, J.G. Retinoschisin gene therapy in photoreceptors, muller glia or all retinal cells in the *RS1h−/−* mouse. *Gene Ther.* 2014, *21*, 585–592.
95. Bennett, J. Immune response following intraocular delivery of recombinant viral vectors. *Gene Ther.* 2003, *10*, 977–982.
96. Li, Q.; Miller, R.; Han, P.Y.; Pang, J.; Dinculescu, A.; Chiodo, V.; Hauswirth, W.W. Intraocular route of AAV2 vector administration defines humoral immune response and therapeutic potential. *Mol. Vis.* 2008, *14*, 1760–1769.
97. Matsuda, T.; Cepko, C.L. Electroporation and rna interference in the rodent retina in vivo and in vitro. *Proc. Natl. Acad. Sci. USA* 2004, *101*, 16–22.
98. Koirala, A.; Conley, S.M.; Naash, M.I. A review of therapeutic prospects of non-viral gene therapy in the retinal pigment epithelium. *Biomaterials* 2013, *34*, 7158–7167.
99. Touchard, E.; Berdugo, M.; Bigey, P.; El Sanharawi, M.; Savoldelli, M.; Naud, M.C.; Jeanny, J.C.; Behar-Cohen, F. Suprachoroidal electrotransfer: A nonviral gene delivery method to transfect the choroid and the retina without detaching the retina. *Mol. Ther.* 2012,

20, 1559–1570.

100. Adijanto, J.; Naash, M.I. Nanoparticle-based technologies for retinal gene therapy. *Eur. J. Pharm. Biopharm.* 2015, *95*, 353–367.
101. Han, Z.; Conley, S.M.; Naash, M.I. Aav and compacted DNA nanoparticles for the treatment of retinal disorders: Challenges and future prospects. *Investig. Ophthalmol. Vis. Sci.* 2011, *52*, 3051–3059.
102. Han, Z.; Conley, S.M.; Makkia, R.; Guo, J.; Cooper, M.J.; Naash, M.I. Comparative analysis of DNA nanoparticles and AAVs for ocular gene delivery. *PLoS ONE* 2012, *7*, e52189.
103. Tamboli, V.; Mishra, G.P.; Mitrat, A.K. Polymeric vectors for ocular gene delivery. *Ther. Deliv.* 2011, *2*, 523–536.
104. Jin, J.; Zhou, K.K.; Park, K.; Hu, Y.; Xu, X.; Zheng, Z.; Tyagi, P.; Kompella, U.B.; Ma, J.X. Anti-inflammatory and antiangiogenic effects of nanoparticle-mediated delivery of a natural angiogenic inhibitor. *Investig. Ophthalmol. Vis. Sci.* 2011, *52*, 6230–6237.
105. Kachi, S.; Esumi, N.; Zack, D.J.; Campochiaro, P.A. Sustained expression after nonviral ocular gene transfer using mammalian promoters. *Gene Ther.* 2006, *13*, 798–804.
106. Kachi, S.; Oshima, Y.; Esumi, N.; Kachi, M.; Rogers, B.; Zack, D.J.; Campochiaro, P.A. Nonviral ocular gene transfer. *Gene Ther.* 2005, *12*, 843–851.
107. Liao, H.W.; Yau, K.W. In vivo gene delivery in the retina using polyethylenimine. *Biotechniques* 2007, *42*, 285–286, 288.
108. Marano, R.J.; Toth, I.; Wimmer, N.; Brankov, M.; Rakoczy, P.E. Dendrimer delivery of an anti-vegf oligonucleotide into the eye: A long-term study into inhibition of laser-induced CNV, distribution, uptake and toxicity. *Gene Ther.* 2005, *12*, 1544–1550.
109. Mitra, R.N.; Han, Z.; Merwin, M.; Al Taai, M.; Conley, S.M.; Naash, M.I. Synthesis and characterization of glycol chitosan DNA nanoparticles for retinal gene delivery. *ChemMedChem* 2014, *9*, 189–196.
110. Mo, Y.; Barnett, M.E.; Takemoto, D.; Davidson, H.; Kompella, U.B. Human serum albumin nanoparticles for efficient delivery of Cu, Zn superoxide dismutase gene. *Mol. Vis.* 2007, *13*, 746–757.
111. Park, K.; Chen, Y.; Hu, Y.; Mayo, A.S.; Kompella, U.B.; Longeras, R.; Ma, J.X. Nanoparticle-mediated expression of an angiogenic inhibitor ameliorates ischemia-induced retinal neovascularization and diabetes-

induced retinal vascular leakage. *Diabetes* 2009, *58*, 1902–1913.

112. Zhang, C.; Wang, Y.S.; Wu, H.; Zhang, Z.X.; Cai, Y.; Hou, H.Y.; Zhao, W.; Yang, X.M.; Ma, J.X. Inhibitory efficacy of hypoxia-inducible factor 1α short hairpin RNA plasmid DNA-loaded poly (d, l-lactide-co-glycolide) nanoparticles on choroidal neovascularization in a laser-induced rat model. *Gene Ther.* 2010, *17*, 338–351.
113. Weleber, R.G.; Francis, P.J.; Trzupek, K.M.; Beattie, C. Leber congenital amaurosis. In *GeneReviews*®; Pagon, R.A., Adam, M.P., Ardinger, H.H., Wallace, S.E., Amemiya, A., Bean, L.J.H., Bird, T.D., Fong, C.T., Mefford, H.C., Smith, R.J.H., et al., Eds.; University of Washington: Seattle, WA, USA, 1993–2017; (last update on 2 May 2013).
114. Redmond, T.M. Focus on molecules: *RPE65*, the visual cycle retinol isomerase. *Exp. Eye Res.* 2009, *88*, 846–847.
115. Koenekoop, R.K. An overview of leber congenital amaurosis: A model to understand human retinal development. *Surv. Ophthalmol.* 2004, *49*, 379–398.
116. Banin, E.; Bandah-Rozenfeld, D.; Obolensky, A.; Cideciyan, A.V.; Aleman, T.S.; Marks-Ohana, D.; Sela, M.; Boye, S.; Sumaroka, A.; Roman, A.J.; et al. Molecular anthropology meets genetic medicine to treat blindness in the North African Jewish population: Human gene therapy initiated in Israel. *Hum. Gene Ther.* 2010, *21*, 1749–1757.
117. Cideciyan, A.V.; Aleman, T.S.; Boye, S.L.; Schwartz, S.B.; Kaushal, S.; Roman, A.J.; Pang, J.J.; Sumaroka, A.; Windsor, E.A.; Wilson, J.M.; et al. Human gene therapy for *RPE65* isomerase deficiency activates the retinoid cycle of vision but with slow rod kinetics. *Proc. Natl. Acad. Sci. USA* 2008, *105*, 15112–15117.
118. Cideciyan, A.V.; Hauswirth, W.W.; Aleman, T.S.; Kaushal, S.; Schwartz, S.B.; Boye, S.L.; Windsor, E.A.; Conlon, T.J.; Sumaroka, A.; Pang, J.J.; et al. Human *RPE65* gene therapy for leber congenital amaurosis: Persistence of early visual improvements and safety at 1 year. *Hum. Gene Ther.* 2009, *20*, 999–1004.
119. Testa, F.; Maguire, A.M.; Rossi, S.; Pierce, E.A.; Melillo, P.; Marshall, K.; Banfi, S.; Surace, E.M.; Sun, J.; Acerra, C.; et al. Three-year follow-up after unilateral subretinal delivery of adeno-associated virus in patients with Leber congenital amaurosis type 2. *Ophthalmology* 2013, *120*, 1283–1291.

120. MacDonald, I.M.; Hume, S.; Chan, S.; Seabra, M.C. Choroideremia. In *GeneReviews®*; Pagon, R.A., Adam, M.P., Ardinger, H.H., Wallace, S.E., Amemiya, A., Bean, L.J.H., Bird, T.D., Fong, C.T., Mefford, H.C., Smith, R.J.H., et al., Eds.; University of Washington: Seattle, WA, USA, 1993–2017; (last update on 26 February 2015).

121. Zinkernagel, M.S.; MacLaren, R.E. Recent advances and future prospects in choroideremia. *Clin. Ophthalmol.* 2015, *9*, 2195–2200.

122. Ma, K.K.; Lin, J.; Boudreault, K.; Chen, R.W.; Tsang, S.H. Phenotyping choroideremia and its carrier state with multimodal imaging techniques. *Retin. Cases Brief Rep.* 2017, *11*, S178–S181.

123. Edwards, T.L.; Jolly, J.K.; Groppe, M.; Barnard, A.R.; Cottriall, C.L.; Tolmachova, T.; Black, G.C.; Webster, A.R.; Lotery, A.J.; Holder, G.E.; et al. Visual acuity after retinal gene therapy for choroideremia. *N. Engl. J. Med.* 2016, *374*, 1996–1998.

124. Barnard, A.R.; Groppe, M.; MacLaren, R.E. Gene therapy for choroideremia using an adeno-associated viral (AAV) vector. *Cold Spring Harb. Perspect. Med.* 2014, *5*, a017293.

125. Johnson, S.; Michaelides, M.; Aligianis, I.A.; Ainsworth, J.R.; Mollon, J.D.; Maher, E.R.; Moore, A.T.; Hunt, D.M. Achromatopsia caused by novel mutations in both *CNGA3* and *CNGB3*. *J. Med. Genet.* 2004, *41*, e20.

126. Greenberg, J.P.; Sherman, J.; Zweifel, S.A.; Chen, R.W.; Duncker, T.; Kohl, S.; Baumann, B.; Wissinger, B.; Yannuzzi, L.A.; Tsang, S.H. Spectral-domain optical coherence tomography staging and autofluorescence imaging in achromatopsia. *JAMA Ophthalmol.* 2014, *132*, 437–445.

127. Kohl, S.; Varsanyi, B.; Antunes, G.A.; Baumann, B.; Hoyng, C.B.; Jagle, H.; Rosenberg, T.; Kellner, U.; Lorenz, B.; Salati, R.; et al. *CNGB3* mutations account for 50% of all cases with autosomal recessive achromatopsia. *Eur. J. Hum. Genet.* 2005, *13*, 302–308.

128. Giblin, J.P.; Comes, N.; Strauss, O.; Gasull, X. Ion channels in the eye: Involvement in ocular pathologies. *Adv. Protein Chem. Struct. Biol.* 2016, *104*, 157–231.

129. Mooy, C.M.; Van Den Born, L.I.; Baarsma, S.; Paridaens, D.A.; Kraaijenbrink, T.; Bergen, A.; Weber, B.H. Hereditary X-linked juvenile retinoschisis: A review of the role of Müller cells. *Arch. Ophthalmol.* 2002, *120*, 979–984.

130. Molday, R.S.; Kellner, U.; Weber, B.H. X-linked juvenile retinoschisis: Clinical diagnosis, genetic analysis, and molecular mechanisms. *Prog. Retin. Eye Res.* 2012, *31*, 195–212.

131. Yu-Wai-Man, P.; Chinnery, P.F. Leber hereditary optic neuropathy. In *GeneReviews®*; Pagon, R.A., Adam, M.P., Ardinger, H.H., Wallace, S.E., Amemiya, A., Bean, L.J.H., Bird, T.D., Fong, C.T., Mefford, H.C., Smith, R.J.H., et al., Eds.; University of Washington: Seattle, WA, USA, 1993–2017; (last update on 23 June 2016).

132. Qi, X.; Sun, L.; Lewin, A.S.; Hauswirth, W.W.; Guy, J. The mutant human *ND4* subunit of complex I induces optic neuropathy in the mouse. *Investig. Ophthalmol. Vis. Sci.* 2007, *48*, 1–10.

133. Ellouze, S.; Augustin, S.; Bouaita, A.; Bonnet, C.; Simonutti, M.; Forster, V.; Picaud, S.; Sahel, J.A.; Corral-Debrinski, M. Optimized allotopic expression of the human mitochondrial *ND4* prevents blindness in a rat model of mitochondrial dysfunction. *Am. J. Hum. Genet.* 2008, *83*, 373–387.

134. Guy, J.; Qi, X.; Koilkonda, R.D.; Arguello, T.; Chou, T.H.; Ruggeri, M.; Porciatti, V.; Lewin, A.S.; Hauswirth, W.W. Efficiency and safety of AAV-mediated gene delivery of the human *ND4* complex I subunit in the mouse visual system. *Investig. Ophthalmol. Vis. Sci.* 2009, *50*, 4205–4214.

135. Zhang, Q. Retinitis pigmentosa: Progress and perspective. *Asia Pac. J. Ophthalmol.* 2016, *5*, 265–271.

136. Hamel, C. Retinitis pigmentosa. *Orphanet J. Rare Dis.* 2006, *1*, 40.

137. Sadeghi, A.M.; Eriksson, K.; Kimberling, W.J.; Sjostrom, A.; Moller, C. Longterm visual prognosis in usher syndrome types 1 and 2. *Acta Ophthalmol. Scand.* 2006, *84*, 537–544.

138. Kimberling, W.J.; Hildebrand, M.S.; Shearer, A.E.; Jensen, M.L.; Halder, J.A.; Trzupek, K.; Cohn, E.S.; Weleber, R.G.; Stone, E.M.; Smith, R.J. Frequency of usher syndrome in two pediatric populations: Implications for genetic screening of deaf and hard of hearing children. *Genet. Med.* 2010, *12*, 512–516.

139. Williams, D.S.; Lopes, V.S. The many different cellular functions of *MYO7A* in the retina. *Biochem. Soc. Trans.* 2011, *39*, 1207–1210.

140. Charbel Issa, P.; Barnard, A.R.; Herrmann, P.; Washington, I.; MacLaren, R.E. Rescue of the Stargardt phenotype in *ABCA4* knockout mice through inhibition of vitamin A dimerization. *Proc. Natl. Acad.*

Sci. USA 2015, *112*, 8415–8420.

141. Congdon, N.; O'Colmain, B.; Klaver, C.C.; Klein, R.; Munoz, B.; Friedman, D.S.; Kempen, J.; Taylor, H.R.; Mitchell, P.; Eye Diseases Prevalence Research, G. Causes and prevalence of visual impairment among adults in the United States. *Arch. Ophthalmol.* 2004, *122*, 477–485.
142. Alexandru, M.R.; Alexandra, N.M. Wet age related macular degeneration management and follow-up. *Rom. J. Ophthalmol.* 2016, *60*, 9–13.
143. Barrangou, R.; Fremaux, C.; Deveau, H.; Richards, M.; Boyaval, P.; Moineau, S.; Romero, D.A.; Horvath, P. CRISPR provides acquired resistance against viruses in prokaryotes. *Science* 2007, *315*, 1709–1712.
144. Makarova, K.S.; Aravind, L.; Grishin, N.V.; Rogozin, I.B.; Koonin, E.V. A DNA repair system specific for thermophilic archaea and bacteria predicted by genomic context analysis. *Nucleic Acids Res.* 2002, *30*, 482–496.
145. Mojica, F.J.; Diez-Villasenor, C.; Soria, E.; Juez, G. Biological significance of a family of regularly spaced repeats in the genomes of archaea, bacteria and mitochondria. *Mol. Microbiol.* 2000, *36*, 244–246.
146. Wiedenheft, B.; Zhou, K.; Jinek, M.; Coyle, S.M.; Ma, W.; Doudna, J.A. Structural basis for DNase activity of a conserved protein implicated in CRISPR-mediated genome defense. *Structure* 2009, *17*, 904–912.
147. Wiedenheft, B.; Sternberg, S.H.; Doudna, J.A. RNA-guided genetic silencing systems in bacteria and archaea. *Nature* 2012, *482*, 331–338.
148. Jinek, M.; Chylinski, K.; Fonfara, I.; Hauer, M.; Doudna, J.A.; Charpentier, E. A programmable dual-RNA-guided DNA endonuclease in adaptive bacterial immunity. *Science* 2012, *337*, 816–821.
149. Pardo, B.; Gomez-Gonzalez, B.; Aguilera, A. DNA repair in mammalian cells: DNA double-strand break repair: How to fix a broken relationship. *Cell Mol. Life Sci.* 2009, *66*, 1039–1056.
150. Oude Blenke, E.; Evers, M.J.; Mastrobattista, E.; van der Oost, J. CRISPR-Cas9 gene editing: Delivery aspects and therapeutic potential. *J. Control. Release* 2016, *244*, 139–148.
151. Chu, V.T.; Weber, T.; Wefers, B.; Wurst, W.; Sander, S.; Rajewsky, K.; Kuhn, R. Increasing the efficiency of homology-directed repair for

CRISPR-Cas9-induced precise gene editing in mammalian cells. *Nat. Biotechnol.* 2015, *33*, 543–548.

152. Dow, L.E. Modeling disease in vivo with CRISPR/Cas9. *Trends Mol. Med.* 2015, *21*, 609–621.

153. Bakondi, B.; Lv, W.; Lu, B.; Jones, M.K.; Tsai, Y.; Kim, K.J.; Levy, R.; Akhtar, A.A.; Breunig, J.J.; Svendsen, C.N.; et al. In vivo CRISPR/Cas9 gene editing corrects retinal dystrophy in the *S334ter*-3 rat model of autosomal dominant retinitis pigmentosa. *Mol. Ther.* 2016, *24*, 556–563.

154. Wu, W.H.; Tsai, Y.T.; Justus, S.; Lee, T.T.; Zhang, L.; Lin, C.S.; Bassuk, A.G.; Mahajan, V.B.; Tsang, S.H. CRISPR repair reveals causative mutation in a preclinical model of retinitis pigmentosa. *Mol. Ther.* 2016, *24*, 1388–1394.

155. Zhong, H.; Chen, Y.; Li, Y.; Chen, R.; Mardon, G. CRISPR-engineered mosaicism rapidly reveals that loss of *Kcnj13* function in mice mimics human disease phenotypes. *Sci. Rep.* 2015, *5*, 8366.

156. Ota, S.; Hisano, Y.; Ikawa, Y.; Kawahara, A. Multiple genome modifications by the CRISPR/Cas9 system in zebrafish. *Genes Cells* 2014, *19*, 555–564.

157. Shah, A.N.; Davey, C.F.; Whitebirch, A.C.; Miller, A.C.; Moens, C.B. Rapid reverse genetic screening using CRISPR in zebrafish. *Nat. Methods* 2015, *12*, 535–540.

158. Wang, S.; Sengel, C.; Emerson, M.M.; Cepko, C.L. A gene regulatory network controls the binary fate decision of rod and bipolar cells in the vertebrate retina. *Dev. Cell* 2014, *30*, 513–527.

159. Cho, S.W.; Kim, S.; Kim, J.M.; Kim, J.S. Targeted genome engineering in human cells with the Cas9 RNA-guided endonuclease. *Nat. Biotechnol.* 2013, *31*, 230–232.

160. Cong, L.; Ran, F.A.; Cox, D.; Lin, S.; Barretto, R.; Habib, N.; Hsu, P.D.; Wu, X.; Jiang, W.; Marraffini, L.A.; et al. Multiplex genome engineering using CRISPR/Cas systems. *Science* 2013, *339*, 819–823.

161. Jinek, M.; East, A.; Cheng, A.; Lin, S.; Ma, E.; Doudna, J. RNA-programmed genome editing in human cells. *eLife* 2013, *2*, e00471.

162. Mali, P.; Yang, L.; Esvelt, K.M.; Aach, J.; Guell, M.; DiCarlo, J.E.; Norville, J.E.; Church, G.M. RNA-guided human genome engineering via Cas9. *Science* 2013, *339*, 823–826.

163. Bassuk, A.G.; Zheng, A.; Li, Y.; Tsang, S.H.; Mahajan, V.B. Precision

medicine: Genetic repair of retinitis pigmentosa in patient-derived stem cells. *Sci. Rep.* 2016, *6*, 19969.

164. Hung, S.S.; Chrysostomou, V.; Li, F.; Lim, J.K.; Wang, J.H.; Powell, J.E.; Tu, L.; Daniszewski, M.; Lo, C.; Wong, R.C.; et al. AAV-mediated CRISPR/Cas gene editing of retinal cells in vivo. *Investig. Ophthalmol. Vis. Sci.* 2016, *57*, 3470–3476.
165. Fu, Y.; Sander, J.D.; Reyon, D.; Cascio, V.M.; Joung, J.K. Improving CRISPR-Cas nuclease specificity using truncated guide RNAs. *Nat. Biotechnol.* 2014, *32*, 279–284.
166. Mali, P.; Aach, J.; Stranges, P.B.; Esvelt, K.M.; Moosburner, M.; Kosuri, S.; Yang, L.; Church, G.M. Cas9 transcriptional activators for target specificity screening and paired nickases for cooperative genome engineering. *Nat. Biotechnol.* 2013, *31*, 833–838.
167. Shen, B.; Zhang, W.; Zhang, J.; Zhou, J.; Wang, J.; Chen, L.; Wang, L.; Hodgkins, A.; Iyer, V.; Huang, X.; et al. Efficient genome modification by CRISPR-Cas9 nickase with minimal off-target effects. *Nat. Methods* 2014, *11*, 399–402.
168. Ran, F.A.; Hsu, P.D.; Lin, C.Y.; Gootenberg, J.S.; Konermann, S.; Trevino, A.E.; Scott, D.A.; Inoue, A.; Matoba, S.; Zhang, Y.; et al. Double nicking by RNA-guided CRISPR Cas9 for enhanced genome editing specificity. *Cell* 2013, *154*, 1380–1389.
169. Hsu, P.D.; Scott, D.A.; Weinstein, J.A.; Ran, F.A.; Konermann, S.; Agarwala, V.; Li, Y.; Fine, E.J.; Wu, X.; Shalem, O.; et al. DNA targeting specificity of RNA-guided Cas9 nucleases. *Nat. Biotechnol.* 2013, *31*, 827–832.
170. Qi, L.S.; Larson, M.H.; Gilbert, L.A.; Doudna, J.A.; Weissman, J.S.; Arkin, A.P.; Lim, W.A. Repurposing CRISPR as an RNA-guided platform for sequence-specific control of gene expression. *Cell* 2013, *152*, 1173–1183.
171. Bikard, D.; Jiang, W.; Samai, P.; Hochschild, A.; Zhang, F.; Marraffini, L.A. Programmable repression and activation of bacterial gene expression using an engineered CRISPR-Cas system. *Nucleic Acids Res.* 2013, *41*, 7429–7437.
172. Gilbert, L.A.; Horlbeck, M.A.; Adamson, B.; Villalta, J.E.; Chen, Y.; Whitehead, E.H.; Guimaraes, C.; Panning, B.; Ploegh, H.L.; Bassik, M.C.; et al. Genome-scale CRISPR-mediated control of gene repression and activation. *Cell* 2014, *159*, 647–661.

173. Gilbert, L.A.; Larson, M.H.; Morsut, L.; Liu, Z.; Brar, G.A.; Torres, S.E.; Stern-Ginossar, N.; Brandman, O.; Whitehead, E.H.; Doudna, J.A.; et al. CRISPR-mediated modular RNA-guided regulation of transcription in eukaryotes. *Cell* 2013, *154*, 442–451.

174. Cheng, A.W.; Wang, H.; Yang, H.; Shi, L.; Katz, Y.; Theunissen, T.W.; Rangarajan, S.; Shivalila, C.S.; Dadon, D.B.; Jaenisch, R. Multiplexed activation of endogenous genes by CRISPR-on, an RNA-guided transcriptional activator system. *Cell Res.* 2013, *23*, 1163–1171.

175. Jung, C.J.; Menoret, S.; Brusselle, L.; Tesson, L.; Usal, C.; Chenouard, V.; Remy, S.; Ouisse, L.H.; Poirier, N.; Vanhove, B.; et al. Comparative analysis of *piggyBac*, CRISPR/Cas9 and TALEN mediated BAC transgenesis in the zygote for the generation of humanized SIRPA rats. *Sci. Rep.* 2016, *6*, 31455.

176. Maeder, M.L.; Linder, S.J.; Cascio, V.M.; Fu, Y.; Ho, Q.H.; Joung, J.K. CRISPR RNA-guided activation of endogenous human genes. *Nat. Methods* 2013, *10*, 977–979.

177. Perez-Pinera, P.; Ousterout, D.G.; Brunger, J.M.; Farin, A.M.; Glass, K.A.; Guilak, F.; Crawford, G.E.; Hartemink, A.J.; Gersbach, C.A. Synergistic and tunable human gene activation by combinations of synthetic transcription factors. *Nat. Methods* 2013, *10*, 239–242.

178. Xiong, X.; Chen, M.; Lim, W.A.; Zhao, D.; Qi, L.S. CRISPR/Cas9 for human genome engineering and disease research. *Annu. Rev. Genom. Hum. Genet.* 2016, *17*, 131–154.

179. Komor, A.C.; Kim, Y.B.; Packer, M.S.; Zuris, J.A.; Liu, D.R. Programmable editing of a target base in genomic DNA without double-stranded DNA cleavage. *Nature* 2016, *533*, 420–424.

180. Hess, G.T.; Fresard, L.; Han, K.; Lee, C.H.; Li, A.; Cimprich, K.A.; Montgomery, S.B.; Bassik, M.C. Directed evolution using dCas9-targeted somatic hypermutation in mammalian cells. *Nat. Methods* 2016, *13*, 1036–1042.

INDEX

I

K

L

M